Ariel Dinar

Free Trade and Agricultural Diversification

Ariel, Dinah.

Free Trade and Agricultural Diversification

Canada and the United States

EDITED BY
Andrew Schmitz

Westview Press
BOULDER, SAN FRANCISCO, & LONDON

Westview Special Studies in Agriculture Science and Policy

This Westview softcover edition is printed on acid-free paper and bound in library-quality, coated covers that carry the highest rating of the National Association of State Textbook Administrators, in consultation with the Association of American Publishers and the Book Manufacturers' Institute.

All rights reserved. No part of this publication may be reproduced or transmitted in any form or by any means, electronic or mechanical, including photocopy, recording, or any information storage and retrieval system, without permission in writing from the publisher.

Copyright © 1989 by Westview Press, Inc.

Published in 1989 in the United States of America by Westview Press, Inc., 5500 Central Avenue, Boulder, Colorado 80301, and in the United Kingdom by Westview Press, Inc., 13 Brunswick Centre, London WC1N 1AF, England

Library of Congress Cataloging-in-Publication Data
Free trade and agricultural diversification: Canada and
  the United States/edited by Andrew Schmitz
    p. cm.—(Westview special studies in agriculture science
and policy)
  Includes bibliographical references.
  ISBN 0-8133-7851-6
  1. Tariff on farm produce—United States. 2. Tariff on farm
produce—Canada. 3. Agriculture and state—United States.
4. Agriculture and state—Canada. 5. Free trade—United States.
6. Free trade—Canada. I. Schmitz, Andrew. II. Series.
HF2651.F27U534 1989
338.1′0973—dc20   89-36173
                                                                   CIP

Printed and bound in the United States of America

∞ The paper used in this publication meets the requirements of the American National Standard for Permanence of Paper for Printed Library Materials Z39.48-1984.

10   9   8   7   6   5   4   3   2   1

Contents

Preface .... *vii*
About the Editors and Contributors... *x*

Chapter 1. **Introduction**... 1
            Andrew Schmitz

Chapter 2. **Agricultural Diversification Strategies:
            Canada and the United States**.................. 8
            Andrew Schmitz

Chapter 3. **Diversification of Prairie Agriculture**............ 37
            William A. Kerr

Chapter 4. **Growth and Development
            of Value-Added Activities** 88
            K. K. Klein and L. Chase-Wilde

Chapter 5. **Freer Trade in the North American
            Beer and Flour Markets**............................. 139
            Colin Carter, Jeffrey Karrenbrock, and William Wilson

Chapter 6. **Irrigation and
            Prairie Agricultural Development** 187
            Surendra N. Kulshreshtha

Chapter 7. **The Adoption of Modern Irrigation Technologies in the United States** 222

Gary Casterline, Ariel Dinar, and David Zilberman

Chapter 8. **Farm Enterprise Size and Diversification in Prairie Agriculture** 249

William J. Brown

Chapter 9. **The Effect of U. S. Farm Programs on Diversification** 303

Richard E. Just and Andrew Schmitz

Chapter 10. **Agricultural Subsidies in Canada: Explicit and Implicit** 329

W. H. Furtan, M. E. Fulton, and K. A. Rosaasen

Index 363

Preface

The majority of the findings presented in this book are part of the results of a study undertaken by the Economic Council of Canada on the future of the Prairie grain economy in Canada. The overall project was referred to the Council by the Prime Minister of Canada in a letter dated March 31, 1987, in which he stated, "I am encouraged to see the Council proposing a significant collaborative effort with federal and provincial governments and the private sector. I am pleased to support this particular study as a vehicle for public debate on a pressing problem which concerns us all—the future of the Prairie grain economy. I expect it to produce an invaluable exchange of information while leaving the Council, as always, to its own independent views, conclusions, and recommendations."

The Economic Council received financial support from the governments of Saskatchewan and Alberta; Agriculture Canada; The Prairie Pools, Inc.; Cargill, Ltd.; and the Royal Bank of Canada. Representatives of these organizations, as well as independent experts, served on the Technical Advisory Committee.

I was asked to serve as Director of Research for the overall study on the future of Prairie agriculture while I held the Hadley Van Vliet Chair at the Department of Agricultural Economics, University of Saskatchewan at Saskatoon. I would like to thank those who supported the appointment as Director of the project.

One of the areas of emphasis, the results of which are contained in this book, deals with diversification of Prairie agriculture and how it was impacted by various policies, including the U. S.-Canada Free Trade Agreement. The other parts of the overall project have been published by the Economic Council of Canada in four volumes: *Grain Market Outlook* (W. H. Furtan, T. Y. Bayri, R. Gray, and G. G. Storey, 1989); *Canada and International Grain Markets: Trends, Policies, and Prospects* (Colin Carter, Alex McCalla, and A. Schmitz, 1989); *Canadian Agricultural Policy and Prairie Agriculture* (Murray Fulton, Ken Rosaasen, and Andrew Schmitz, 1989); and *Canadian Prairie Farming, 1960 Through 2000* (Lou

Auer, 1989). The chapters contained in this book by Schmitz; Brown; Kulshreshtha; Furtan, Fulton, and Rosaasen; Kerr; and Klein and Chase-Wilde report the findings of the diversification phase of this research project.

Carter, Karrenbrock, and Wilson were doing parallel work focusing specifically on the flour milling and the brewing industries and the effect the U. S.-Canada Free Trade Agreement has on these industries. Their findings are reported in a separate chapter. Also, I was able to add some perspectives from the United States through research conducted by Just, Zilberman, and others. Just and I present findings on the impact of U. S. farm programs on U. S. diversification; and the Casterline, Dinar, and Zilberman chapter on U. S. irrigation and diversification parallels the study by Kulshreshtha.

In terms of the above, the usual caveats are made that the findings, conclusions, and recommendations are those of the authors. They are not those of the Economic Council of Canada.

Many people have contributed to the compilation of this volume, and it is impossible to acknowledge everyone involved. However, there are several people I would like to acknowledge specifically. One is Judith Maxwell, Chairman of the Economic Council of Canada. Another is Caroline Pestieau of the Economic Council, who contributed significantly to the research content of various chapters of this book and other reports already published by the Council; her critical evaluation of the research was invaluable. To Dr. Harold Bjarnason, Associate Deputy Minister, Agriculture Canada, my thanks for strongly supporting this research. My thanks are extended also to the Technical Advisory Committee. I would like to thank my wife, Carole, for editing many of the chapters of this book and the previous manuscripts. Also, I thank the Word Processing Unit of the Department of Agricultural and Resource Economics, University of California at Berkeley, for editing and typing the final manuscripts. Specific thanks go to Ikuko Takeshita, Shirley Ramacher, Amor Nolan, Angela Erickson, Roberta Bregger, and Caren Oto.

Finally, early in the compilation of the four volumes mentioned and the chapters for this book, it became apparent that the research results should be published in a high-quality format to ensure that the material would be widely distributed and read. A large number of the four volumes previously listed are already in circulation across North America and other parts of the globe and have found their way not only into academic circles but also into the policy arena. It is hoped that this book will have the same success. Because of the intended audience, the writing style of these chapters and the previous publications is somewhat different than is normally the case for academic authors. We have attempted to write these manuscripts so that they are readable by the general public. Wherever possible, we have either completely deleted the mathematics and highly theoretical treatments or have put them in appendices.

Again, as Director of Research for the project on the future of Prairie agriculture for the Economic Council of Canada, I hope that the results from this research will make a positive and significant contribution to solutions in dealing not only with Prairie agriculture but with the entire North American agricultural complex. The Free Trade Agreement clearly indicates that U. S. and Canadian agriculture is highly interconnected and, hopefully, that trade problems in those areas where Canada and the United States compete directly abroad can be resolved by forming stronger relationships with each other. My special thanks to the Economic Council of Canada for financial support for this volume and to Premier Grant Devine of Saskatchewan for providing the initial thrust for the overall study undertaken by the Economic Council of Canada on the future of Prairie agriculture.

Andrew Schmitz
Berkeley, California

About the Editor and Contributors

ANDREW SCHMITZ is a graduate of the University of Saskatchewan, Department of Agricultural Economics, and of the University of Wisconsin, Department of Economics. His area of emphasis is international economics and agricultural policy. He is Chairman of the Department of Agricultural and Resource Economics, University of California at Berkeley where he holds the George W. and Elsie M. Robinson Endowed Chair. He is also an Adjunct Professor, Department of Agricultural Economics, University of Saskatchewan in Saskatoon. Schmitz is a Fellow of the American Agricultural Economics Association. He has published numerous papers and books and has received several awards for his research, including the Harold Groves Doctoral Dissertation Award for the best thesis in the Department of Economics at the University of Wisconsin, five awards from the American Agricultural Economics Association on research discovery and quality, and an award for a book on grain export cartels. In 1986, he was the first recipient of the Department of Agricultural Economics, University of Saskatchewan, Hadley Van Vliet Chair and also received the University of Saskatchewan Outstanding Graduate Award for the College of Agriculture's 75th Anniversary. Schmitz has consulted for numerous organizations, including many law firms which focus on antitrust and price-fixing considerations, as well as for the World Bank, Canada Department of Agriculture, U. S. Department of Agriculture, the U. S. Central Intelligence Agency, the National Grain and Feed Association, the International Trade Commission, the Potash Corporation of Saskatchewan, and the U. S. Office of Technology Assessment.

WILLIAM J. BROWN is an Associate Professor with the Department of Agricultural Economics at the University of Saskatchewan where he teaches Farm Business and Financial Management at the Vocational and Undergraduate level. He received his B.S.A. and M.Sc. in Agricultural Economics from the Universities of Manitoba and Alberta, respectively. His publications include: "A Risk Efficiency Analysis of Crop Rotations in Saskatchewan," "Irrigation and Farm Level Risks: A Case Study of the South Saskatchewan River Project," and others dealing with farm level economics and risks.

COLIN CARTER obtained a B.A. in Economics and an M.Sc. in Agricultural Economics from the University of Alberta. His studies continued at the University of California at Berkeley, where he obtained an M.A. in Economics and, in 1980, a Ph.D. in Agricultural Economics. He is Professor of Agricultural Economics at the University of California-Davis. His published work has covered such topics as import tariffs in the world wheat market, grain export cartels, futures market efficiency, U. S. wheat policy, China's trade in grains, and Japanese trade policy. He recently held a 3-year fellowship from the Kellogg International Fellowship Program in Food Systems.

GARY L. CASTERLINE is a student in the doctoral program at the Department of Agricultural and Resource Economics at the University of California at Berkeley. He received his B.S. in Agricultural Economics from Montana State University. He has contributed to projects with the Electric Power Research Institute and the U. S. Department of Agriculture concerning irrigation technology. His Ph.D. dissertation is on the diffusion of irrigation technologies in the United States.

LINDA CHASE-WILDE is a Market Economist with Alberta Agriculture, Edmonton, Alberta. From 1984 to 1989, she was Assistant Professor of Agricultural Economics in the Department of Rural Economy at the University of Alberta, Edmonton, Alberta. She received a B.A. degree in Comparative Religions from Carleton University, Ottawa, in 1972; an M.Sc. degree in Agricultural Economics from the University of Alberta in 1979; and a Ph.D. degree in Agricultural Economics from Michigan State University in 1984. Most of Dr. Chase-Wilde's research has been in the area of international grain trade.

ARIEL DINAR is a Visiting Researcher in Resource Economics at the University of California-Riverside and the San Joaquin Valley Drainage Program. His Ph.D. in Agricultural Economics is from The Hebrew University of Jerusalem, Israel. He has written on economics of water quality, economics of irrigation with respect to water-soil-crop, salinity, and drainage related problems.

MURRAY FULTON is Associate Professor with the Department of Agricultural Economics and holds a position in the Centre for the Study of Co-operatives at the University of Saskatchewan. Professor Fulton received his undergraduate degree in Agricultural Economics at the University of Saskatchewan in 1977. He received his Masters degree in Agricultural Economics from Texas A&M University in 1978 and his B.A./M.A. in Philosophy, Politics, and Economics from Oxford University in 1980. He earned his Ph.D. in Agricultural Economics from the University of California at Berkeley in 1985. Dr. Fulton's teaching and research interests are in the areas of agricultural policy, industrial organization, and cooperatives. His recent research includes a review of agricultural policies affecting Canadian prairie agriculture, mergers in the multinationals in international grain markets, and the impact of cooperatives in oligopolistic industries.

W. H. FURTAN is Professor of Agricultural Economics at the University of Saskatchewan. He received his B.S.A. and M.Sc. at the University of Saskatchewan and his Ph.D. at Purdue University. His research is in the area of agricultural policy and technological change. He has published numerous articles in such journals as the *Journal of Public Economy* and the *Quarterly Journal of Economics*.

RICHARD E. JUST is Professor of Agricultural and Resource Economics at the University of Maryland. He received his graduate training and was on the faculty for many years at the University of California at Berkeley. He has written extensively in the areas of quantitative agricultural policy and economic welfare analysis. His publications include "Applied Welfare Economics and Public Policy "(1982) and numerous articles published in such journals as *American Economic Review, Econometrica, International Economic Review, Quarterly Journal of Economics, American Journal of Agricultural Economics, Journal of Economic Theory, Journal of Econometrics, Journal of Development Economics, Economic Development and Cultural Change,* and *Oxford Economic Papers*. He has served as Editor of the *American Journal of Agricultural Economics* and on the editorial board of several other journals. He is a Fellow of the American Agricultural Economics Association and is named in *Who's Who in Economics* as one of the major economists from 1700-1984.

JEFFREY D. KARRENBROCK is an Economist at the Federal Reserve Bank of St. Louis. He received his B.S. in Agricultural Economics from the University of Missouri-Columbia and his M.S. in Agricultural Economics from the University of California-Davis. He contributes to two of the bank's publications—*Review* and *Pieces of Eight*, a regional economics publication. His research interests, in addition to international trade, include agribusiness and natural resource industry analysis.

WILLIAM A. KERR is Professor of Agricultural Economics in the Department of Economics at the University of Calgary, Calgary, Canada. He received a B.A. in International Relations from the University of British Columbia, an M.A. in Trade and Development Economics from Simon Fraser University, and a Ph.D. in Agricultural Economics from the University of British Columbia. Dr. Kerr has written extensively on agricultural trade issues with works having appeared in the *American Journal of Agricultural Economics*, the *Journal of World Trade*, the *Journal of Economic Studies,* and the *Canadian Journal of Agricultural Economics*. He has dealt extensively with the U. S.-Canada Free Trade Agreement and Canada-EEC trade disputes and is currently is researching Canada's trade with the Pacific Rim countries.

KURT K. KLEIN is a Professor in the Department of Economics at the University of Lethbridge, Lethbridge, Alberta. He received a Diploma in Vocational Agriculture in 1961, a B.S.A. degree in Agriculture in 1970, and an M.Sc. in Agricultural Economics in 1972—all at the University of Saskatchewan. Professor Klein received a Ph.D. degree in Agricultural Economics from Purdue University in 1976. Prior to joining the faculty at the University of Lethbridge, he was a research economist at the Agriculture Canada Research Station, Lethbridge, Alberta. Dr. Klein has done extensive research on the economics of agricultural technology on western Canadian farms and agricultural models. He has published the results of his research in the *Canadian Journal of Agricultural Economics* and in several biological sciences journals.

SURENDRA N. KULSHRESHTHA is Professor of Agricultural Economics at the University of Saskatchewan, Saskatoon. He studied at the Agra University, India, and at the University of Manitoba where he received his Ph.D. in agricultural economics. His publications include a number of articles and monographs dealing with irrigation, value of water, water use, an input-output model, and econometric models. He is presently Associate Editor of the *Canadian Water Resources Journal*.

KEN A. ROSAASEN is Professor of Agricultural Economics at the University of Saskatchewan. A native of Saskatchewan, he received his B.S.A. and M.Sc. from the University of Saskatchewan. He has written extensively in the area of agricultural policy, in particular, on issues dealing with Saskatchewan agriculture. He currently teaches agricultural marketing at the University.

WILLIAM WILSON is Associate Professor of Agricultural Economics at North Dakota State University. He received his Ph.D. in Agricultural Economics from the University of Manitoba in 1980 and was a Visiting Scholar at the Food Research Institute at Stanford University in 1987-88. He has written extensively in the areas of industrial organization in the grain marketing industries and international competition, with articles appearing in numerous economic and agricultural journals as well as three chapters in books. He is currently a member of the advisory committee for the Federal Grain Inspection Service and was involved in the Office of Technology Assessment project on grain quality. In addition, he serves on numerous other academic and industry committees and is a consultant to a number of domestic and international agricultural businesses and organizations.

DAVID ZILBERMAN is Professor of Agricultural Economics at the Department of Agricultural and Resource Economics at the University of California at Berkeley. He was born in Israel and received his B.A. in Economics and Statistics at Tel Aviv University. His Ph.D. was earned at the University of California at Berkeley. Dr. Zilberman's areas of specialty are irrigation and pesticide economics, environmental regulations, economics of technological change, economics of natural resources, microeconomic theory, and agricultural policy. He has consulted for the U.S. Department of Agriculture, the World Bank, the

U. S. Environmental Protection Agency, the California Air Quality Control Board, Public Interest Economics, the Chicago Board of Trade, and Koor Trade and Galil Technologies in Israel.

1

Introduction

Andrew Schmitz

The 1970s were boom years for the grain economy, both on the Prairies in Canada and in the midwestern United States. The world's cry for more grain (including wheat) during the 1970s led to a rise in both grain prices and land values. Generally, because grain enterprises became more profitable than livestock operations, there was a large sell off of cattle herds during the 1970s both in Canada and the United States. The high grain prices led farmers, at least on the Prairies, toward specialization in grain with the major crop being wheat. Wheat became king in the Prairie region. The 1980s saw a reversal of the 1970s. Grain prices plummeted and input costs soared. Farm bankruptcies in North American agriculture rose significantly. Farmers were caught in a cost-price squeeze. The problem was even worse than might have been the case since many farmers became specialized in grain production and no longer operated diversified enterprises. This was in sharp contrast to the 1950s, for example, when many grain farmers also had livestock enterprises.

To highlight the focus on grains and the associated problems with specialization, the *Toronto Globe and Mail* did a study on the region of Central Butte, Saskatchewan, where I was raised on a farm and currently operate a wheat and cattle operation with my brother. The study report is entitled "Harvest of Despair: How Wheat Has Wrecked the Prairie Economy" (Nikiforuk). The author of the report contends that increased specialization in wheat production was financially devastating and that it dealt rural communities irreversible hardships.

This book addresses some of the issues surrounding the debate over whether or not it is in the best interest in the long run to pursue strategies which leave agriculture specialized and thus in a highly vulnerable position and the extent to which past programs have either added to or detracted from diversification activities. As history teaches, commodity prices are very volatile. In this situation it is very difficult for farmers to adjust to a price collapse in a specific set of commodities if they produce those commodities with highly specialized inputs including management skills. From a government policy standpoint, is it more economically prudent to provide incentives to farmers which encourage a diversified farm portfolio? Even broader than this, from an individual management standpoint, farmers have to consider a portfolio which does not include only agriculture but has additional elements, e.g., investments in nonfarm activities such as bonds. As the saying goes, it is prudent not to put all one's eggs in a single basket.

Value-Added Activities

Related to the above is the issue of value-added activities for Prairie agriculture and that in the Midwest. It has often been stated that Canada and the United States export agricultural products in raw form and that this must change. Value-added products are the key to economic growth and development. Over 60 percent of U. S. exports are shipped out in an unprocessed form increasing from 62 percent in 1972 to 68 percent in 1981. In 1972, 27 percent of exports were semiprocessed, but this percentage decreased to 22 percent in 1981. However, only 10 percent of exports were truly processed. In Canada over the period 1982-83, exports averaged over $9 billion. Of this amount, roughly 70 percent was exported in unprocessed form as primary products. However, in the early 1960s, exports of raw products were only roughly 65 percent of the total. Thus, exports of raw products have become relatively more significant.

Even though both the United States and Canada export goods largely in unprocessed form, one difference is that Canada's level of primary exports as a proportion of the total has always been higher. In 1967 the United States was shipping 65 percent of total unprocessed exports while Canada was shipping 68 percent. Similarly, in 1981, the United States shipped 68 percent in raw form while Canada exported 71 percent. The data for both Canada and the United States are disturbing relative to exports from European countries. For example, in 1980 France exported 52 percent of its agricultural products in primary form while semiprocessed exports constituted 33 percent and highly processed exports constituted 15 percent of the total. The corresponding figures for Sweden were 56 percent, 21 percent, and 23 percent.

This book puts the issues on diversification and value added in perspective. Several chapters focus on barriers to diversification and value-added activities. In this context, for example, case studies are made of the brewing industry and the flour milling industry. The implications of the U. S.-Canada Free Trade Agreement are presented. Also, a broader range of studies are included which focus on such activities as livestock and farm machinery manufacturing.

In promoting diversification and value-added activities, governments have to be careful that the use of instruments, such as subsidies, does not create an international trading environment in which the trading partner imposes countervailing duties to offset the effect of production subsidies.

This book analyzes numerous border disputes in agriculture between the United States and Canada and suggests means by which such disputes can be avoided. This is extremely important because of such policy initiatives as the Western Diversification Initiative in Canada. Here an attempt is being made to broaden the economic activities of the Prairie region. Tied into this strategy is the promotion of international trade since many of the activities which are being promoted in the Prairie region involve exports to other nations.

Diversification

There is no single definition of diversification. One of the authors defines diversification as a process whereby more and more activities are added either to a single farm enterprise or to a region. For example, through irrigation, diversification occurs if more and more production activities are added to the resource base through time. In turn, if livestock activities are added to grain activities as a result of irrigation, then irrigation has brought about economic diversification. Another author, on the other hand, looks at diversification differently. The argument is made that diversification only occurs if it brings about more income stability than would otherwise be the case. For example, if a region has four activities, all of which are correlated in the same manner, then the region is not diversified since, when prices rise, all of the industries are profitable; but when prices fall, all of the industries simultaneously become unprofitable. Therefore, just because a region has many activities, it does not follow that the region is diversified. To be truly diversified, some of the economic activities have to be negatively correlated. That is, when one industry prospers, the other industry will have the opposite characteristics.

In the above context, international trade impacts the degree of diversification. Generally, trade theory argues that the more a region becomes engaged in exports, the greater will be the degree of specialization in that region. Therefore, there may appear to be a contradiction in that expanded international trade may lead a region away from diversification. On the other hand, many policymakers view diversification as a plus. This leads to the ultimate question: "Is diversification necessarily a desirable goal?" It is difficult to answer this question, but it is important to keep in mind that, even though a region may experience a great deal of income instability over time, this may be preferable to a time path where incomes are relatively stable. One also has to be concerned about the mean income through time of a region. It may well be that the mean income coupled with high instability is higher than the mean income associated with stability. This separation is important since, if a region exhibits a high degree of instability but higher mean incomes than would be the case if stability through diversification were attempted, then governments should pursue income stabilization schemes. Thus, income stabilization schemes associated with specialization may be preferable to diversification in the absence of stabilization schemes. One has to keep in mind that, for the Prairie region and for the Midwest generally, international trade is vital; but along with international trade comes price volatility, even though the gains from trade are positive. One does not want to choke off trade only to promote income stability. Clearly, in this case, lower levels of income will result.

A focus of this book is on the impact of the U. S.-Canada Free Trade Agreement. It extends the analysis of Schmitz and Carter. Will free trade add to diversification or cause increased specialization in agriculture? How can trade disputes be resolved in an easy and costless manner so that the maximum economic gains from trade liberalization can be achieved?

Portfolio and Farm Enterprise Selection

In terms of on-farm diversification, the debate has always taken place about whether or not farmers should diversify their activities and thus avoid the high degree of income volatility associated with agriculture. It was clearly the focus of the *Toronto Globe and Mail* report referred to earlier. For example, at one time many farmers in the Prairies had both grain and livestock operations. It was generally felt that, when grain prices were

Introduction

high, cattle prices were low and vice versa, and thus over time, farmers could achieve income stability by having more than one enterprise. There is an excellent chapter in this volume which deals with this issue. As the results show, a farmer does not necessarily have to diversify his/her portfolio only within agriculture. Clearly, diversification opportunities are available through investing in such activities as stocks, bonds, and bank certificates. In addition, the risk-averse farmer may choose a different form of portfolio investments than a risk-taking entrepreneur. The results show the different enterprise combinations for different types of risk-taking attitudes.

It has generally been found in agriculture that significant economies of scale exist in any given farm enterprise. That is, as output increases on a given farm, per-unit costs fall. Thus, an individual farmer is left with the problem of whether to specialize or diversify since, by diversifying into more than one enterprise, the farmer may not be able to achieve the economies of scale necessary to survive. In other words, for a single farm family, it is extremely difficult to achieve economies of scale in three or four enterprises. This is because a single family is incapable of managing this many enterprises, at least at the scale necessary to achieve minimum per-unit costs. Thus, it may well be that farm families have correctly chosen the specialization path and attempted diversification through off-farm activities.

The above raises the interesting question as to whether each individual farmer should diversify his/her operation or whether each individual farmer should have to specialize in farm enterprises in order for the region to become diversified. For example, for true diversification to occur, it may well be that some farmers have to specialize in grain production while others specialize in other activities including livestock production. It is generally inferred that, for a region to be diversified, an individual farmer should have to diversify his/her operation. This may well not be the case. For many individual farms, it may be impossible to carry out successful diversified operations as is done, for example, by the Hutterites in Canada. Many of these operations produce several combinations of commodities including pork, beef, and grain. One has to recognize that these colonies are sufficiently large such that they can achieve economies of scale in each of their enterprises through management specialization in each activity. Such management and laborer supplies, however, are not available for the average family farm in either the Prairie region or the United States.

Subsidies—Explicit and Implicit

Economic diversification and value-added activities are influenced by the degree to which governments support such activities. Controversy has arisen in Canada over the degree to which various farm sectors are subsidized. Western agriculture, because of its export orientation, receives visible support from governments. Examples include the 1988-89 drought payment and the earlier special Canadian Grains Program. The latter was in response to the lowering of the loan rate in the U. S. 1985 Farm Bill. Eastern Canada, however, is generally dominated by supply management. In these cases, a farmer's income is supported through various means such as import and production quotas. Also, an important component is the government expenditure on research and development (R&D) activities. These expenditures may or may not be neutral across the Canadian agricultural region. One study examines the nature of farm subsidies in Canadian agriculture by comparing income transfers to eastern agriculture with those made to Prairie farmers. It is generally found that the support is higher in the East than in the Prairie region. In addition, the largest component of R&D is in horticultural crops where production is the greatest in Eastern Canada.

The degree of federal and provincial assistance influences the value-added activities forthcoming from agriculture. It is generally found that the diversification activities for eastern agriculture are greater than those which occur in the western region. The data and the findings from this chapter are of significance to policymakers in Canada if they have neutrality of farm programs as one objective. In addition, as the study illustrates, programs designed for eastern agriculture are much more permanent than those pertaining to the West. It was only since the 1980s that large income transfers were made to Prairie agriculture.

The analysis which deals with the level of government transfers between the Prairie region and Eastern Canada is not intended to create divisiveness between Eastern and Western Canada. The opposite is intended, which is to create harmony between the various agricultural sectors in Canada. Policies and programs should be implemented which do not favor any given agricultural sector or region.

In Canada, controversy grows over the future direction of agricultural policy and what impact specific policies will have on economic diversification. One such policy deals with transportation. Under the current arrangement, the Crow rate is paid to the railways. This is a payment made by the federal government to make up the difference between current transportation costs and that which was agreed to and set forward as the Crow rate. Some organizations, especially livestock producers, are arguing that the payment should be made to producers rather than to the railways. They contend that this would add significantly to the livestock sector and the various value-added activities related to this important sector. They argue that paying the producers rather than the railways will change the relative prices in favor of using grains for domestic use rather than for shipment to export markets. This is one example where the outcome will significantly impact the direction and future of the Prairie agricultural region. In addition, the method of payment has significant effects for eastern agriculture as well.

In conclusion, several issues related to agricultural diversification are not fully developed in this book. For example, one argument often made is that, for agricultural diversification to occur, a significant manufacturing sector has to be present. This adds to the off-farm employment opportunities available. An analysis of this dimension of diversification will be possible as more data become available. Two such projects—Excel's beef slaughtering and processing plant in Alberta and the Cargill, Inc., proposed fertilizer plant in Saskatchewan—will provide the needed data.

References

Nikiforuk, Andrew. "Harvest of Despair." *Report on Business Magazine* in *The Globe and Mail* (April 1988):36-47.

Schmitz, Andrew, and Colin Carter. "Sectoral Perspective: Agriculture." In *Perspectives on a U. S.-Canadian Free Trade Agreement,* edited by R. N. Stern, P. H. Treziese, and John Whalley. Washington, D. C.: Brookings Institution, 1987.

2

Agricultural Diversification Strategies: Canada and the United States

Andrew Schmitz

Background

Governments have provided and will continue to provide assistance to the agricultural and nonagricultural sectors of the economy. The Western Canadian Diversification Initiative aimed at economic diversification is one such example. It is important, however, when providing such assistance, that such funding does not bring about charges of unfair trade practices by trading partners. Because of unfair trade practices, at least as alleged by the countries undertaking unfair trade practice litigation, the General Agreement on Tariffs and Trade (GATT) has been a forum whose purpose is to provide both a broad framework to achieve freer trade in agricultural products and to provide a dispute settlement mechanism. In addition, individual countries have taken up their own lawsuits aimed explicitly at a trading partner who is being accused of violating GATT rules.

This chapter discusses border disputes between Canada and the United States in the context of GATT. Even with the free trade agreement between the United States and Canada, trade irritants remain. In the future, if the trend toward a diversified agriculture is to be continued, border trade disputes should be avoided. To do this, the background which leads to trade irritants has to be understood.

In the past few years, the number of agricultural trade actions between Canada and the United States has increased, demonstrating that there are serious frictions in the agricultural trade relationship between the two countries. Although agriculture represents only 4 percent of total bilateral trade, most of the recent countervailing duty and antidumping investigations between the two countries have been in agricultural products. If fish and forest products are included under the definition of agriculture, then the percentage of trade cases involving agricultural products becomes even greater. During the 1980-1986 period, there were 13 U. S. countervailing duty investigations of Canadian exports. Of those, 9 involved agricultural products where the most significant cases centered around hogs and pork, Atlantic groundfish, softwood products, and softwood lumber. During the 1980-1985 period, 6 of the 10 U. S. antidumping investigations of Canadian exports involved agricultural or processed food products [U. S. International Trade Commission (ITC), *Annual Report*, various issues]. In contrast, during the 1980-1987 period, there was only one Canadian countervailing duty investigation which involved U. S. exports of corn into Canada.

In these trade disputes, subsidies—either direct or through production stabilization programs—cause the most important problems since the contention is made that export subsidies have trade-distorting effects and, therefore, should be prohibited. However, as discussed later, the impact that subsidies have is not clear. A subsidy can be put in place which need not have any effect on output or trade. Regardless of their effects, the extensive use in agriculture of domestic or general subsidies is an important policy instrument for alleviating poverty, addressing unfair imbalances of income, sustaining employment, eliminating regional industrial economic and social disparities, and promoting economic growth.

The Subsidies and Countervailing Duties Code, signed as a result of the Tokyo Round in 1979, explicitly prohibits the use of export subsidies—that is, government programs directed at exports are contingent on export performance—except those on certain primary products such as agricultural products. In Article 11 of the Code, however, the signatories recognized that domestic subsidies may be used for beneficial policy purposes and should not be disciplined except where they have trade-distorting or injurious effects (agreement on interpretation and application of Articles VI, XVI, and XXIII of the GATT, 1979).

Under the GATT and the Subsidies and Countervailing Duties Codes, domestic administration of countervailing duty laws is sanctioned as part of the international system for disciplining the use of trade-distorting subsidies. As Jackson states, "Even nations with very similar economic systems such as two industrial country market economies can find that minor variations in their economic systems can create situations which have the appearance of unfairness. These situations may have arisen almost completely by accident. . . . When two societies with even minor economic differences desire to work together, frictions or misunderstandings can occur unless there is an interface mechanism" (Jackson, p. 123).

It would appear that, after the recent experience of the countervailing duty cases involving Canadian softwood lumber and U. S. corn, one can reasonably observe that the domestic trade law systems are not working as an interface mechanism. There is substantive disagreement between the two countries on some of the rules—for example, the definition of a countervailable subsidy. Also, controversy exists over the impact subsidies have. Unfortunately, there also appears to be a corresponding lack of confidence in the ability of the trade remedy systems to apply fairly or interpret the rules as they are understood. This is evidenced by the fact that Canada and the United States, as a result of the softwood lumber and corn investigations, complained to the GATT that the investigations were not conducted in accordance with GATT rules. In the softwood lumber case, the GATT investigation was formally ended by the negotiation of the memorandum of understanding between the governments of Canada and the United States which was signed on December 30, 1986, by which the two countries agreed to resolve differences with respect to the conditions affecting trade in softwood lumber products. The dispute settlement mechanism will not work unless there is substantive agreement on the rules. For an international dispute settlement mechanism to be effective, the parties involved must have confidence that the mechanisms they have selected to resolve the disputes are designed to apply fairly. The softwood lumber case, in particular, demonstrated that Canadians not only disagree with the U. S. Department of Commerce (DOC) interpretation of the rules but also that they have lost confidence in the ability of the U. S. system to be fair.

According to Hudec, there are many reasons for the decline of confidence and the lack of utilization of these procedures. The U. S. administration, particularly since 1984, has expressed its lack of confidence in the GATT system. It has pursued a unilateral policy of reducing trade barriers by negotiating bilateral and multilateral agreements and by aggressively attacking unfair trade practices through the use of U. S. trade laws. This policy was enunciated in the U. S. Trade and Tariff Act of 1984. Since 1979, The United States has provided its domestic industries with private rights to initiate antidumping and countervailing duty investigations in an administrative system that is mandatory and time limited. This system, which was established under the U. S. Trade Agreements Act of 1979, is generally procedurally biased in favor of domestic complaints. Canada has adopted an essentially similar set of rules and procedures in the Special Import Measures Act (Horlich, Bello, and Talbot).

The antidumping and countervailing systems of both the United States and Canada were specifically designed to provide fast, effective relief to domestic industries threatened by imported goods. According to Hudec, on examination of the trade bill passed by Congress, the United States has gone beyond the intentions of the signatories to the antidumping, subsidies, and countervailing duties codes in its attempts to reduce trade barriers and to attack unfair trade practices throughout the world. The United States, through administrative and judicial interpretation in specific trade cases and progressive legislative amendments has created a new set of rules that are administered unilaterally by U. S. agencies and courts.

As a result of these recent U. S. actions, there is now a major disagreement between Canada and the United States over the rules of the game. Also, there is a corresponding lack of confidence in the fairness of their domestic trade regulatory systems to resolve disputes. There appears to be a basic difference of perception about what is fair and what is unfair in international trade. An important area of disagreement is in the definition and measurement of countervailable or actionable subsidies. As a result of the DOC preliminary determination in the softwood lumber case and recent proposals reviewed by Congress, the U. S. interpretations of specific or countervailable subsidies are becoming increasingly broader.

Although the United States objected strongly to Canada's ruling on U. S. corn (including initiation of formal procedures in the GATT), Canada's Department of National Revenue, Revenue Canada, has generally been more lenient than the United States concerning rulings on countervailing duty cases. It found only 4 of 64 programs investigated to be countervailable. In its determinations to date, Revenue Canada has applied a narrower definition of specific targeted or countervailable subsidy than has the DOC. One reason appears to be that Canada's Special Import Measures Act contains a provision stating that, in considering the definition of subsidy or subsidization, Revenue Canada is to take fully into account Articles 9 and 11 of the Subsidies and Countervailing Duties Agreement.

The primary purpose of the Code is to bring subsidies under more effective international discipline and reduce or eliminate those which are trade distorting and cause injury to domestic producers. Article 9 prohibits signatories from granting export subsidies on products other than certain primary products while Article 11 states that the signatories recognize that domestic subsidies are widely used as important instruments for the promotion of social and economic policy objectives. The signatories agreed that in designing their government assistance programs they would seek to avoid causing injury to a domestic industry of another signatory. This point is extremely important when considering government assistance aimed at diversifying an economy's resource base. Also, many elements of farm programs contain government transfers to producers, but these are generally not trade distorting.

The Free Trade Agreement Between the United States and Canada

The negotiation of a free trade area agreement provides an opportunity for Canada and the United States to develop common rules for regulating the use of subsidies and other trade-distorting measures. However, the question remains as to whether or not these two countries can devise a better mechanism than the current domestic trade remedy systems for dealing with their trade disputes. One option would be to replace existing countervailing duty laws with a new binational system for regulating subsidy practices. However, as Hudec points out, problems remain. First, it is extremely difficult to develop clear legal principles for determining which subsidies will have trade-distorting effects. This is a topic of ongoing debate in the GATT round. No agreement seems to be forthcoming on whether or not programs in Canada such as the Western Grain Stabilization Act or the Special Canadian Grains Program create trade distortions. Second, any general statutory definition, of necessity, will be overly broad and subject to interpretation by the agencies or authorities charged with administrating the agreement. A better proposal (Hudec) might be to include a provision similar to Article 92 of the European Community (EC) Treaty which declares that government assistance in any form that distorts or threatens to distort competition insofar as it affects trade between member states is incompatible with the Common Market. The EC Treaty also provides lists of categories of subsidy programs that are deemed to be or may be considered to be compatible with the Common Market. Under the EC system, the European Commission is required to review all existing government assistance programs and to monitor the development of new ones. The Commission has the authority to review existing or proposed programs and to order a member country to abolish or alter a program.

Where a member country does not comply with a ruling of the commission, that ruling may be enforced in the European Court of Justice.

In preparing lists of categories of programs that are already acceptable and those that may be acceptable, the Canadian and U. S. negotiators might also be guided by Articles 9 and 11 of the Subsidies and Countervailing Duties Code. The basic purpose of any rule that is developed to regulate the use of subsidies should be to eliminate government programs that distort or that threaten to distort competition insofar as they affect trade between the two countries (Steger).

Adoption of a binational system for regulating the use of subsidies could be phased in gradually over time where the two countries agree initially to create or not to create any new trade-distorting programs and to phase out objectionable programs over a period of time sufficient for adjustment to take place. In the transitional period, the domestic agencies responsible for administrating trade laws could continue to investigate cases using domestic laws but supplementing these by the rules contained in the new agreement.

Work is underway to address the question of which types of agricultural subsidy programs should be prohibited, which should be reviewable, and which should be acceptable. Professor Warley, for example, has developed a red-, green-, and amber-light approach to the problem of categorizing subsidy programs in light of their trade-distorting effects (Warley). In addition, current work is being done by Schmitz and Vercammen for the Office of the U. S. Trade Representative which ranks programs not only according to their associated subsidy levels but also in terms of efficiency characteristics.

The second aspect of this proposal would be establish a binational agency responsible for administering the rules. A free trade area agreement will include new obligations and rules concerning subsidies, dumping, restrictive business practices, protection of intellectual property, government procurement, product standards and technical regulations, trade and services, and investment. It is crucial that an appropriate joint mechanism be established to administer the agreement, to monitor legislative and policy developments as well as the activities of private firms, and to assist in resolving disputes (Hudec).

The free trade agreement should also include a transparency requirement where countries should be required to notify the binational agency of their existing programs and any new programs it proposes to implement. The agency could be given the authority to review proposed domestic programs; to consult with the two governments; and, where necessary, to issue orders declaring certain programs incompatible with the objectives of the agreement (Hudec). This would help minimize market distortions created by government programs. However, from an economic perspective, a great deal more work is needed on subsidy definitions and their effects. As Schmitz and Sigurdson and Just, Hueth, and Schmitz point out, subsidies do not necessarily lead to increased output and trade distorting effects. To the extent an agreement contains extensive new rules governing the use of subsidies, a permanent quasi-judicial tribunal could be established. This would be part of a binational agency for resolving certain types of disputes, particularly those affecting private parties.

The GATT Experience[1]

The GATT is an international agreement that establishes rules of behavior for governments in the area of international trade policy. It contains an adjudication procedure that permits member countries to bring lawsuits about violations of those rules. Since the GATT came into force in 1948, over 136 GATT lawsuits have been initiated (Hudec). Table 2.1 lists the number of appearances that each GATT member has made in GATT lawsuits from 1960 to 1988, listing separately the number of appearances as a defendant and the number of appearances as a plaintiff. Table 2.2 lists the parties who were suing and being sued by the five most active GATT litigants—the EC, the United States, Canada, Japan, and Australia. The following observations can be made:

1. Almost one-third of the GATT lawsuits during this period were lawsuits between the United States and the EC.
2. In addition to the 26 lawsuits between the United States and the EC, 45 of the remaining 54 lawsuits involved either the United States or the EC as one of the parties. Only 9 of the 80 lawsuits involved neither.
3. The EC and the United States litigated more frequently with each other than with others. The United States accounted for 26 percent of the complaints filed against other GATT countries, but 53 percent of the complaints were filed against the EC. The EC accounted for only 11 percent of the complaints filed against others, but 56 percent of the complaints were filed against the United States. The same disproportionate shares show up when one examines the activity of the United States and the EC as defendants. The EC was the target of

TABLE 2.1

Appearances as Defendant and Plaintiff Under Gatt

Defendants		Plaintiffs	
EC[a]	33[b]	U. S.	31
U. S.	15[c]	EC[d]	15
Japan	9	Canada	7
Canada	7	Australia	6
U.K.	3	Brazil	3
Spain	2	Japan	2
Greece	1	Chile	2
Denmark	1	Hong Kong	2
Norway	1	India	2
New Zealand	1	Uruguay	1
Finland	1	Israel	1
Switzerland	1	Korea	1
Jamaica	1	Argentina	1
Brazil	1	Poland	1
Chile	1	Nicaragua	1
		South Africa	1
		Finland	1
TOTAL	78[e]	TOTAL	78[f]

[a] Includes complaints both against the EC and against member states: EC itself, 27; France, 3; Italy, 1; Belgium, 1; and the Netherlands, 1.

[b] 17 by United States.

[c] 8 by the EC.

[d] There were no separate complaints by individual EC member states.

[e] Plus one case with 10 defendants and one case submitted jointly by the two parties.

[f] Plus one case with 15 plaintiffs and one case submitted jointly by the two parties.

Source: Hudec.

TABLE 2.2

Opposing Parties (GATT)

	As plaintiff, sued:		As defendant, sued by:	
EC	U. S.	8	U. S.	17
(49[a])	Canada	3	Australia	5
	Chile	1	Canada	3
	Finland	1	Chile	2
	Switzerland	1	Korea	1
	Japan	1	Brazil	1
			Hong Kong	1
			Argentina	1
			Japan	1
			(10 countries)	1
		15		33
United	EC	17	EC	8
States	Japan	5	Canada	3
(47[a])	Canada	3	Japan	1
	Greece	1	India	1
	Denmark	1	Poland	1
	Jamaica	1	Nicaragua	1
	Spain	1		
	U. K.	1		
	Brazil	1		
		31		15
Canada	U. S.	3	U. S.	3
(14)	EC	3	EC	3
	Japan	1	South Africa	1
		7		7
Japan	U. S.	1	U. S.	5
(11)	EC	1	Australia	1
			Canada	1
			India	1
			EC	1
		2		9
Australia	EC	5		
(6)	Japan	1		
		6		

[a]Total includes one case submitted jointly by the United States and the EC.

Source: Hudec.

30 percent of the complaints filed by other GATT countries but was the defendant in 57 percent of the U. S. complaints. The United States was the target of only 13 percent of the complaints filed by others but 53 percent of the EC complaints were brought against the United States.
4. The U. S.-EC litigation is not distinctive in terms of each party's role as plaintiff and defendant.
5. With or without the EC, the United States was responsible for initiating a very large share of the GATT lawsuits during this period. It accounted for 40 percent of all complaints since 1961 with the EC a distant second at 21 percent. Excluding all lawsuits between the United States and the EC, the United States still accounted for 26 percent of the rest with Canada next at 13 percent.
6. With or without the United States, the EC has been the target of a very large share of the lawsuits during this period. Excluding suits between the EC and the United States, the EC was the defendant in 30 percent of the remaining cases with Japan next at 17 percent. Table 2.3 gives a breakdown of the disputes according to the product area affected by the trade policy measure in question. Disputes are divided into three broad categories: (1) complaints about measures that affect trade in agricultural and fishery products, (2) complaints about measures that affect trade in industrial and mining products, and (3) complaints about trade measures of a general character that have no particular product focus or effect in the case at hand. According to Hudec,

   a. Of all the GATT lawsuits from 1960 to 1985, 54 percent of the complaints involved trade in agricultural products.
   b. The EC has been the target of 58 percent of all lawsuits involving agricultural products.
   c. The percentage of agricultural complaints and litigation against the EC is twice as high as the percentage in litigation against other GATT countries. Litigation against other GATT countries involved agriculture trade measures only 38 percent of the time while litigation against the EC, on the other hand, involved complaints about agricultural trade in 73 percent of the cases.
   d. Lawsuits over agricultural trade measures are clearly the reason why the EC is the GATT's most frequent defendant. In disputes not involving agricultural products, the EC has been sued no more than often than the United States—nine times each.

TABLE 2.3

Disputes Affecting Trade, by Product Area

Cases[a]	Agricultural/ fishery		Industrial/ mining		General	
	No.	%	No.	%	No.	%
All vs. EC (33)	24	73	5	15	4	12
U. S. vs. EC (17)	13	76	1	6	3	18
Other vs. EC (16)	11	69	4	25	1	6
All vs. U. S. (15)	6	40	6	40	3	20
EC vs. U. S. (8)	3	38	3	38	2	25
Other vs. U. S. (7)	3	43	3	43	1	14
All vs. Other (30)	12	40	16	53	2	7
EC vs. Other (7)	2	25	4	63	1	12
U. S. vs. Other (14)	6	43	7	50	1	7
Other vs. Other (9)	3	33	6	67	0	0
Unclassified (2)	2	100	b			

[a]"Other" means all other litigants except the EC and/or the United States.
[b]Blanks indicate no data available.
Source: Hudec.

    e. The volume of agricultural lawsuits against the EC is not due to any peculiarity of U. S.-EC litigation. Other governments accounted for 44 percent of the agricultural lawsuits against the EC. Agricultural complaints represented an almost equally high percentage of their total complaints against the EC, as did the agricultural complaints of the U. S.—69 percent to 77 percent. It appears that everyone has special problems with EC agriculture.

In summary, the data suggest that the litigation-generating effect of the Common Agricultural Policy (CAP) in general is not merely a peculiar U. S. reaction. Table 2.4 gives a breakdown according to the type of trade policy measure being complained about. These disputes are divided into

TABLE 2.4

Trade Disputes by Policy Measures

	Trade barriers						Discriminatory element[d]	
	Subsidy[a]		Tariff[b]		Nontariff[c]			
	No.	%	No.	%	No.	%	No.	%
All cases (78)	16	21	12	15	50	64	18	29
All vs. EC (33)	13	39	3	9	17	51	6	30
U. S. vs. EC (17)	8	47	1	6	8	47	1	11
Other vs. EC (16)	5	31	2	13	9	56	5	45
All vs. U. S. (15)	2	13	3	20	10	67	5	38
EC vs. U. S. (8)	2	25	2	25	4	50	1	17
Other vs. U. S. (7)	0	0	1	14	6	86	4	57
All vs. Other (29)	1	3	6	21	22	76	7	25
EC vs. Other (6)	0	0	2	33	4	67	0	0
U. S. vs. Other (14)	1	7	2	14	11	79	4	30
Other vs. Other (9)	0	0	2	22	7	77	3	33

[a]Includes both export subsidies and domestic production subsidies.

[b]Includes EC variable levies.

[c]Includes border and internal measures and quantitative and tax-type measures.

[d]The number and percentage of trade barriers which contain a clear legal attack on a discriminatory element.

Source: Hudec.

(1) complaints about subsidies, including both export subsidies and domestic production subsidies; (2) complaints about tariffs, including EC variable levies; and (3) complaints about nontariff barriers, including both border measures and internal measures and both quantitative and tax-type measures. The following points can be noted.

1. Subsidies accounted for an important share of litigation against the EC (13 out of 33 complaints) but were only a negligible factor in the lawsuits against other GATT countries during this period (3 out of 45 complaints). Of the 13 subsidy complaints against the EC, 10 involved agriculture.
2. Subsidy complaints were an even larger share of U. S. complaints against the EC—8 out of 17.
3. The United States suffered almost exactly the same percentage of complaints as did the EC against what might be called antisubsidy measures. Four complaints against the United States involved legal objections to some aspect of the U. S. countervailing duty law. A fifth complaint involved the U. S. export subsidy on sales of wheat flour to Egypt—an act of retaliation against the EC wheat flour subsidy. Finally, a sixth complaint involved the U. S. corporate income tax law regarding an export subsidy.
4. Of the 64 trade barriers complained about during the period, 51 were nontariff trade barriers.

The United States—EC Litigation

Legal Actions in the 1960s (Example)

To give an example of such actions, a brief review of the celebrated Chicken War Dispute is given. The Chicken War arose when the United States declined to accept the compensation offered for withdrawing the West German binding on poultry and announced that they would retaliate by withdrawing some concessions of their own on $44 million worth of EC trade. The EC contested the size of the retaliation and this collateral dispute was submitted to the GATT panel. The United States then retaliated by withdrawing concessions in the amount determined by the panel—$26 million. The real purpose of the Chicken War retaliation was to give a dramatic expression of U. S. displeasure with the level of protection adopted in the CAP on poultry. It appears that the retaliation had no effect

Diversification Strategies 21

at all since the EC went about finishing the CAP and setting its support prices much as before, with levels of protection uniformly higher than the United States thought appropriate.

Legal Actions in the 1970s (Examples)

One such involvement was a rather long skirmish over EC tariff preferences on citrus products in favor of Mediterranean suppliers. The United States repeatedly protested the illegality of such preferences but never formally demanded legal ruling. The matter was eventually settled in 1973 with an agreement adjusting the seasonal tariff rates in a manner favorable to U. S. exporters.

In 1974, the U. S. Executive Branch went to the U. S. Congress to obtain new negotiating authority for the Tokyo Round trade negotiations. One of the conditions that Congress added was a new procedure known as Section 301. It was designed to compel the U. S. Executive Branch to enforce U. S. legal rights more vigorously than had been done in the past. The pressures generated by Section 301 yielded seven GATT lawsuits by the United States during the years of the Tokyo Round negotiations (1975 through 1979). Three were against Japan, two against the EC, and one each against Canada and Spain. Both complaints against the Community were filed in 1976 and both involved short-term trade measures in the agricultural sector taken to alleviate surplus situations in product sectors benefitting from CAP price supports. One case involved a minimum import price regime for imported tomato products and the other involved a mixing regulation requiring the use of surplus dairy products with imported animal feeds.

Legal Actions in the 1980s (Examples)

In 1979 the U. S. Congress passed major legislation approving and implementing all the trade agreements negotiated in the Tokyo Round. There was a promise that the Executive Branch would even more rigorously enforce GATT legal rights in the future. Congress strengthened Section 301. Between December, 1981, and July, 1982, the United States filed seven GATT lawsuits against the EC. Five concerned subsidies in agricultural products, one involved the renewal of the 1970-1973 complaint about preferential tariffs on citrus products, and one involved an industrial trade problem. The complaints included: (1) export subsidy on wheat flour (subsidy codes), (2) export subsidy on pasta (subsidy codes), (3) export subsidy on poultry (subsidy codes), (4) production subsidy on canned fruit and raisins (Article XXIII), (5) export subsidy on sugar (subsidy codes), and (6) preferential tariff on citrus products (Article XXIII). In many of the legal cases, calls were made for changes in EC policy that ranged from extremely difficult to politically impossible.

According to Hudec, the legal claims in three of the agricultural subsidy cases attacked critical aspects of the CAP. The wheat flour case sought to establish legal limits on the amounts of CAP surplus production that could be disposed of via subsidized exports. The pasta complaint attempted to prevent or at least seriously hinder the Community from giving subsidies in exports of processed foods. The canned fruit complaint would have caused similar effects on food processors selling within the EC because the complaints sought to bar subsidies to domestic producers for products whose EC tariffs had been bound in previous GATT negotiations. All of these issues and decisions fell short of satisfactory resolution. For example, the wheat flour panel was unable to reach a decision on whether EC wheat flour exports had exceeded the equitable share standard of Subsidies Code Article 10. The panel in the pasta case issued a 4:1 divided report in which the majority found that the export subsidy on pasta products was prohibited by Article 9 of the Subsidies Code.

The United States—Canada Litigation

As illustrated earlier, there have been several cases involving agricultural trade disputes not only between the United States and the EC but between the United States and Canada as well. In addition, lawsuits have prevailed which include commodities of significant importance to the Prairie Region (such as potash) but are generally not included under the narrow definition of agricultural products. In this section, comments are made on three areas—pork, potash, and corn. In the first two cases, the United States sued Canada for using subsidies to distort export trade between the two countries. In the latter case, Canada initiated the investigation against the United States for alleged dumping of corn into the Canadian market. These cases are discussed briefly in order to expand on the points made earlier concerning the use of government programs to both expand output and diversify the resource base of a region.

Canadian Hog Exports to the United States

Background and Outcome. The National Pork Producers Council (NPPC) of the United States, on November 2, 1984, brought a formal complaint to the ITC against the Canadian exportation of live hogs, pork, and processed pork products to the United States. The NPPC contested that imports from Canada were being subsidized because of production subsidies given producers in Canada by both the provincial and federal governments. The NPPC contended that these subsidies caused injury to the U. S. industry because it expanded Canadian exports to the United States and also caused prices to drop below a competitive equilibrium situation. The complaint arose largely because of the increase in Canadian exports to the United States. For example, from 1974 to 1984, pork exports to the United States increased from 85,850,000 pounds to 573,630,000 pounds.

The formal complaint by the NPPC to the ITC was referred to the DOC, which ruled that hog exports to the United States were subsidized through production subsidies. In December, the ITC reviewed these findings and, in the preliminary determination, ruled that U. S. producers and processors were injured as a result of the subsidies. In June of 1985, the ITC made its final ruling. They ruled that only Canadian live hog exports were subsidized and it was this subsidy which resulted in injury to U. S. producers. Exports of pork and pork products, while found to be subsidized, did not result in injury. As a result, tariffs were imposed on Canadian live hog exports to the United States.

The first question which arose in this litigation was whether or not the subsidies in Canada encourage production sufficiently to cause significant damage to the U. S. hog industry. If a program is subsidized by government, it could be viewed as an unfair trade practice but only the government portion of the program should be viewed as an unfair benefit. Also, the question remains as to whether or not government contributions to an industry cause significant harm to the importing nation and, if so, what size of duty is required to restore prices. Whether exports should be countervailable depend on the extent to which the increased production due to subsidies causes a significant price decline. The extent to which a government program results in an increase in production depends on whether a shift, a movement, or no response to the supply curve occurred in the exporting country. In considering the case for countervailing duties, simply establishing that a subsidy exists is not sufficient reason for proceeding with a duty. It needs to be shown that the subsidy is the type which results in either a shift or a movement of the supply curve. Once it is established that a shift or movement has occurred, it also has to be determined whether or not the increased production resulted in a significant price decline in the importing country. To do this, the price elasticity of the importing demand curve and the exporting supply curve need to be considered. The more price elastic the importer's demand curve, or the more inelastic the exporter's supply curve, the smaller will be the price decline as a result of the subsidy. Also, in this particular case, it was necessary to determine whether or not the subsidy on production of the primary product was passed through the processors in the exporting country. As Schmitz and Sigurdson show, the processor in the exporting country receives no pass-through benefit from the production subsidy since the margin remains unchanged. Since there is no subsidy passed through to the processors in the exporting country, no countervail duty should be applied.

It is necessary in this case to examine the international and U. S. trade laws on countervailing duties. The primary objective of GATT in the area of countervail law was to provide a mechanism by which member countries could establish a duty to offset the impact of imports which were the result of subsidies. The intention of GATT appears not to have been one of dictation to member countries on how to transfer funds between sectors in their country; rather, its intention was to ensure that these transfers do not injure individuals in other countries. This injury would be as a result of imports which were a direct result of subsidies in the exporting country.

Article 6.B.1. of the GATT deals with countervailing duties. Of particular interest are those sections which define injury and the size of a countervailing duty. In determining injury, the Agreement states that a determination of injury shall involve an objective examination of both the volume of subsidized imports and their effect on prices and that both are important when investigating the market for like products. Within this statement there are two key principles. The first is that, in determining injury, a country shall concern itself only with subsidized imports. The article does not directly state that a country should address only the incremental increase of imports which is the result of the subsidy. It is inferred, however, that this is what is meant. Second, in determining injury, a country can only consider the effect it has on like products. Therefore, a subsidy on an input or raw product cannot be taken into account in determining the damage from the finished products. In determining injury from imports, the Agreement also states, in Article 6.4, that injury must be the result of the subsidized imports, and not because of other factors that at the same time are causing injury to the industry. This point is extremely crucial.

A further point of interest is the size of duty applied once a countervailing duty has been deemed as justified. The Agreement gives the authority of setting the duty to the importing country. The Agreement only mandates that the duty not be more than the amount of the subsidy.

There appear to be important differences between U. S. definitions and interpretations of trade laws, both written and in practice, and those laid out in Article 6 of the GATT (Grey; Langford; Rugman). The most important difference appears to be the linkage which is made between subsidized imports and injury. In determining injury, the U. S. trade law requires that the volume of imports be evaluated in determining the effect on price (Stowell). Unlike Article 6 of the GATT, there is no mention of subsidized imports. The United States requires that there only needs to be evidence of a subsidy, and the necessity of establishing the linkage between the subsidy and the increased imports is not clearly stated or practiced. The U. S. trade law is not explicit in stating that the injury to the industry must be the result of the imports and not other factors which existed. The law directs the ITC to examine the impact imports had on the effected industry by observing industry indices such as output, sales, market shares, profits, and other related indicators. The law does not direct the ITC to ensure that they separate the injury and consider only that which was the result of the subsidized imports. This is in sharp contrast to Canadian trade law where causality on the linkage between the subsidized imports and injury must clearly be established.

It is interesting that in the final ruling by the ITC only three of the five commissioners ruled on the case. Two of the five commissioners declined because of conflict of interest. The final ruling, as stated earlier, placed duties on live hog exports to the United States but products remained duty free. It was also interesting that one of the commissioners argued that tariffs should not be placed on either live hogs or pork products, one commissioner contended that only duties be placed on live hogs, and the other commissioner ruled that duties should be placed on both live hogs and products. It appears that the commission took the average of each commissioner's finding and thus concluded to impose a tariff only on live hogs. From this, one can conclude that there was no strong evidence that subsidization programs in Canada were causing injury to U. S. pork producers.

The ITC commissioners, according to procedures set down by the U. S. trade law, could not rule on the subsidy issue; hence, they had to take the DOC subsidy ruling as given. The task of the ITC commissioners was to determine whether or not these subsidies resulted in injury to U. S. pork producers and related sectors. The ITC, therefore, ruled that subsidies on live swine caused injury to the U. S. industry and the subsidy on processed pork did not cause injury. The case under investigation seems to have posed a conflict between law and economics. The legal definition of a product allowed the DOC to consider live hogs and products as one good; therefore, it only had to show that a subsidy existed which then, by definition, applied to both live hogs and products. The ITC, which considered them as separate products, had to take the DOC ruling that a subsidy existed on live hogs and processed pork and pork products even though no investigation was undertaken to see if this was true. If one used economics rather than law, the conclusion likely would have been that there was not a subsidy on processed pork. The ITC, by default, ruled correctly because it felt that processed pork exports did not cause injury to U. S. producers.

In 1989, once again, the pork case drew international attention. In the ITC final ruling, duties were imposed on both live hogs and products and the duty on live hogs was higher than previously. Clearly, trade irritants between the two countries have not gone away.

Nature of Programs. From a Canadian perspective, additional points are of interest. The ITC ruling that Canadian stabilization programs conferred a subsidy on producers certainly appears to have been consistent with the economic analysis. However, the ITC ruling that these subsidies also injured the U. S. industry was not consistent with economic arguments. First, the U. S. hog industry was experiencing difficult times as evidenced by a decline in hog production and low to negative profits at the farm level. Second, the ITC concluded that there is a relationship between U. S. and Canadian prices and prices are related to pork production. Third, they found that Canadian exports to the United States and their share of the market have been increasing for the period examined. These factors, together with the fact that Canadian producers were being subsidized, led the ITC to conclude that Canadian exports of hogs was the cause of the decrease in hog production and profits in the U. S. industry. In their ruling, the ITC did not examine the effect the stabilization subsidy payment had on Canadian production and hence the effect on exports. Their ruling appears to have lacked economic content.

To highlight the economics of the case, one point is stressed. There is a strong argument as to why the size of the subsidy, measured ex post, may overstate the subsidy in Canada. Producers do not know for certain at the time they join the program if they will receive a subsidy. Stabilization schemes are viewed by producers as income insurance schemes just as is the case, for example, with crop insurance. Additional factors suggest that the actual measurement and amount of an export subsidy or an ex post subsidy should be discounted heavily in order to determine the true ex ante subsidy. The analysis by Schmitz and Sigurdson suggests that Canadian production and exports of hogs due to subsidies had not caused injury to the U. S. hog and pork sectors. This supports the hypothesis that producers do not respond the same way to subsidies granted through stabilization programs as they do to a direct price increase through some form of target price mechanism. In addition, the ITC, upon determining that the Canadian hog program provided a subsidy to Canadian producers, set the duty at the level of the Canadian subsidy as calculated by U. S. authorities. This ruling, though perhaps consistent with the U. S. trade law, raises questions. Depending on the type of subsidy, the duty necessary to restore prices is generally less than the actual subsidy. In the case of a subsidy given after the time when production decisions are made (like the Canadian programs), no duty is required as a subsidy does not result in a production response.

What would have been the impact on this case had the U. S.-Canada Free Trade Agreement been in place? As Schmitz and Sigurdson argued, it is questionable whether the outcome would have been any different. The agreement in Article 1902 gives each country the right to apply its own countervailing duty law as has been done in the past. Thus, the United States could still have applied its definition of injury as set out by law in past cases. If the free trade agreement was to have an influence on the ruling, it would have to be because of Article 1904 which allows a country to appeal the final determination to a binational panel review rather than to a judicial review. The binational panel would consist of two members from each country plus an equally acceptable fifth member. The purpose of this review panel is to allow impartial reviews of the final determination and to remove the influence of political pressures. In the hog case, an impartial panel may have ruled that the size of the two countries is such that Canada's exports could have no influence on the U. S. industry.

Potash

Saskatchewan is a major producer of potash, and a large percentage of its production goes to the U. S. market. For several years prior to 1987, the price of potash sold in the U. S. market was insufficient to cover total costs of operating the Canadian potash industry. In this case, because prices were below those necessary to cover costs of production, the Canadian industry was vulnerable to antidumping charges on the part of the United States. On February 10, 1987, two American firms filed a lawsuit against several Canadian producers for dumping potash on the U. S. market at prices which, they argued, were at least 40 percent below the cost of production. The two firms were Lundberg Industries of Dallas, Texas, and New Mexico Potash Corporation of Memphis, Tennessee. The two U. S. firms at that time accounted for roughly 15 percent of the potash consumed in the United States. The suit was filed through the DOC and the ITC. To convict a country for dumping a product in the United States, one of three actions had to be proven: (1) the Canadian potash price in the United States was less than the price sold for in Canada, (2) the Canadian potash price in the offshore market was less than in the U. S. market, and (3) the Canadian potash was sold at less than the cost of production. The United States perceived that Canadian producers were price cutting to increase their market share in the United States. The Canadian market share increased from 76.7 percent in 1984 to 82.6 percent in 1986. The ITC agreed on April 3, 1987, that there was indication of unfair price discrimination. On August 21, 1987, the DOC announced preliminary duties on potash. The margin percentages were set at different levels for the different potash firms. This was in large part because the costs of production vary by firm. The duties ranged from $9.14 per tonne for IMC to $85.2 per tonne for Noranda.

The provincial government of Saskatchewan responded by limiting potash production to increase the world price of potash. If the Saskatchewan producers exceeded their production limits, they faced a fine of $1 million. The Potash Corporation of Saskatchewan also announced in September, 1987, that it was increasing its price by 60 percent to $93 (U. S.) per tonne from $58 (U. S.) per tonne. In December of 1987, the Saskatchewan producers agreed to stop shipping potash to the United States at prices that undercut U. S. prices.

As a result, on January 8, 1988, the Canada and U. S. dispute case was suspended for five years by an agreement between Saskatchewan and New Mexico producers. The duties on potash are to be removed and the agreement suspended until 1993 as long as, within the five-year period, Saskatchewan firms do not sell their product below the fair market value. There is no actual minimum price set in the agreement. The agreement defined the minimum price for each producer as a price equal to the current fair value of the potash minus 15 percent of its preliminary dumping margin. The fair value for a producer is (1) the constructed value of its potash if it is selling below cost of production, (2) its Canadian price if it has 5 percent or more of total sales in Canada or above its cost of production, or (3) its offshore price if it is selling at or above cost of production and has insufficient sales in Canada. On January 25, 1988, the Potash Corporation of Saskatchewan announced that it would decrease its price by 15 percent, according to the fair market value definition.

Corn

In the mid-1980s, the Ontario corn growers brought a lawsuit against American producers for dumping corn into the Ontario market. This was the first such litigation ever brought against a large agricultural trade exporting sector such as U. S. corn. It essentially challenged the U. S. Farm Bill. Under the Bill, the loan rate was lowered to drop world prices below the minimum previously set. In addition, the Export Enhancement Program added additional export subsidies in world markets.

This case was heard by the Canadian Import Tribunal. In the final ruling, the Canadians imposed high duties on U. S. corn imports exceeding $1.00 per bushel. However, in a later review, the duty was dropped to roughly 50 percent below the initial ruling. In light of this particular case, it is interesting that similar lawsuits have not been brought against the United States for agricultural commodity dumping in world markets. However, this case apparently has been used by the EC, for example, only to suggest that the United States is not pure when it comes to the area of agricultural subsidies.[2]

An Overview

These cases are extremely important to review since they highlight several important features of litigation in agricultural trade matters.

1. They generally involve subsidies. In the pork and potash cases, subsidies were involved. The pork stabilization programs entailed a government transfer to producers while in the potash cases, at least for the Potash Corporation of Saskatchewan, transfers were made from the

provincial government in order to keep the potash industry afloat. In the corn case, the subsidies were to U. S. producers from the U. S. government.
2. Subsidies are generally measured using accounting rather than economic principles.
3. Subsidies, by themselves, do not cause legal actions in agricultural trade areas. Someone has to be the irritant in order to persuade governments to act. In other words, someone has to be suing someone, and generally the one who is doing the suing initially is not the government. For example, in the hog case, the U. S. hog producers, through their respective organizations sued Canada; in the potash case, two private potash firms sued the Canadian industry; and in the corn case, the Ontario corn producers sued the United States. Producer groups filed these particular lawsuits through government agencies, but the government agencies by themselves were not instigators.[3]
4. The question which remains is, why weren't these particular cases handled directly through GATT? It appears that, if GATT is a true legal body, these lawsuits could have been handled through that mechanism. However, as indicated earlier, serious questions can be raised concerning the effectiveness of GATT as a judicial body. In addition, it may well have been that in all of these particular cases the litigants felt that they could do much better by suing the other country through their own governments than they could by going through the GATT process.
5. What is of interest in all of these cases is that they involve producers who are competing against goods being shipped in from abroad. In terms of an export sector, which is being injured because of unfair trade practices on the part of importers, there is really no legal court, except GATT, to handle these particular cases. For example, if an exporter in country A is being treated unfairly by an importer in country B, then country A cannot hear the case in its own country nor could it hear the case in the country where the importers reside. This observation becomes apparent since those cases reviewed earlier (those where the injury is done to an exporter) are handled, if at all, in the GATT process.
6. Other studies have been done on the impact of freer trade in minerals other than potash. In the uranium case, for example, the results for

free trade are positive, and attempts by the United States to restrict uranium imports have generally failed.
7. Canadian beef exports to the United States have also been under scrutiny by U. S. producers, but action by the United States has not been taken even in view of Canadian beef stabilization programs.
8. Transportation subsidies through the Western Grain Transportation Act have also been an issue in the free trade agreement and will remain a point of contention.
9. Institutions have been challenged. For example, the U. S. Wheat Associates have contended that the Canadian Wheat Board gives Canada an unfair advantage in exporting grain from Canada to both the United States and abroad.
10. There is a great deal of arbitrariness in deciding the outcomes of the various cases directly involving the United States and Canada and cases where GATT is directly involved.
11. Agricultural trade between the United States and Canada is important; and freer trade is in the best interest for, at least, western agriculture. However, since grain exports are of major importance, the role of GATT is important as a forum for freer trade. This is because the major subsidy issues involve countries other than the United States and Canada as importers. The United States and Canada compete largely in third markets. Japan, for example, is a large wheat importer, importing from both the United States and Canada. It is also a significant importer of canola from Canada and soybeans from the United States. The impact of freer trade in all markets around the world is much larger than for freer trade between the United States and Canada, even though trade in grains and oil crops between the United States and Canada would increase.
12. Interestingly, in the Canada-U. S. trade disputes, processed products have faired better than those of an unprocessed nature.
13. This report has not delved into the question of nontariff barriers and their impact on diversification strategies. As shown in *Canada's Agriculture in a World Trade Context* (Schmitz), nontariff barriers in agricultural trade are generally far greater than tariff barriers. Nontariff barriers may even increase as a substitute for reduced tariff protection. In this regard, it is important to avoid these impediments to trade by adopting standard labeling practices and the like, as used in the United States. For example, a barrier to trade in beef was overcome by

modifying the rib cut. Small changes in labeling, packing, etc., can greatly reduce nontariff barriers to trade.

Guidelines

There are several important features in the above discussion that provide a framework for allocating money from the federal government to promote diversification in the Prairie region.

1. In spite of the free trade agreement between Canada and the United States, it is highly unlikely that, at least in the short term, a binational tribunal will be established to handle trade disputes between the two countries. As a result, the near future will witness more legal cases; and they likely will be handled in a similar manner to those which have existed in the past. As illustrated, the irritant is usually a producer or producer group in the importing country. Given this fact, money should be allocated into areas which do not provide subsidies to produce a commodity that is in direct competition with producers in the actual commodity importing country.
2. From the results, one can conclude that the majority of cases are against the use of production subsidies. Interestingly, cases do not arise because governments use money to promote, through advertising or other means, a product in export markets. For example, if a firm, through the assistance of government, creates a new product in an export market, then clearly the importing country has no basis for filing a lawsuit. By definition, if a new product is introduced, then there are no domestic producers who are competing against this particular product line. Also, even if one can shift the demand for a product by somehow differentiating it from other products abroad—for example, if the Canadian pork industry, through advertising, can differentiate Canadian pork from that produced in America and this perception is firmly in mind by the consuming public—then this type of promotion does not constitute the basis for legal action. As a result, money spent, either for generating new products or expanding the market demand for existing products by modifying them through promoting product differentiation, is legitimate grounds for allocating money for the purposes of diversification. Therefore, product classification is

important, and a distinction should be made between existing products and new product lines.
3. The distinction between U. S.-Canada trade and Canada's trade to non-U. S. markets is very critical in designing long-term strategies. From the record, it is important that the Canadians avoid competing in the U. S. market through the use of production subsidies and/or through dumping. Part of the emphasis should be on non-U. S. markets. It is interesting, for example, that the hog industry in Japan has not brought countervailing duty action against the exportation of Canadian pork to that market. It was the United States that responded to the stabilization programs in Canada, not the Japanese. As a result, if Canada wants to proceed further with traditional products that are exported to both the United States and to areas such as Japan, it may want to pursue some form of differentiated pricing—perhaps along the lines of the U. S. Export Enhancement Program.
4. Because of the nature of the litigation, it is important that Canada continue to expand its exports into commodity lines where little competition exists. Even if these commodities were produced with various types of government assistance, at least initially, the Canadian industry is unlikely to receive court actions on the part of importers. For example, the Japanese are unlikely to bring a lawsuit against Canada if brine shrimp were to be exported to Japan from Canada where the Canadian industry received government support through Western Diversification initiatives.
5. It is important to keep in mind that the past cases involving agricultural litigation were brought about on an ad hoc basis and ruled on with arbitrary procedures and standards. As a result, in making decisions concerning the allocation of money, one has to essentially predict what the consequences will be in terms of legal actions if grants are given to increase production of existing commodities exported to the various markets including the United States. Unfortunately, however, there are few, if any, guidelines upon which to base one's predictions. For example, in the U. S.-Canadian hog dispute, there was no justification for the imposition of duties in the first place. Canadian stabilization programs did not cause injury to U. S. producers. Output can increase, but there can be little or no injury because of the small market share Canada realizes. Also, the duty was placed on live hogs but not products—a ruling which was unpredictable. This suggests that further processing of Canadian hogs may be possible without being

subjected to countervailing duty law. (However, in 1989, the United States is also seeking for duties on processed parts.) Therefore, in terms of past experiences, one avenue to pursue further is the entire area of red meat processing. Here, it appears likely that further expansion is possible with government assistance without being subject to countervailing duties. Certainly, if one uses the pork case as a precedent, then Canadians would be exempt from this endeavor.

6. In terms of No. 5 above, it is important to stress that value-added activities in Canada, even backed with government support, are likely to find a better chance of entering the United States duty free than are primary product exports which are encouraged through government assistance in Canada. However, this may not be true for trade with Japan where, for example, canola oil from Canada is met with tariffs while the raw product enters relatively duty free. The degree of protection depends, in part, on the lobbying efforts of the special interest groups being affected. For example, crushers in Japan are very well organized (which is not the case for U. S. pork slaughtering and processing firms).

7. In terms of new product lines, there are many areas which are unlikely to receive a great deal of scrutiny by the United States. For example, there are many areas in the manufacturing of specialty bakery products (e.g., wheat crackers) and the like, which would receive little attention in the U. S. market in terms of countervailing duty action. The development of niche markets (generally, in new product areas) will likely receive very little attention from the United States. Items such as pork, potash, corn, and the like receive attention because they are major products.

8. A good example of one direction which is not countervailable is the Western Diversification's granting of money to the potash industry for developing export markets in China. In this particular case, because the money is spent for export development, countervailable duty action will not be forthcoming. Also, because of the nature of the country, actions are essentially not forthcoming regardless of the nature of the source of export expansion. In this particular case, potash is not a new product; therefore, demand expansion essentially comes by promotion of the product to higher use levels. This, again, is a demand expansion activity promoted through the federal government and does not entail production subsidies in producing Canadian potash.

Diversification Strategies 35

9. In terms of No. 8 above, another area is the promotion and development of export of red meat to the Pacific Rim. For example, even though the Japanese import demand for red meat is growing, Canada's export share to that market has fallen. Using money for export market development promotion does not provide the basis for border dispute litigation.
10. Providing assistance for basic research and development (e.g., biotechnology) and applying this to new product lines also is generally not regarded as a production subsidy. This is evident from the calculation of the many producer subsidy measures. This line of assistance should be encouraged.

Notes

[1] A large part of this material was obtained from Hudec.

[2] Based on my involvement in all three of the cases discussed (hogs, potash, and corn), politics plays an important role. It is my opinion that persuasive economic arguments were down played in reaching the final rulings.

[3] In countervailing duty cases, if country A sues B, the level of subsidies in country A does not enter as part of the litigation. The economic arguments for not having them included are unclear.

References

Grey, R. C. *United States Trade Policy Legislation: A Canadian View*. Montreal, Quebec: Institute for Research on Public Policy, 1982.

Horlich, Gary N., Judith Heppler Bello, and James J. R. Talbot. "A Manual of U. S. Trade Laws" *International Business Lawyer* (June 1985):249.

Hudec, R. E. "Legal Issues in US-EC Trade Policy: GATT Litigation 1960-1985." Chapter 2, pp. 17-58: In *Issues in US-EC Trade Relations*, edited by R. E. Baldwin, C. B. Hamilton, and A. Sapir. Chicago, Illinois: University of Chicago Press, 1988.

Jackson, John H. "Achieving a Balance in International Trade." *International Business Lawyer* (April 1986):123.

Just, R. E., D. Hueth, and A. Schmitz. *Applied Welfare Economies and Public Policy.* Englewood Cliffs, New Jersey: Prentice-Hall, Inc., 1982.

Langford, J. R. "American Investigation Involving Temporal Perishable Agricultural Commodities." Winnipeg, Canada, 1985. Mimeographed.

Rugman, A. M. "United States Protectionism and Canadian Trade Law." *Journal of World Trade Law* (July-August 1986):363-380.

Schmitz, A. "Canada's Agriculture in a World Trade Context." Paper prepared for Agriculture Canada, Regina, Saskatchewan, 1984.

Schmitz, Andrew, and Dale Sigurdson. *Stabilization Programs and Countervailing Duties: Canadian Hog Exports to the United States.* Calgary, Alberta: University of Calgary Press, forthcoming.

Schmitz, Andrew, and J. Vercammen. "PSE's and the Efficiency of Farm Programs." University of California, Berkeley. September, 1989. Mimeographed.

Steger, Debra, P. "The Impact of U. S. Trade Laws on Canadian Economic Policies." *Policy Harmonization: The Effects of a Canadian-American Free Trade Area.* Toronto, Ontario: C. D. Howe Institute, 1986.

Stowell, A. M. *U. S. International Trade Laws.* Washington, D. C.: Bureau of National Affairs, Inc., 1987.

U. S. International Trade Commission. *Annual Report.* Washington, D. C.: U. S. International Trade Commission, various issues.

Warley, T. K. "Implications for Canadian Agricultural Policy." Paper presented at the Conference on U. S.-Canada Trade in Agriculture: Managing the Disputes. University of Guelph, Guelph, Ontario, April, 1987.

3

Diversification of Prairie Agriculture

William A. Kerr

Introduction

The economy of the Canadian Prairies is characterized by resource-based production which is surplus to regional requirements. As a result, prosperity is dependent to a considerable degree upon the ability to secure outlets for production and the prices received in external markets. These markets can be either international or in other regions of Canada. Given that resource-based production, whether agricultural or nonagricultural, has large linkages to both the manufacturing and service sectors located in the Prairies, changes in the external trading environment will generate significant "ripple effects." When prices are rising and trade barriers are receding, prosperity will increase. On the other hand, falling prices and protectionist trends create economic hardship. Over the long term, both prices and protectionism have exhibited sufficient fluctuation to characterize the Prairie region as a "boom and bust" economy. In the agricultural sector these external forces may be exacerbated by the vicissitudes of weather. While the effects of such fluctuations are largely borne by the residents of the region, the federal government bears considerable responsibility for mitigating the impact of economic downturns. Of course, it also benefits from additional revenues in times of economic expansion. Such fluctuations, however, lead to wasted investment and wasted human resources. Hence, all segments of Canadian society have an interest in the economic performance of the Prairie region. While calls for reform are seldom heard during periods of prosperity, in times of economic distress the search for alternatives to the existing system will be intensified. This chapter identifies alternatives which may arise from changes originating in the international trading environment for food and agricultural products.

An economy which is dependent on international markets will always be less secure than one which largely serves a domestic market. The ability of government to influence the international trading system will always be less than its ability to influence the performance of the domestic economy. The means available to a country to influence the international environment within which it trades are a function of its relative economic power, its ability to form cooperative alliances with other nations, and the degree to which it can successfully apply moral suasion to induce trading partners to abide by existing and agreed-to rules of international trade. The ability to influence the course of international events has two facets. First, securing access to external markets for products surplus to domestic requirements is a function of the degree to which protectionist forces in foreign markets can be countered. This has aspects pertaining to both the absolute quantities of products moving to foreign markets and the value added which they internalize. Second, little control can be exercised over prices received in external markets and, hence, it is difficult to counter the undesirable consequences arising from being a "price taker."

With respect to the long-term economic performance of Prairie agriculture, five options appear available: (1) no change to the current policy regime, (2) improvement and expansion of income stabilization programs, (3) reduced interaction with foreign markets, (4) broad-based diversification out of agriculture, and (5) diversification within agriculture. The current investigation will concentrate on the latter. This seems the logical point of departure. There are few, if any, possible changes in current policy that will reduce the historical dependence of the agricultural sector upon world markets unless there are identifiable changes in the international trading environment which would result in significant diversification within the agricultural sector itself. Of course, it is always possible to commit additional fiscal resources to income stabilization. Further, if major policy-induced opportunities for this diversification within the agricultural sector exist, then such diversification is likely to be accomplished with less resources or economic costs than the remaining two alternatives. Diversification out of agriculture will mean major commitments of resources, both in terms of productive investment and social policies, to aid in the transition. Reduced interaction with foreign markets without diversification out of agriculture would mean a major rationalization of the Prairie economy, wasted resources, and considerable population effects. While these issues will not be addressed directly, the prospects for diversification within agriculture will determine whether a choice will have to be made between long-term continuation of the "boom and bust" cycle, or increased budgetary expenditure on income stabilization, and serious consideration of either diversification out of agriculture or reduced foreign market interaction.

The chapter begins with a definition of diversification which has an economic interpretation and which is empirically measurable. This will provide a means by which the impacts of changes in the trading environment upon diversification can be assessed. A discussion of the economic ramifications of trade liberalization or increased protectionism will then be presented. An historical base case which allows comparison with various alternatives involving changes to the international environment is developed at this point. This base case is also used to identify the direction in which the agricultural economy should evolve if diversification is to be enhanced. Thereafter, specific alternatives regarding changes to the trading environment are developed and compared to the base case to determine if they will lead to increased diversification. The specific cases investigated are: (1) the effect of the U. S.-Canada Free Trade Agreement (FTA), (2) the effect of a significant reduction in agricultural trade barriers arising from the current round of General Agreement on Tariffs and Trade (GATT) talks, and (3) the effect of a failure at the GATT resulting in increased European Community (EC) protectionism.

Diversification: A Definition

It is a common perception that the degree of diversification is a function of the number of products produced by a region. The larger the number of products the more diversified the economy. While this is one possible definition of diversification, it is not an operational one and its unidimensional character may actually be misleading. In a naive way, this definition of diversification relies on the "law of large numbers" in a global general equilibrium framework. The underlying assumption is that, by producing a large number of products, the probability of having positively correlated intermarket variations in prices is reduced. In some sense then, diversification is defined by its objective reduction in the regional variability of income. Positively correlated intermarket variations in prices increase the variability of regional income whereas negatively correlated variations will reduce it. For example, assume that a region produces two products. If the prices of the two products move together (i.e., when the price of the first good rises, the price of the other good also rises and, when the price of the first falls, the price of the second also falls) then the entire economy's income increases or decreases at the same time. On the other hand, if prices move in an offsetting manner (i.e., when the price of the first good rises, the price of the second tends to fall), then the regional income will exhibit more stability. Of course, some gains can be made from movements within the positively correlated range. If the mix of outputs moves from being concentrated in highly and positively correlated commodities to those with positive but lower degrees of correlation, then variation will be somewhat reduced.

When the range of products produced in a region is limited (an absence of large numbers), adding to the number of products may increase the variability of regional returns. This would happen if the price variations for the additional products were positively correlated with the price variations of the existing range of goods produced and if the variance in the price of new products were larger than the variance of the prices of existing products. Hence, the notion that the production of a wider variety of goods is desirable because it will reduce the variation in regional income cannot be sustained. As a result, this definition of diversification would appear questionable. If more goods are not unequivocally better, then such a definition has no economic interpretation. Clearly, in terms of reducing the variability of income, a region might well be better off with a smaller number of negatively correlated outputs. Of course, the variability of regional income will also be affected by the share of income arising from negatively correlated outputs.

For the purpose of this chapter, agricultural diversification will be defined by its objective reduction in the variability of income. Hence, a region will be considered more diversified if, as a result of a change in the economic constraints (in this case, the international trading environment), the variability of income is reduced. This also provides an operational definition for quantitative comparison.[1]

Changes in the international trading environment will, through price adjustments and subsequent resource responses, alter the share of total output contributed by any individual activity. Hence, it is possible to compare the expected variance of gross returns arising from a change in the international trading environment with those returns which existed prior to the change. The relative size of the variance would determine whether the change was diversification enhancing or diversification retarding. Of course, this relative measure abstracts from changes in levels of gross revenue which would arise from the change in the international trading environment. A change in the trading environment could lead to an increase in the level of gross returns and, at the same time, be diversification retarding.

The analysis developed here is based on gross revenues. While net revenue would be a more desired measure, it is not tractable or meaningful at the level of aggregation for which data are available. Net revenue must account for costs. However, given the well-known variation in cost structures among farms, no meaningful measure of net revenue can be developed. This is better left to analyses of individual representative farms where models can be tested for alternative cost configurations. The major loss arising from the use of the gross revenue approach is in the area of livestock production where feed grains are inputs to the production process. When livestock prices are increasing and feed prices are decreasing, profits are rising. The gross revenue approach does not take into account the interactive effects of the two markets on the welfare of the sector. Given this limitation, the use of the gross revenue approach can still provide considerable insights into the problems of diversification since output prices remain a major source of agricultural instability.

The effects of changes to the trading environment on diversification may not be the only facets of trade liberalization which are of interest. The effects on the level of income and the value added accruing to the region are also important. Examination of the effects of changes to the trading environment concentrate on long-run adjustments. Hence, information on income level is of limited use because any sustained increase tends to become capitalized into fixed resources. Thus, any result relating to the income level should be interpreted very carefully as the actual benefit to regional welfare may be overestimated. The relative measure of income level presented should be viewed in this light.

Of more interest may be the change in the composition of output arising from new trading environments. Protectionist measures are often designed so as to maximize the value added accruing to the importing country. Trade liberalization will tend to reverse this process. An increase in value added will increase the impact of the sector on the regional economy. At the same time, however, adding to the value added may or may not reduce the variation in income. Again, this depends on the price correlations and the covariance terms.

The quantitative measure of diversification suggested may also provide considerable policy insights. For example, those combinations of activities whose expansion will contribute most to reducing the variance in gross revenues can be identified. Then, if the objective is indeed to reduce the variance of gross returns, expansion of these industries could be encouraged through policy initiatives. In a similar vein, commodities with a positive covariance could be discouraged or, at least, exempted from policy measures. Further, those commodities with volatile prices and a positive covariance coefficient might be targeted for coordinated price stabilization policies to aid in reducing fluctuations in regional revenues.

The Effect of Changes to the Trading Environment on Diversification

Changes to the international trading environment can take many forms. They can, however, be loosely divided into two categories: those measures which alter the price at which goods cross the border and those measures which directly affect the quantity of commodity traded. Tariffs, variable levies, export subsidies, transportation subsidies, and input subsidies are examples of the former. Import quotas, licensing requirements, voluntary export restraints, and most nontariff barriers (such as health regulations, inspection procedures, and consumer-protection legislation) affect the quantities traded directly.

The effect of changes to price-distorting policies can be illustrated by the removal of a tariff. The removal of a tariff will lead to changes in the relative prices of tradable goods and, through supply responses, alter the regional output mix. This can be illustrated for the case of U. S.-Canada trade by Figure 3.1.

Figure 3.1(a) illustrates the case where Canada is an exporter of the commodity. Canadian demand and supply curves are depicted as DC and SC. At any price above where DC = SC (i.e., P1), Canada will have product available for export. For example, at PC the export supply equals QS - QD. Assuming the "small exporting country" model, Canada can sell

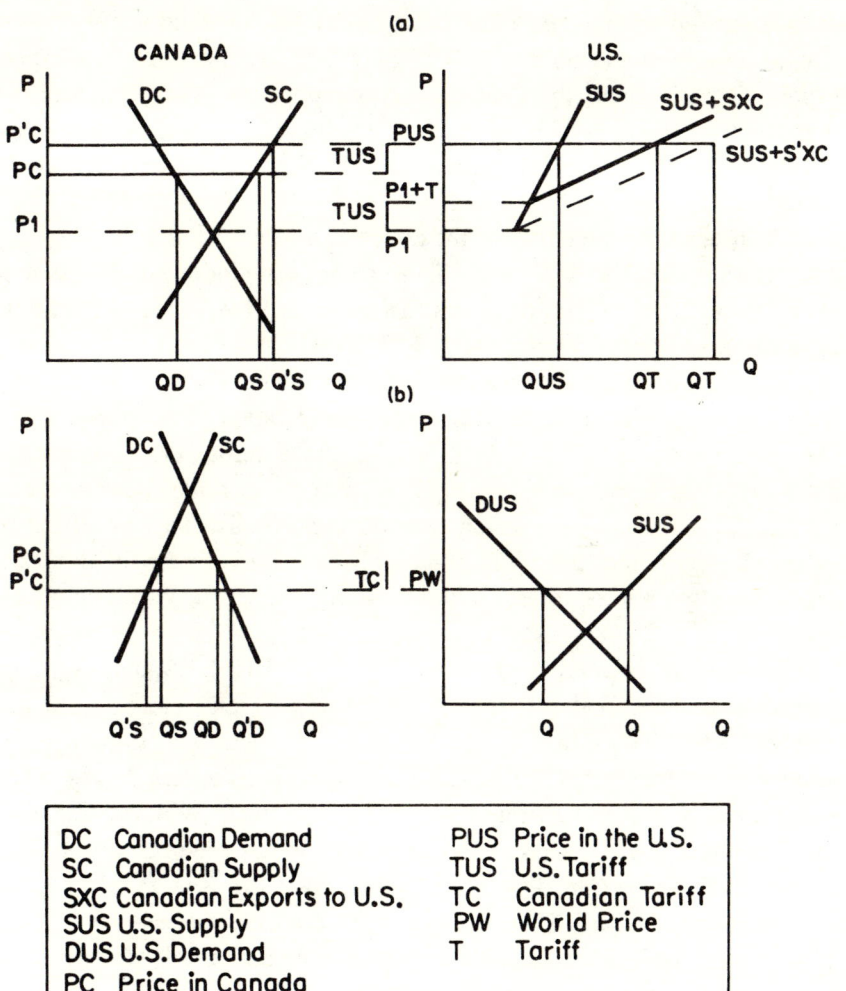

FIGURE 3.1. Tariff Removal

DC Canadian Demand
SC Canadian Supply
SXC Canadian Exports to U.S.
SUS U.S. Supply
DUS U.S. Demand
PC Price in Canada
PUS Price in the U.S.
TUS U.S. Tariff
TC Canadian Tariff
PW World Price
T   Tariff

all that it wishes at the U. S. import price, PUS. Changes in the quantity of Canadian exports will not affect PUS.

With the tariff (TUS) in place, supplies of Canadian product will become available in the United States at P1 + T. Hence, Canadian supplies must be added to U. S. supplies at any price so that total supply in the United States equals SUS + SXC where SXC is the Canadian export supply. To determine the Canadian export supply at any U. S. price, one must subtract the unit value of the tariff from the U. S. price to determine the Canadian price. For example, at price PUS, the Canadian price would be PC.

The removal of the tariff has the effect of shifting the total U. S. supply from SUS + SXC to SUS + S'XC as Canadian product now becomes available at P1 in the United States instead of at P1 + T. The effect is to raise the price in Canada from PC to P'C. Total change in Canadian output is Q'S - QS. This represents the increase in the contribution to total regional output of the product as a result of trade liberalization. The removal of a foreign export subsidy simply increases the external price, causing a movement along the domestic supply curve. The removal of an input subsidy in a market into which Canada exports also has the effect of increasing the border price. The share of the commodity in the domestic production mix will be reflected in a movement along the domestic supply curve.

The case of imports is presented in Figure 3.1(b). Again, domestic Canadian demand and supply are represented by DC and SC. At any external price below that where DC = SC, Canada will import. It is assumed that the United States is in a net export position at PW where PW is the internationally determined price of the product. With the Canadian tariff in place, the effective import price for Canada becomes PW + TC = PC where TC is the Canadian tariff. Again, given the usual assumptions of the small country case, Canada can purchase all that it wants at PW. In other words, the level of Canadian imports has no discernible effect on PW.

At PC, Canada is willing to import QD - QS. If the tariff is removed, the price in Canada falls to P'C = PW and Canada imports Q'D - Q'S. The decrease in the contribution to regional output of this commodity becomes QS - Q'S as a result of trade liberalization. Of course, this discussion abstracts from transportation costs and the costs imposed by other border measures. The effects of trade liberalization are also partial equilibrium in nature, and general equilibrium aspects are ignored. In other words, the shifts in resources between inefficient and efficient industries, which one would expect from trade liberalization, have not been included. Modeling these general equilibrium aspects is normally perceived as intractable ex ante. To the extent that these adjustments are ignored, the analysis will provide biased estimates.

The information requirements for such cases appear quite manageable. Information is required on the size of the tariff reduction; the current quantity, QS; the current price level; and the domestic elasticity of supply for each commodity. In the case of the U. S.-Canada trade agreement, the tariff reductions are available from the supplementary tariff schedules of the FTA. In the optimistic and pessimistic GATT cases, the future trading environment is less clear and assumptions regarding future levels of trade restrictions must be made. Quantities of output and prices are generally available for tradables. Domestic supply elasticities are available for a number of commodities while, for others, some assumptions will have to be made regarding supply elasticities. These can be varied to determine the sensitivity of the variance in regional income to these assumptions. As the prices of tradables are assumed to be exogenous in both cases, removal of the tariffs will have no effect on the variability of prices. The expected supply responses are a result of the reduction in tariffs. This allows the before-and-after variance in regional gross revenues to be compared. Variations in quantities supplied in response to such short-run price changes remain. Hence, the measure developed does not account for lagged short-run supply responses to price variations. It should be pointed out that, as the actual variation in prices remains unchanged, the contribution of these short-run supply responses will also remain unchanged. They will simply take place around the level of output represented by the new total share of output. As almost all agricultural commodities exhibit a lag between changes in price and the supply response, the direct variation in annual returns attributable to exogenous price changes can be calculated at the new share level. Of course, the imposition of, or increases in, tariff levels will have the opposite effect. Changes in levels of quantitative restrictions, such as import quotas, can be treated in a similar fashion.

Income Variability in the Current Trading Environment

The case to be used as a basis of comparison of the impacts on gross income variability for the various trade alternatives developed in later sections is presented. The base case itself can provide considerable insights into the problems of income variability manifest in Prairie agriculture.

Commodities and Data

Based on their importance to the international trade of the Prairie region, 27 commodities or commodity groupings were selected for examination. In a few cases—such as fresh and processed fruit—where imports were significant but production in the Prairies was insignificant, the products were excluded. The analysis was done on a final product basis so inputs to other agricultural activities were not included. For example, forage is almost entirely consumed by domestic livestock. Therefore, to prevent double counting, it was not included as a separate category. Although not heavily traded, the major supply-management commodities—dairy, chicken, turkey, and eggs—were included because they do represent significant components of Prairie farm income. To ignore these commodities would have considerably underestimated the variability in regional income and reduced the policy information available. Finally, other minor nontradables were excluded largely due to problems of acquiring complete data series. Speculation regarding exports of new products was not undertaken as prices (and hence price variations) in such products do not exist. Other studies in the series examined such opportunities. Given these exclusions, the estimates of income variability will be biased to the extent that such excluded commodities would contribute to the variability. The list of commodities examined, however, does include the major part of Prairie agricultural production. The product divisions generally conform to the classifications of Canadian international trade statistics. This facilitates the matching of tariff rates to commodities for the various trade cases. A complete list of the commodities included is presented in Table 3.1.

The price data required to calculate the gross variance of income were collected for each of the selected commodities. In some cases the trade classifications represented aggregates of product groups for which only the prices for the individual components of the aggregate were available. In these cases a weighted average of the individual component prices was used

TABLE 3.1

Selected Commodities

Feeder cattle (F Cattle)	Rye
Slaughter hogs (S Hogs)	Wheat
High quality beef (HQ Beef)	Hard spring wheat flour (HSW Flour)
Manufacturing beef (LQ Beef)	Durum, semolina flour (DS Flour)
Pork	Fresh tomatoes (FR Tomato)
Processed pork (Proc Pork)	Other fresh vegetables (Fresh Veg)
Chicken	Processed vegetables (Proc Veg)
Turkey	Sugar beets (S Beets)
Dairy	Linseed oil-cake-meal (Proc Lin)
Eggs	Canola oil-cake-meal (Proc Cano)
Honey	Certified seed (Seeds)
Barley	Flaxseed
Oats	Canola
	Wool

in the calculation of income variance. The prices were collected on an annual basis for the years 1977 to 1986. In those years where prices were reported on a crop-year basis, the price was applied to the year that the crop was grown (e.g., the 1986-87 crop-year price was considered to be the 1986 price). All prices were converted to 1986 dollars using the Price Index for Gross Domestic Product (dollar figures in this chapter are Canadian dollars). This provides for the calculation of income variance with inflationary trends removed. The year 1986 was chosen so that the real dollar value of the price variation could be standardized to the quantity data. The latest year for which there was a complete set of production data available was 1986.

Where possible, all quantities were adjusted for further value added to prevent double counting. For example, as the analysis is conducted on a final product basis, actual production quantities of pork were adjusted at the appropriate rate to reflect that portion of production which was further processed. In a similar fashion, barley was adjusted to reflect that portion which was used for animal feed within the Prairie region, wheat for flour production, canola for the proportion crushed, etc. Hence, the variance of incomes calculated represented those of the agricultural sector rather than farm level incomes. This method would seem more appropriate to the study of changes to the trading environment as they did have effects on the degree of value added internalized in traded goods as well as on the mix of farm-level commodities produced.

The Variance of Income

Income variance is only important as a relative measure. Given the squared and cross-product terms, the absolute values tended to be very large. This was particularly true when large aggregates, such as Prairie agricultural production, were being used. What is important is the relative contribution of the elements of the variance-covariance matrix to the total variance. For simplicity, all values reported in the text were standardized to the 18th decimal (e.g., 9.356 would be 9.356E+18). Of course, positive covariances add to the total variance while negative covariances reduce it. What determines whether the covariance is positive or negative is the relationship between the two commodity prices. If the prices are positively correlated over time, the covariance will add to the total variance. The opposite is true when prices are negatively correlated. Hence, the price correlation matrix can provide considerable information regarding the likely ability of an economy to diversify. The price correlation matrix is reported in Table 3.2.

The values on the diagonal were all 1.000, reflecting the perfect correlation of own prices. The positively correlated prices outweighed, to a considerable degree, those which were negatively correlated. Leaving aside the diagonal, the total number of negatively correlated combinations was 115 while there were 236 positively correlated pairs. The negatively correlated prices were concentrated in a few commodities. Thus, a broadly based diversification strategy will be of only limited success. In addition, the only export commodities which were consistently negatively correlated

TABLE 3.2

Correlation Matrix of Prairie Region Prices, 1977-1986

	F Cattle	S Hogs	HQ Beef	LQ Beef	Pork	Proc Pork	Chicken
F Cattle	1.000						
S Hogs	0.383	1.000					
HQ Beef	0.891	0.394	1.000				
LQ Beef	0.889	0.210	0.967	1.000			
Pork	0.288	0.961	0.249	0.072	1.000		
Proc Pork	0.517	0.539	0.264	0.166	0.602	1.000	
Chicken	0.366	0.795	0.544	0.409	0.751	0.191	1.000
Turkey	0.526	0.877	0.649	0.497	0.826	0.409	0.954
Dairy	0.455	0.294	0.694	0.630	0.190	-0.119	0.745
Eggs	0.165	0.563	0.456	0.352	0.527	-0.005	0.893
Honey	0.687	0.767	0.811	0.695	0.687	0.359	0.786
Barley	0.662	0.169	0.885	0.881	0.073	-0.053	0.551
Oats	0.420	0.426	0.689	0.589	0.348	-0.061	0.710
Rye	0.683	0.393	0.860	0.839	0.331	0.044	0.717

(Continued on next page.)

TABLE 3.2—continued.

	Turkey	Dairy	Eggs	Honey	Barley	Oats	Rye
F Cattle							
S Hogs							
HQ Beef							
LQ Beef							
Pork							
Proc Pork							
Chicken							
Turkey	1.000						
Dairy	0.660	1.000					
Eggs	0.829	0.732	1.000				
Honey	0.891	0.583	0.689	1.000			
Barley	0.550	0.818	0.557	0.684	1.000		
Oats	0.656	0.800	0.643	0.728	0.821	1.000	
Rye	0.713	0.794	0.668	0.825	0.943	0.850	1.000

(Continued on next page.)

TABLE 3.2—continued.

	F Cattle	S Hogs	HQ Beef	LQ Beef	Pork	Proc Pork	Chicken
Wheat	0.691	0.534	0.902	0.835	0.393	0.035	0.768
HSW Flour	-0.267	-0.708	-0.176	-0.080	-0.704	-0.314	-0.709
DS Flour	-0.223	-0.678	-0.127	-0.048	-0.653	-0.155	-0.610
FR Tomato	-0.381	-0.476	-0.487	-0.369	-0.496	-0.399	-0.681
Fresh Veg	-0.143	-0.704	-0.215	-0.099	-0.733	-0.211	-0.505
Proc Veg	-0.653	-0.525	-0.446	-0.443	-0.612	-0.586	-0.526
S Beets	0.804	0.557	0.866	0.837	0.506	0.286	0.696
Proc Lins	0.556	0.662	0.741	0.676	0.599	0.110	0.896
Proc Cano	0.619	0.659	0.787	0.729	0.569	0.130	0.874
Seeds	0.593	0.632	0.428	0.353	0.700	0.725	0.488
Flaxseed	0.678	0.589	0.840	0.756	0.473	0.105	0.822
Canola	0.301	0.735	0.345	0.246	0.713	0.070	0.858
Wool	0.809	0.603	0.943	0.865	0.501	0.350	0.760

(Continued on next page.)

TABLE 3.2—continued.

	Turkey	Dairy	Eggs	Honey	Barley	Oats	Rye
Wheat	0.777	0.814	0.666	0.869	0.892	0.869	0.935
HSW Flour	-0.637	-0.390	-0.379	-0.355	-0.077	-0.295	-0.277
DS Flour	-0.531	-0.272	-0.244	-0.333	-0.025	-0.275	-0.242
FR Tomato	-0.721	-0.667	-0.744	-0.643	-0.489	-0.575	-0.571
Fresh Veg	-0.595	-0.006	-0.457	-0.704	-0.137	-0.333	-0.374
Proc Veg	-0.593	-0.296	-0.327	-0.591	-0.337	-0.301	-0.561
S Beets	0.760	0.603	0.542	0.875	0.823	0.704	0.930
Proc Lins	0.876	0.724	0.816	0.891	0.762	0.804	0.914
Proc Cano	0.865	0.734	0.784	0.896	0.765	0.792	0.908
Seeds	0.603	0.208	0.301	0.652	0.252	0.393	0.465
Flaxseed	0.813	0.862	0.687	0.858	0.822	0.888	0.905
Canola	0.767	0.599	0.632	0.629	0.343	0.605	0.579
Wool	0.837	0.737	0.643	0.904	0.862	0.783	0.908

(Continued on next page.)

TABLE 3.2—continued.

	Wheat	HSW Flour	DS Flour	FR Tomato	Fresh Veg	Proc Veg	S Beets
Wheat	1.000						
HSW Flour	-0.368	1.000					
DS Flour	-0.364	0.944	1.000				
FR Tomato	-0.500	0.280	0.054	1.000			
Fresh Veg	-0.386	0.182	0.285	0.267	1.000		
Proc Veg	-0.414	0.556	0.469	0.553	0.374	1.000	
S Beets	0.889	-0.387	-0.378	-0.454	-0.498	-0.688	1.000
Proc Lins	0.910	-0.501	-0.467	-0.592	-0.566	-0.597	0.895
Proc Cano	0.928	-0.505	-0.472	-0.589	-0.513	-0.602	0.892
Seeds	0.382	-0.452	-0.356	-0.653	-0.460	-0.884	0.581
Flaxseed	0.958	-0.499	-0.462	-0.641	-0.353	-0.536	0.838
Canola	0.630	-0.842	-0.847	-0.473	-0.477	-0.607	0.584
Wool	0.937	-0.377	-0.295	-0.615	-0.378	-0.559	0.927

(Continued on next page.)

TABLE 3.2—continued.

	Proc Lins	Proc Cano	Seeds	Flaxseed	Canola	Wool
Wheat						
HSW Flour						
DS Flour						
FR Tomato						
Fresh Veg						
Proc Veg						
S Beets						
Proc Lins	1.000					
Proc Cano	0.990	1.000				
Seeds	0.535	0.551	1.000			
Flaxseed	0.902	0.927	0.528	1.000		
Canola	0.767	0.758	0.524	0.756	1.000	
Wool	0.881	0.899	0.557	0.903	0.536	1.000

Sources: Statistics Canada, Cat. No.22-003, 22-007, 22-201, 23-001, 23-007, 23-202, 23-205, 23-210, 32-228, 65-202, and 65-203; Agriculture Canada, *Livestock Market Review*, 1977-1986.

were the relatively minor flour commodities. The other goods with consistently negative correlations were vegetables. As vegetables are generally imported and receive tariff protection, any move toward trade liberalization will lead to a contraction of output and a decrease in the contribution the commodity can make to the reduction in the variance in gross income.

The correlation coefficients can, however, cover up many short-term advantages. For example, while the correlation between feeder cattle and barley is .662 over the decade, between 1984 and 1986 they have moved in opposite directions. Feeder cattle prices were strengthening while barley prices fell. The strong feeder cattle prices reflect the opportunities in cattle feeding arising from poor barley prices. Of course, strong feeder cattle prices translate into increased incomes for cow-calf operations as feedlot operators bid up the price of feeders to take advantage of low barley prices. When such windows of opportunity arise, they can be capitalized upon by farmers and on-farm diversification can help reduce the variability of the operation's income. Other such opportunities may also present themselves.

Given that the major export commodities are all positively correlated and imported commodities are negatively correlated, the likelihood that significant reductions in variance will arise from trade liberalization seems remote. This is because changes in the mix of outputs arising from trade liberalization will only change the distribution among positively related prices rather than move the region into a mix of outputs which contain a greater proportion of negatively correlated prices. Given the existing pattern of price correlations, no opportunities for such diversification appear to be available. Still, this does not mean that the variance of income cannot be reduced considerably by expanding away from commodities which have a heavy weighting and which were highly positively correlated into those that were important but less positively correlated. These opportunities are explored below.

To facilitate this examination, the variance-covariance matrix for the base case is presented in Table 3.3. The values indicate the relative contribution of the various components to the variability of income. Those on the diagonal represent the variance arising from the movement of the prices of commodities directly. The other terms represent the interactive impacts of the 27 commodities.

TABLE 3.3
Variance/Covariance Matrix of Prairie Region Prices-Base Case

	Cattle	Swine	HQ Beef	LQ Beef	Pork	Proc Pork	Chicken
Cattle	8.821						
Swine	1.206	0.281					
HQ Beef	138.338	10.929	683.681				
LQ Beef	23.443	0.990	224.545	19.712			
Pork	16.056	9.551	121.872	5.995	87.866		
Proc Pork	3.128	0.582	14.040	1.500	11.492	1.037	
Chicken	3.533	1.369	46.228	5.903	22.883	0.632	2.643
Turkey	1.752	0.521	19.049	2.474	8.688	0.467	1.741
Dairy	2.693	0.310	36.120	5.570	3.545	-0.242	2.411
Eggs	0.914	0.557	22.228	2.913	9.211	-0.010	2.708
Honey	2.896	0.577	30.081	4.376	9.146	0.519	1.813
Barley	186.770	8.485	2,197.018	371.554	65.410	-5.128	85.099
Oats	20.477	3.708	295.948	42.919	53.550	-1.021	18.948
Rye	16.357	1.681	181.335	30.047	24.996	0.362	9.400

(Continued on next page.)

TABLE 3.3—continued.

	Turkey	Dairy	Eggs	Honey	Barley	Oats	Rye
Cattle							
Swine							
HQ Beef							
LQ Beef							
Pork							
Proc Pork							
Chicken							
Turkey	0.315						
Dairy	0.738	0.992					
Eggs	0.868	1.360	0.870				
Honey	0.710	0.824	0.912	0.504			
Barley	29.300	77.387	49.371	46.098	2,255.913		
Oats	6.043	13.079	9.841	8.486	639.903	67.385	
Rye	3.227	6.380	5.024	4.720	361.190	56.297	16.256

(Continued on next page.)

TABLE 3.3—continued.

	Cattle	Swine	HQ Beef	LQ Beef	Pork	Proc Pork	Chicken
Wheat	734.268	101.205	8,435.846	1,326.963	1,319.195	12.680	446.631
HSW Flour	-6.163	-2.912	-35.656	-2.759	-51.198	-2.479	-8.945
DS Flour	-0.162	-0.088	-0.810	-0.052	-1.495	-0.039	-0.242
FR Tomato	-0.018	-0.004	-0.200	-0.026	-0.073	-0.006	-0.017
Fresh Veg	-0.666	-0.585	-8.812	-0.691	-10.770	-0.337	-1.288
Proc Veg	-0.512	-0.073	-3.081	-0.519	-1.514	-0.158	-0.226
S Beets	14.963	1.850	141.887	23.303	29.729	1.826	7.094
Lin Oil	1.975	0.420	23.167	3.589	6.718	0.134	1.741
Rape Oil	27.346	5.193	306.222	48.188	79.321	1.963	21.138
Seeds	36.842	7.016	234.476	32.852	137.392	15.450	16.597
Flaxseed	42.452	6.578	463.090	70.747	93.481	2.250	28.157
Rapeseed	43.453	18.951	439.100	53.133	325.321	3.479	67.929
Wool	0.096	0.013	0.990	0.154	0.189	0.014	0.050

(Continued on next page.)

TABLE 3.3—continued.

	Turkey	Dairy	Eggs	Honey	Barley	Oats	Rye
Wheat	155.901	289.997	222.184	220.569	15,150.118	2,551.578	1,348.691
HSW Flour	-2.771	-3.016	-2.741	-1.952	-28.338	-18.808	-8.656
DS Flour	-0.073	-0.066	-0.055	-0.058	-0.295	-0.551	-0.238
FR Tomato	-0.006	-0.010	-0.011	-0.007	-0.364	-0.074	-0.036
Fresh Veg	-0.523	-0.009	-0.668	-0.784	-10.227	-4.290	-2.361
Proc Veg	-0.088	-0.078	-0.080	-0.111	-4.221	-0.653	-0.597
S Beets	2.673	3.763	3.170	3.891	244.959	36.237	23.510
Lin Oil	0.588	0.862	0.911	0.756	43.298	7.898	4.407
Rape Oil	7.217	10.875	10.880	9.458	540.426	96.707	54.447
Seeds	7.088	4.339	5.873	9.693	250.703	67.536	39.272
Flaxseed	9.617	18.103	13.516	12.837	823.365	153.669	76.918
Rapeseed	20.951	29.028	28.696	21.734	793.482	241.689	113.576
Wool	0.019	0.029	0.024	0.026	1.643	0.258	0.147

(Continued on next page.)

TABLE 3.3—continued.

	Wheat	HSW Flour	DS Flour	FR Tomato	Fresh Veg	Proc Veg	S Beets
Wheat	32,000.805						
HSW Flour	-510.070	15.043					
DS Flour	-15.885	0.894	0.015				
FR Tomato	-1.400	0.017	0.0001	0.00006			
Fresh Veg	-108.330	1.107	0.055	0.003	0.614		
Proc Veg	-19.568	0.570	0.015	0.001	0.077	0.017	
S Beets	997.111	-9.418	-0.289	-0.022	-2.447	-0.570	9.826
Lin Oil	194.822	-2.322	-0.068	-0.006	-0.531	-0.094	3.356
Rape Oil	2,470.045	-29.172	-0.857	-0.069	-5.980	-1.182	41.619
Seeds	1,428.882	-36.736	-0.910	-0.107	-7.542	-2.443	38.141
Flaxseed	3,612.530	-40.785	-1.189	-0.106	-5.824	-1.491	55.378
Rapeseed	5,483.335	-159.012	-5.035	-0.180	-18.212	-3.901	89.060
Wool	6.727	-0.059	-0.001	0.000	-0.012	-0.003	0.117

(Continued on next page.)

TABLE 3.3—continued.

	Lin Oil	Rape Oil	Seeds	Flaxseed	Rapeseed	Wool
Wheat						
HSW Flour						
DS Flour						
FR Tomato						
Fresh Veg						
Proc Veg						
S Beets						
Lin Oil	0.358					
Rape Oil	8.810	55.355				
Seeds	6.699	85.764	109.564			
Flaxseed	11.377	145.351	116.511	111.082		
Rapeseed	22.333	274.685	267.226	387.621	592.408	
Wool	0.021	0.268	0.234	0.382	0.524	0.0004

Sources: Statistics Canada, Cat. No. 22-003, 22-007, 22-201, 23-001, 23-007, 23-202, 23-205, 23-210, 32-228, 65-202, and 65-203; Agriculture Canada, *Livestock Market Review*, 1977-1986.

The key issues can be discussed with reference to Figure 3.2. It displays a scatter diagram of the price correlations against the covariance values where the latter is plotted on a logarithmic scale. In general, the plot runs from the bottom left-hand to the top right-hand corner; that is, it has a positive slope. Furthermore, the log scale, which is used for ease of illustration, distorts the actual configuration as the very large (and mostly positive) covariances are de-emphasised. Those commodities with large covariances also tend to be highly positively correlated. A well-diversified economy would have the opposite configuration. In other words, the distribution would slope downward from left to right. Those commodities with large covariances would also be those with prices that are strongly negatively correlated. Positively correlated prices would be concentrated in the commodities with small covariances. The top left-hand and bottom right-hand quadrants are very sparse for Prairie agriculture as presently constituted. This means there are no obvious avenues for resource shifting which will lead to significant diversification. Of course, any shift in the mix of outputs which moves the orientation of the distribution closer to one which slopes downward from left to right is likely to reduce the total variance of gross income in the Prairie region.

The total variance of gross income in the Prairie region (as defined by the 27 selected commodities) is 9.539. This figure will be used to evaluate the impact of all subsequent cases.

The Base Case Extended

To reduce the variance in income for the Prairie region, what is required is that resources be moved from outputs, whose prices are highly correlated and have a large covariance, into a mix of products, whose prices are less highly correlated and have a smaller covariance. The case developed allows for a movement of resources out of wheat production into the production of additional processed pork, fresh pork, canola, and processed canola products. This alteration of the configuration of Prairie agriculture arises from the price correlations and covariances observable in the base case. It also provides a feasible combination of outputs given Canadian resources.

The movement out of wheat production is logical given its contribution to the total variance of returns. As the most important Prairie agricultural commodity, it will have a heavy weighting in the covariance calculation. Unfortunately for diversification, wheat prices are also highly positively correlated with the prices of most other major Prairie commodities. Of the

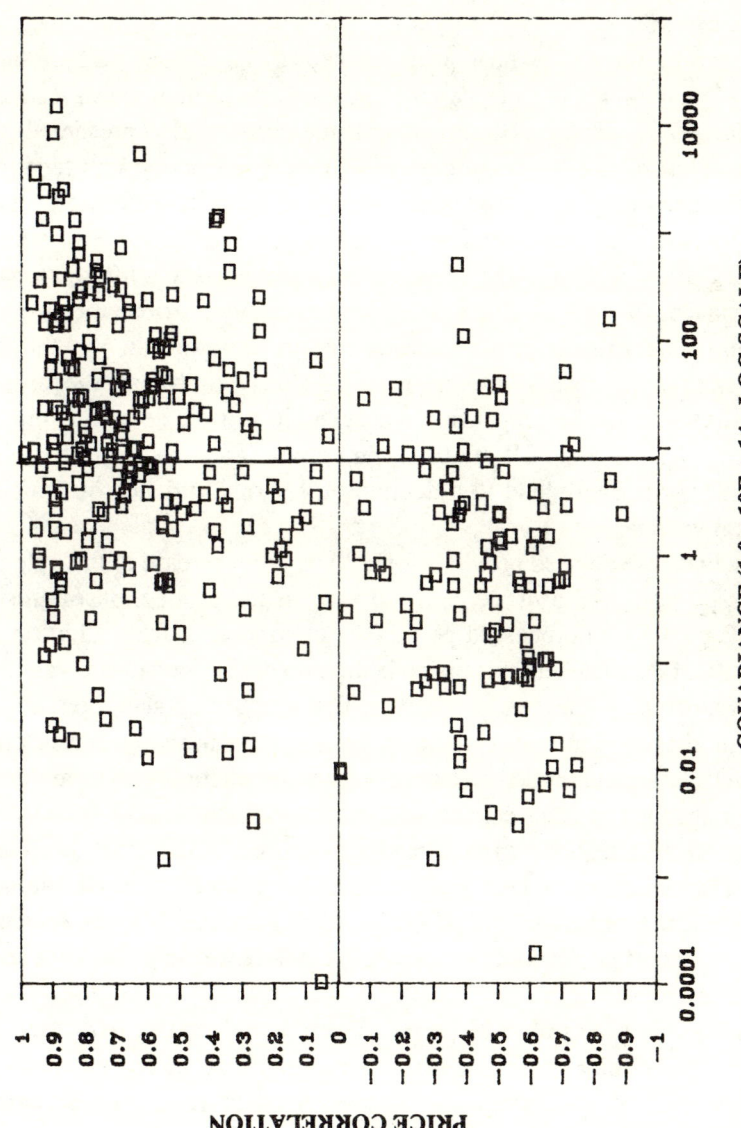

FIGURE 3.2. Price Correlation Versus Covariance - Prairie Commodities, 1977 - 86

positively correlated pairs, only feeder cattle, the three pork industry classifications, eggs, canola, and certified seeds have a coefficient of less than .700. Table 3.4 presents the distribution of price correlations for selected commodities.

The correlation with wheat price exceeds .70 for over half of the remaining 26 commodities. Further, the covariance values for wheat tend to be very large. The distribution of covariance terms is also presented in Table 3.4. Hence, moving resources out of wheat production will tend to have a considerable effect on the total variance of income. Processed pork products, on the other hand, tend not to be as highly correlated with other commodity prices although the tendency to positive correlation remains. The covariances for pork products are also more evenly distributed between fairly equal positive and negative values. Again, however, the weighting remains skewed to positive covariances. Relative to wheat, the other commodities selected for expansion also exhibit a lower degree of price correlation. As these are all relatively important, an expansion of these commodities will likely lead to a decline in the total variance of Prairie agricultural income.

The actual case devised provides for a 20 percent increase in the output of processed pork and a 40 percent increase in the production of fresh pork. The increase in processed pork is assumed to arise from additional animals so that the total herd expansion is the sum of the animal equivalents of the expansion of the two products. The number of slaughter hogs utilized in these calculations remains constant because the cases are constructed on a value-added basis; hence, intermediate products are not included in the total variance calculation.

The extra feed required to produce these animals is expected to come from land diverted from wheat production to the production of additional barley. Thus, the quantity of wheat produced will decline. The movement of resources takes into account the appropriate differences in yields between commodities.

The case also allows for a 25 percent increase in the quantity of canola produced. This may represent an upper bound for increased canola production. All of this increase is also assumed to arise from acreage diverted from wheat production. Two-thirds of this additional canola production is further processed.[2]

TABLE 3.4

Distribution of Price Correlations and Covariances Selected Commodities

Correlation	Price correlations				
	Wheat	Proc pork	Canola	Pork	Proc canola
.90 to 1.00	6	0	0	1	4
.80 to .89	6	0	1	1	5
.70 to .79	2	1	6	3	7
.60 to .69	3	1	4	2	2
.50 to .59	1	2	5	5	2
.40 to .49	0	1	0	1	0
.30 to .39	2	2	3	3	0
.20 to .29	0	2	1	2	0
.10 to .19	0	5	0	1	1
.00 to .09	1	3	1	2	0
- .10 to -.01	0	3	0	0	0
- .20 to -.11	0	2	0	0	0
- .30 to -.21	0	1	0	0	0
- .40 to -.31	3	2	0	0	0
- .50 to -.41	2	0	2	1	1
- .60 to -.51	0	0	0	0	3
- .70 to -.61	0	0	1	2	1
- .80 to -.71	0	0	0	2	0
- .90 to -.81	0	0	2	0	0
-1.00 to -.91	0	0	0	0	0

Covariance size	Covariance									
	Wheat		Proc pork		Canola		Pork		Proc canola	
	+	-	+	-	+	-	+	-	+	-
10^{18}	1									
10^{17}	9				1		1		1	
10^{16}	9	2			8	1	3		4	
10^{15}	1	1	4		10	1	9	2	10	1
10^{14}	1	1	6	3	1	2	7	2	5	2
10^{13}			6	3	1	1	1		1	1
10^{12}			1	1			1			1
10^{11}				2						

The new variance-covariance matrix for this case can be found in Kerr (1989, Appendix 3) and is denoted "base case extension." The total variance of gross income for this case is 9.353. This represents a 2 percent decline relative to the base case. While the decline suggests this scenario would represent some progress toward diversification, the variance in income remains very large. This simply indicates the difficulties associated with diversification within the agricultural sector alone. Since prices are so interrelated, targeting sectors for diversification programs may not yield significant improvements. It also suggests that the effects on diversification arising from trade liberalization are not likely to be large. The changes expected from the U. S.-Canada FTA are developed in the next section.

U. S.-Canada Free Trade

The most significant trade policy issue in recent Canadian history is the free trade agreement signed by the United States and Canada. Previously, it was claimed that limited access to U. S. markets had prevented Canadian firms from achieving economies of scale. While this may be the case for the agricultural processing industry, restrictions on access to the U. S. market had not been the reason why some Prairie farmers had not realized all of the economies of scale available. If trade liberalization leads to economies of scale in the processing sector, it is likely to change both the mix of products produced in the Prairies as well as the degree of value added which will accrue in the Prairies. To the extent that these changes reduce total income variance, they will be diversification enhancing. Ex ante, however, it is not possible to determine whether the changes brought about by the FTA will be diversification enhancing or diversification retarding. In addition, even if no significant economies of scale can be realized, enhanced and more secure market access to the United States may alter trade patterns and provide a stimulus to alter resource commitments in the Prairies.

The Agricultural Provisions

The FTA provides for the elimination of all tariffs over a period of 10 years, with the exception of fresh vegetables and fruits. For these commodities, Canada can reimpose existing seasonal tariffs for a period of 20 years. These "snapback" provisions can only be activated under a limiting set of circumstances and, thus, will have little impact except under very adverse market conditions. In general, however, the tariff levels are small and not perceived to be the major restricting influences on agricultural trade.

Both countries currently utilize countercyclical meat import policies which allow the restriction of imports if traded quantities exceed formula-established trigger levels. Of course, the law simply backs up a system of negotiated "voluntary" export restraints which are the actual means used to constrain imports. While these provisions are aimed primarily at offshore imports, they have meant that Canadian exports to the United States, at times, have been cut off in November or December. The value of trade lost when these restrictions are imposed can be considerable, but more worrisome are the security of supply concerns which have been created for U. S. customers. Therefore, Canadian beef exporters have found it difficult to develop and maintain consistent market channels by which to move products to the United States. The provisions of the FTA now exempt both countries from their respective meat import acts which should result in Canadian processors having more secure access to U. S. markets.

The FTA also contains provisions by which the two countries will undertake to harmonize technical regulations. These are particularly sensitive in the area of health regulations. In the past, there have been instances, on both sides, when differences in health regulations have been used to restrict trade—particularly in meat and livestock. Further, border inspections have been used as short-term inhibitors to trade. The frequency of such inspections will now be reduced. Other nontariff barriers will also be eliminated. The success of all such reductions to these barriers will depend on the effectiveness of the dispute settlement mechanism. This can only be assessed once it begins operation.

The Canadian Wheat Board (in conjunction with the Department of External Affairs) has restricted the import of grains from the United States through an import license requirement. This allows the Board to better carry out its responsibilities relating to Canada's international grain trade because it removes any instability which would arise from grain moving into the Wheat Board area. To prevent retaliation by the United States for this trade barrier, exports to the United States have been voluntarily restricted. Under the FTA, these import licenses are to be eliminated.

The activation of these provisions is contingent upon the alignment of Canadian and U. S. subsidy levels. These provisions are not likely to come into effect in the near future. This is because U. S. subsidies on wheat are approximately $50 per tonne higher than those in Canada. Subsidies on barley and oats are also higher. A working group on grains and oilseeds will continue negotiation in this area. Given the current confrontational attitude of the United States toward the subsidy policies of the EC, any significant reduction in U. S. subsidies seems unlikely in the near future. In fact, unless there is significant progress at the current round of the multilateral trade talks, U. S. subsidy levels may well increase over time.

The FTA sets out the methodology for determining equivalent subsidies in great detail. A long list of U. S. programs is included for purposes of calculation, including all aspects of the 1985 U. S. Food Security Act and related export subsidies. In addition, some less-direct forms of subsidization are included such as the U. S. Army Corps of Engineers, Inland Waterways Program; research expenditures; and state agricultural budgetary allocations which are specific to particular grains. Canadian programs included are Western Grain Stabilization payments, Canadian Wheat Board pool deficits, the Special Canadian Grains program, provincial stabilization schemes, research expenditures, branch line rehabilitation, crop insurance, and cash advance programs to name a few. Many of the subsidies are tied to market prices so that, for example, in times of drought the level of the subsidy declines. In years of high yields, the subsidies increase. Given the complexity of the subsidies involved and their year-to-year volatility, any consistent alignment would seem problematic at best, making the opening of cross-border trade a relatively remote possibility.

In addition, the agreement allows each party to introduce contingency protection measures or to reintroduce import restrictions on wheat, oats, barley, and their products (such as flour) if imports increase significantly as a result of a change to agricultural programs. This would seem to considerably restrict the ability of governments to introduce policies aimed at encouraging exports as a means of increasing the degree of diversification.

While not directly a result of the FTA, low erucic acid rapeseed (canola in Canada) oil was recently granted "generally regarded as safe" (GRAS) status by the United States. By the terms of the FTA, the label, "canola," can now be substituted for the term, rapeseed, when the product is marketed in the United States. This should greatly improve Canadian opportunities to expand into U. S. markets for canola and its processed derivatives. The major U. S. import-restriction policies for sugar and sugar products remain in force, effectively eliminating most Canadian products from the U. S. market. The only concessions relate to food products which have a sugar content of less than 10 percent.

Clearly, the provisions on grain allow for the maintenance of the major Canadian policy instruments in this area and, hence, have more to do with the third objective of the federal government than they do with improving market access. The other area where the retention of policy instruments took precedence over trade considerations was in the commodities where supply-management marketing boards are in place. Canadian poultry and egg producers, in particular, perceived that they had a great deal to lose from trade liberalization. In addition, one of the major pillars upon which the last 20 years of Canadian agricultural policy had been built would have to have been abandoned. Neither the federal nor provincial governments were willing to take that step. Of course, abandonment of the supply-management system would have, after the long-run adjustment, a destabilizing effect on the prices of poultry commodities. To the extent that supply management isolates the Canadian market, it reduces price variability. As Canadian prices for these commodities are now considerably higher than prices in the United States, import quotas are required to prevent large quantities of U. S. product from flowing into Canada and reducing the market share of domestic producers. Under the terms of the FTA, imports will be held at levels equal to the average of the previous five years. Little emphasis is placed on the dairy sector in the FTA, suggesting that there was little pressure for change on either side of the border. As a result, the ability of the Boards to restrict supply is maintained. Furthermore, if required, new supply-management boards can be implemented.

The elimination of tariffs on supply-managed products has the potential to affect processors and food manufacturers who produce derivative products. This is because the prices of raw product input into TV dinners, chicken pies, etc., are higher in Canada as a result of supply management. Some of these products are not on the import control list (viz., chicken/turkey cordon bleu, chicken/turkey TV dinners, or chicken/turkey Kiev). There are provisions in the FTA which will allow the Canadian government to add these products to the import control list if the industry is being damaged by large increases in U. S. exports. In the dairy industry, a considerable number of products already have been added to the import control list. Hence, it would appear that there is little likelihood of any significant change from the status quo for the supply-management commodities or their derivatives as a result of the FTA. While this has meant that the Canadian government has been able to achieve the objective of maintaining the main elements of Canadian agricultural policy intact, it also means that major changes will not likely arise as a result of the FTA.

The Conservative Scenario

The basic assumptions of this case are that the only response to the FTA from Prairie agriculture results from the incentive provided by the removal of the tariffs (i.e., all changes to quantities produced take place from movements along existing supply functions). This implies that no market development is undertaken. This may be due to other barriers to trade (such as technical barriers) which are not removed in the future because of an inability to agree to their form, or it may be the result of an inability or unwillingness by Canadians to exploit opportunities as they arise. It is further assumed that there is no alignment of subsidies, meaning that the institutional status quo of the grains sector remains in force. Trade in canola and canola products is expected to increase. The supply-management commodities, in conformance with the FTA provisions, also experience no change. The trade regime for sugar is assumed to remain unchanged.

The commodities included in the group of 27 selected above can be divided into two groups: net export commodities and net import commodities. These classifications were determined from the Statistics Canada trade figures. The commodities on the net export list include feeder cattle, slaughter hogs, high-quality beef, low-quality beef, fresh pork, processed pork, honey, processed linseed products, processed canola products, certified seed, flax, and canola. The net import list includes fresh tomatoes, fresh vegetables, processed vegetables, and wool.

To determine the effect of the removal of tariffs, the elasticity-of-supply estimates were combined with the published tariff removal schedules included in the FTA. An attempt was made to utilize the most up-to-date supply elasticities available (Kerr 1989, Appendix 4). To this end, individuals in Agriculture Canada were contacted and their current estimates requested. Where no estimates were available from Agriculture Canada, a search of the relevant literature was conducted and the most recent estimates were utilized. In a few cases, no published estimates could be found and an elasticity of 1.00 was assumed. As new trade equilibriums should reflect long-run adjustments, long-run elasticities were used wherever possible.

The tariff rates used in the calculations are found in the volumes included with the FTA. The applicable rates for the net export commodities were the U. S. import tariffs. These were converted to Canadian dollar equivalents to reflect the true price effect for Canadian exporters. For net import commodities, the Canadian tariffs were utilized. In those cases where there were commodities with different tariff rates subsumed within one of the 27 commodity groupings, a weighted average of the applicable rates was adopted. All tariffs were then converted so that they conform to the calculation units of the 27 commodities (Kerr 1989, Appendix 5). The tariffs are scheduled to be reduced according to various timetables. All adjustments are assumed to have taken place at the point when all tariffs are removed.

The changes in tariffs and elasticities were applied to 1986 prices and quantities. Where ad valorem tariffs are in place, they were calculated as a percent of the 1986 price. In the cases of live hogs, unprocessed pork, and wool, no tariffs exist. Regarding live hogs, however, the current U. S. countervailing duty was assumed to remain in force. This assumption is made because the Canadian subsidies which have been found countervailable are still in effect and no motions have been made to remove them. Some additional assumptions regarding the movement of resources between commodities were also required. The production of canola and flax is increased due to the reduction of the U. S. tariffs. This has two components, the direct increase due to the reduction of the tariff on the raw product and the indirect increase due to the reduction of the tariff on the processed derivatives. It is assumed that the total increase in production arises from the transfer of resources out of wheat production. Wheat production is reduced by the appropriate adjustment factor for each crop. Given the importance and high positive price correlations between wheat and other commodities, spreading the resource shifts among the other grains would tend to reduce the effect on income variance. The additional production of animals will require feed. All additional feed requirements are assumed to come from the diversion of barley which would have been sold out of the Prairie region. In other words, it is moved into higher value-added production and the direct contribution of barley to income variance is reduced.

The detailed results of the changes to the variance-covariance matrix arising from the conservative free trade case can be found in Kerr (1989, Appendix 3). This can be compared to the base case. The total variance of income for the Prairie region which arises from this scenario is 9.550. This represents a 0.12 percent increase in variance relative to the base case.

An almost unchanged variance should not have been unexpected. As the tariff rates are generally small relative to the price of the products, the changes in quantities generated from the elasticity estimates tended to be modest. Further, as the major grain crops and supply-management commodities remain untouched by the FTA, much of the production mix remains as it was before the agreement. In the case of grains, only indirect trade effects arise from the movement of resources between commodities. Any reductions in the variance of income which arise from the increase in exports are partially offset by the reduction in the production of imported commodities. As noted above, the prices of these commodities tend to be negatively correlated. Thus, any reduction in their production will tend to increase the total variance of gross income. This case further highlights the problems associated with the diversification of Prairie agriculture. Trade liberalization, in the absence of additional measures which actively channel resources in a diversification-enhancing direction, will not lead to a significant reduction in the variance in regional income.

Sensitivity analyses were also conducted. The results (presented in Kerr 1989, Appendix 3) were not particularly sensitive to the elasticities assumed. This conclusion should not have been unexpected given the offsetting interactions between the commodities indicated by the price correlation matrix. Further, as the elasticities increase or decrease, the supply response of the imported commodities will move to offset, to some extent, any changes in exports.

The assumptions embodied in these cases are, of course, very conservative since they do not allow any market expansion and only allow limited movement of resources among sectors. A less restrictive case will now be developed.

The Optimistic Scenario

Due to general equilibrium problems relating to the ex ante estimation of the effects of trade liberalization, few quantitative estimates have been made of the expected impacts of the trade liberalization process resulting from the FTA. The liberalization process concerns not only the removal of tariffs and quotas but also the removal of nontariff barriers and other less formal impediments to market development. Such trade-inhibiting practices can span the range of actions from restrictions on business travel to the abuse of health standards. Provisions exist in the FTA to address such trade irritants once the agreement comes into force or at least to promote a negotiated solution to contentious issues in the period after implementation.

The red meat sector has received the greatest attention in terms of these issues for two primary reasons. First, given the perishability and potential health risks associated with meat, the myriad of existing health regulations provide considerable scope for abuse in aid of market protection. Second, there is a large potential market perceived to be available in California (Kerr 1986). The ability to tap this market has, in part, been limited by the application of nontariff barriers by the United States (Gillis *et al.*). If all of the changes in the FTA are implemented and further negotiations are successful in aligning technical standards and removing other trade irritants, considerable opportunities should present themselves for the exports of red meat out of the Prairie region (Bruce and Kerr). This, of course, also depends upon whether or not Canadian producers and processors are willing to exploit the opportunities which arise.[3]

The other major area where opportunities are likely to be present is canola and canola products. As mentioned above, the combination of the changes brought by the FTA and the granting of GRAS status for this oilseed should provide considerable market potential for Canadian product.

Deloitte, Haskins, and Sells did a study on the expected effects of trade liberalization on the Canadian Prairies. This study developed a number of alternative cases for the year 1995 of the changes in Prairie agriculture which would result from trade liberalization. The most optimistic of these cases will be used here as the basis for the optimistic U. S.-Canada FTA scenario. The Deloitte study assumed no major changes to the grain economy but assumed maximum access for red meat and canola products. This would appear to conform fairly closely to the provisions of the actual FTA.

The Deloitte study also took into account the changes in West-East trade which were expected to arise out of trade liberalization with the United States. In particular, while there was to be considerable expansion of beef exports into the California and Pacific Northwest markets of the United States, there was also to be some reduction in the movement of beef to markets in Central Canada. Thus, the results for beef are net of these West-East changes. In addition, barley production was expected to decline as a result of imports of U. S. corn and other feeds into Central Canada. This was partially offset by an increase in the export of malting barley to the United States.[4]

The net changes in Prairie output derived in the Deloitte study were: beef, 12 percent increase; pork, 2 percent increase; barley, 6 percent decrease; canola, 3 percent increase; flax, 1 percent increase; and rye, 1 percent increase. In our case, all of the increase in canola production is assumed to be in processed form for export to the United States. The pork increase is divided between fresh and frozen pork in the ratio of base case production. All additional beef production is assumed to be in the form of high quality beef. The additional feed required is assumed to be reallocated from existing barley marketings outside of the Prairie region. All other commodities enter at the values used in the conservative U. S.-Canada FTA scenario and, hence, are assumed to realize only the gains or reductions arising from the removal of the tariffs.

The variance-covariance matrix for this case can be found in Kerr (1989, Appendix 3). The total value of the variance in total gross revenue is 9.630. This represents an almost 1 percent increase over the base case. The positive effect on total variance in this case is greater than in any of the conservative cases. The major reason for this increase is the expansion of high quality beef production. Certainly, such an expansion would have a positive effect on the value added accruing in the Prairie region and, hence, lead to an increase in the absolute level of income. However, the effect on diversification will not be significant because beef prices are relatively highly correlated with the prices of other commodities. As is well known, beef prices are also highly variable. Hence, the increased contribution of high quality beef to the total value of output also tends to add considerably to the variance of income.

The effect of trade liberalization on the diversification of Prairie agriculture is somewhat disappointing. While the income effects of the FTA are likely to be of considerable benefit on average, it would seem that the post-FTA era will leave the Prairie agricultural sector as vulnerable to large swings in income as it was in the past. Of course, if the FTA provides opportunities for expansion into entirely new export lines, the variance of income could be reduced. This will depend upon the absolute size of the price variance and how the price movements of any new goods are correlated with the prices of existing commodities.

The Multilateral Agricultural Trade Environment

What should the effect be of removing trade barriers under the GATT?[5] Basically, liberalization will have to involve not only the EC but Japan, the United States, other developed economies, and the developing countries as well. It is not possible for any one of these economic units to abandon their agricultural trade policies without reciprocation from other major producers. Hence, it is only meaningful to talk about trade liberalization between Canada and the EC, for example, within a framework of mutual multilateral reductions in levels of farm support. This requires a multicountry model which can account for the interactive impacts of a general reduction in protectionism.

The confrontations and recriminations which the problems in agricultural trade have caused led to a high priority being given to agriculture at the talks. However, the outcome of the negotiations is far from clear. While there are considerable pressures for reform, little real progress has apparently been made to date. The proposals of the major protagonists remain far apart. Given that the total resources committed to agricultural production considerably exceed those required to supply needs at undistorted prices, complete abandonment of all trade-distorting agricultural policies would mean massive, and politically unacceptable, alterations to the patterns of resource use in agriculture. Hence, even if the Uruguay Round could be considered a success at its conclusion, only modest changes in agricultural policies should be expected. They will also be multinational in nature. Of course, the multilateral talks could fail and the agricultural trading environment could continue to deteriorate. These considerations will form the basis of multilateral alternatives developed below.

The Optimistic Multilateral Scenario

Concerned over the deteriorating international trading environment, the Council of Ministers of the Organization for Economic Cooperation and Development (OECD) requested a major study of these problems in May, 1982. This endeavor took almost five years to complete (it was submitted in late April, 1987). The study provides a comprehensive multinational examination of agricultural policies and includes, as an important element, an examination of "the market impacts and economic consequences of a gradual and balanced reduction in assistance to producers" (OECD, p. 3). The basis of this analysis is a:

> ". . . near linear, medium-term partial equilibrium comparative static model of agricultural production, demand and trade. . . . The model system is built around individual country models which are linked through trade between countries" (OECD, p. 137).

This model was chosen for a number of reasons. First, it provides the framework for a multilateral reduction in agricultural trade and farm support policies. Second, it has a Canadian submodel from which quantitative estimates are available. Third, it is one of the most disaggregated models available. This is important as it allows for commodity interactions in its estimates. In all, 14 tradable commodities and 11 economic blocks are explicitly modeled. Further, the model allows for changes to both border measures (such as tariffs and quotas) and domestic policies affecting prices and quantity of output.

The OECD assumed a 10 percent reduction in agricultural support. As even those who perceive that progress is possible at the multinational negotiations realize, "domestic political realities . . . appear to preclude true trade liberalization in agriculture by phasing out most or all domestic agricultural programs" (Hathaway, p. 142). In this light, such an apparently modest degree of success could be considered a realistic projection.

The projections of the OECD model are used as the basis for our optimistic multilateral scenario. The study assumed:

". . . assistance on all commodities in all countries is reduced by 10 per cent. Policies measured on a volume basis (United States Grains Farmer Owned Reserve and Set Aside, Japanese Paddy Field Reorientation Programme, USA and EEC dairy stock changes) were also reduced by 10 percent"[6] (OECD, p. 144).

The reductions in trade restrictions and subsidies in the multilateral model produced a number of effects on international prices. Reference prices for livestock products increased while prices for products used for feeding livestock tended to decline. Prices of beef rose 1.5 percent. Prices of wheat and coarse grains fell slightly. Pork and poultry prices rose slightly.

As the published OECD reports did not include estimates of the changes in trade volumes, the research section of the OECD in Paris was contacted and the actual quantities secured. The OECD model indicated that Canadian exports of beef would increase by 46 percent, canola by 0.59 percent, and pork by 0.11 percent. Wheat and barley exports declined by 1.05 percent and 1.95 percent, respectively. Imports of chicken increased 11 percent while imports of wool decreased 2.08 percent. The OECD study used the years 1979-1981 as the basis for its estimates. Given that the level of farm support internationally has increased considerably since then, these changes represent conservative values. To be consistent with the base case in this study, the percentage changes are applied to 1986 Canadian exports. As surplus milk production in the Prairie provinces pertains more to the management of seasonal milk supply than a structural problem, increases in milk product exports were assumed to come from surplus production in Central Canada. All changes in beef, pork, wheat, barley, and canola exports were credited to Prairie production. In the cases of poultry and wool, the new quantities were apportioned across the country according to the current share of national production. The changes in animal feed requirements were assumed to be the result of the diversion of production surplus to the Prairie region.

The variance-covariance matrix for this case is reported in Kerr (1989, Appendix 3). The value of the total gross variance in Prairie agricultural income is 9.565. This represents a 0.27 percent increase relative to the base case. This slight rise results from the movement of resources into beef production with its highly variable prices. While the small size of the total change may be surprising, it is more a comment on the limited possibilities available for multinational trade liberalization than the possibilities for diversification as:

> "A reduction in assistance of 10 percent means that PSE/CSEs[7] are reduced by 10 percent. The decline in effective prices will generally be less. If, for example, the PSE is 50 percent of the effective price, a 10 percent reduction in PSE will mean a 5 percent initial reduction in the effective producer prices" (OECD, p. 144).

The adjustments which would result from the multinational alterations in national agricultural policies could have a significant effect on the interrelationships between prices of various commodities. Hence, more than for any other alternative developed, the assumption that the interprice movements remain the same as in the past can be questioned and the results interpreted with increased caution. This is particularly true because trade liberalization is likely to reduce the variability of world prices. Export subsidies (especially those of the EC) and variable levies are responsible for increasing the variability of world prices. Stable international prices may induce increased resource commitments in exporting countries which could lead to larger trade effects than are accounted for in the OECD model. Still, the total effect on world prices for the commodities utilized in this scenario (as predicted by the OECD model) ranged between +1.46 percent for beef to -0.35 percent for barley (OECD, p. 32).

The Pessimistic Multilateral Scenario

The reduction in levels of domestic agricultural support and the subsequent trade effects assumed above are based on the premise that significant progress will be made at the current GATT negotiations. As the results of these discussions are far from clear at this point, failure remains a distinct possibility. This could lead to a further deterioration in the international trading system for agricultural commodities. In this case, it will be assumed that the U. S.-Canada FTA is in place and that access to the U. S. market is assured. Levels of Japanese protectionism remain as

Diversification of Prairie Agriculture 79

presently constituted. It is assumed, however, that the EC increases its protectionist measures in aid of self-sufficiency. However, no increase in the level of commodities moving out of the EC under the various export subsidy programs is assumed. These would seem to be reasonable assumptions, given the current direction of Common Agricultural Policy (CAP) policies. Attempts are being made in Brussels to stabilize or reduce output through reduced prices or quantitative controls. These policies are having some limited success. Hence, the quantities of EC commodities moving into world markets will likely stabilize. The implied reduction in output, however, will free up European agricultural resources. Currently, CAP policies are being initiated to encourage the movement of these freed-up resources into the production of commodities for which the EC is less than self-sufficient. This is particularly true in the case of animal feeds and edible oils. As these commodities now comprise most of Canada's exports to the EC, in a deteriorating trade environment such exports would likely be at risk.

It will be assumed for this "worst case" scenario that imports of major Canadian commodities into the EC are eliminated. While this assumption is unrealistic, particularly in the case of high-quality Canadian hard wheat which serves special culinary needs, it does represent the worst possible case. The quantity and value of exports to the EC for major Canadian commodities in 1986 are presented in Table 3.5.

It is further assumed that 50 percent of these exports find alternative markets offshore. For the other 50 percent, it is assumed that prices decline sufficiently so that there is an incentive to transfer the resources used to produce these exports into the production of barley for animal feed. This lower priced barley allows for an expansion of red meat and livestock output. The additional available feed is apportioned among the various livestock commodities—high quality beef, pork, and processed pork—according to the current ratios of production. As market access to the United States is assured through the FTA, the additional production is assumed to be exported to U. S. markets.

The total value of the variance of gross income is 10.010 (Kerr 1989, Appendix 3). This represents a 4.94 percent increase relative to the base case. Such a change is clearly diversification retarding. The major reason for this is the movement of resources into beef production with its relatively high price correlations and variable prices. This case represents the largest

TABLE 3.5
Canadian Exports to the European Community, 1986

Commodity	Classification	Quantity	Value
		tonnes	thousand dollars
Barley	06119	143,673	12,541
Oats	06131, 06133, 06139	41,314	3,777
Wheat	06164, 06165, 06167, 06168, 06169	1,495,060	296,961
Rapeseed oil, cake, and meal	15351	22,973	3,394
Rapeseed	21240	60,418	12,612

Source: Statistics Canada, *Exports: Merchandise Trade*, Cat. No. 65-202.

deviation from the base case of all the analyses done in this study. Despite this, the change in the total variance of gross income is still less than 5 percent. Thus, it would appear that, no matter what changes in the international trading environment come about, the effects on the diversification of Prairie agriculture are not likely to be significant.

Summary and Conclusions

Twenty-seven commodities were suggested for the investigation. From the outset, the limitations to the diversification of Prairie agriculture were apparent. Prices over the last decade tended to be positively correlated for all of the important Prairie commodities. In addition, those commodities whose contribution to total Prairie income is large tended to exhibit the highest levels of positive correlation. This indicated that feasible resource substitution possibilities would not be particularly diversification enhancing. For example, movements out of export grain production into livestock would not yield significant gains because both sets of prices were generally highly positively correlated with the prices of most other commodities.

In all, a base case and seven additional cases were developed. These other alternatives were all compared to the base case. The first case was constructed using information from the base case price-correlation matrix and variance-covariance matrix. The purpose of this scenario was to identify commodities with weakly positively correlated prices which also made large contributions to total income. Once these commodities were identified, it was assumed that policies were implemented to encourage the transfer of resources into the production of these commodities and out of the production of those commodities with high positive-price correlations. This was the benchmark against which the trade cases could be judged. This scenario yielded only a 2 percent reduction in the total gross-income variance. The limited success from this direct targeting suggested that changes in the trading environment would also have only a limited effect on diversification enhancement. This is borne out by the fact that none of the trade cases was as diversification enhancing as the "base case extension" scenario.

Based on the changes expected from the U. S.-Canada FTA, four alternatives were developed. Three of these could be considered conservative cases because they assume that the only response to the FTA arises from reactions to changes in the tariff levels. The first of these three used estimates of supply elasticities and reductions in tariff values to determine the new mix of outputs. Two additional cases were developed to test the sensitivity of the results to the assumed elasticities. It was determined that the results were not materially affected by changes to the elasticities. Two of these three scenarios were found to be diversification retarding and one to be diversification enhancing. The effects were very small, however, being within 1.5 percent of the base case.

An optimistic FTA scenario was also developed. It assumed that nontariff barriers are removed and that Canadians aggressively seek to exploit market opportunities in the United States. The results of a previous study were used to provide information on the expected changes. Major expansions were expected in the areas of red meat and canola. This alternative proved to be somewhat diversification retarding. This was due largely to the fact that beef prices are highly variable and positively correlated with most other prices. Still, the increase in variance was less than 1 percent greater than the base case.

Two cases relating to changes in multinational trade were also developed. The first assumes that the Uruguay Round of GATT discussions is successful and that reductions in agricultural support programs and trade restrictions are manifest in the future. The results of a major OECD study on trade liberalization were adapted for our purposes. This alternative was found to increase the total gross variance of income slightly. The second alternative assumed that the GATT talks fail and, as a consequence, the EC continues to follow policies to increase its level of self-sufficiency. Canadian exports are reduced commensurate with these policies. This scenario led to an almost 5 percent increase in the total gross variance in income and, hence, a reduced degree of diversification. A summary of the results is provided in Table 3.6.

With the usual limitations of empirical works in mind, the results of this study would appear to indicate that little alteration in the diversification of Prairie agriculture can be expected from changes in the trading environment investigated. Of course, other markets (such as Japan, China, or Russia) might provide more diversification-enhancing markets, but investigations of these markets was beyond the terms of reference of this study. This result stems from the relationship between the movement of prices among commodities and the constraints imposed by the existing resource base on substitution possibilities. These are forces largely beyond the control of policymakers. Hence, if a significant increase in the stability of Prairie income is desired, then it will probably mean that diversification of agriculture must be pursued in combination with nonagricultural activities. As a result, other activities must be found to augment Prairie regional income.

While changes in the trading environment will not lead to a significant decrease in the variability of regional income, trade liberalization is likely to have other desirable effects. Although there is a tendency for increases in income to be capitalized over the long run all of the trade liberalization cases resulted in increases in gross regional income (Kerr 1989, p. 71). This is the expected result from trade theory. Hence, trade liberalization would seem to provide for improvements in income while having little effect on diversification and, thus, can be viewed as a positive step.

TABLE 3.6
Comparison of Scenarios

Scenario	Total variance of gross income[a]	Percent change from base case
Base case	9.539	
Base case extension	9.353	-1.95
Conservative U. S.-Canada FTA scenario	9.550	+ .12
Optimistic U. S.-Canada FTA scenario	9.630	+ .95
Optimistic multilateral scenario	9.565	+ .27
Pessimistic multilateral scenario	10.010	+ 4.94

[a]All values are 10^{18}.

It is also important to note that all of the trade alternatives will lead to an increase in the percentage of value-added production in the Prairie region (Kerr 1989, p. 3). This increase is largely the result of movements out of primary grain sales and into more livestock and red meat production as well as oilseed processing. Given that trade restrictions tend to be devised so as to maximize the ability of the importing country to capture value added, this may be the major benefit to Prairie agriculture arising from trade liberalization.

These results do not mean that trading opportunities will not arise in the future for new products. Such opportunities may arise from more inventive methods of processing and presenting the traditional products of the Prairie region. This would also increase the value added accruing to the Prairies. It should be remembered, however, that a simple expansion in the number of products produced in a region will not necessarily lead to a reduction in the variance of income for the region. That will depend upon the way in which the prices of the new products are correlated with each other and with existing products.

Trade liberalization is likely to increase incomes and value added. Still, given the configuration of the prices of major commodities and the fact that trade liberalization is not likely to alter them significantly, the variation in agricultural income is likely to remain a fact of life for Prairie producers. As any effects of trade liberalization on diversification are likely to be marginal at best, the income and value-added opportunities from trade-induced specialization should be pursued. As a result, incomes should rise. Policy effort could then be channeled into designing and refining non trade-distorting, decoupled stabilization programs.

The results, however, do not mean that individual farmers could not reduce the variability of their operation's incomes through alterations to their mix of outputs. Considerable opportunities would seem to present themselves.

While the results of this examination may appear disappointing, they do suggest that, if increased diversification for the purpose of reducing income variability in the Prairies is desired, it will not be sufficient to trust changes in the trading environment to bring it about. Other policy avenues will have to be actively explored.

Notes

[1]The quantitative measure which will be used is the total variance of gross revenues from major commodities for the prairie region. If c_i, price, is random and a normally distributed variable and x_i, output, is constant, then, in the n variable case, the total variance of gross revenues from tradable commodities can be formulated as:

$$Var\ (cx) = x'DX$$

where x = column vector of production quantities
x' = x transpose
c = row vector of expected prices
D = variance-covariance matrix of expected prices

or

$$Var \sum_{i=1}^{n} c_i x_i = \sum_{i=1}^{n} \sum_{j=1}^{n} x_i vc_i c_j x_j$$

where $vc_i c_j = vc_i^2$ when $i = j$, e.g., vc_i^2 = variance (c_i) and $vc_i c_j$ = covariance $(c_i c_j)$.

If $vc_i c_j$ is negative, it acts to reduce the variance of gross returns according to its contribution to total output. If it is positive, it will add to the total variance of gross returns.

[2] The resource use changes are assumed to have little or no effect on prices.

[3] For a discussion of the perceived attitudes and aptitudes of Canadians, see Schmitz.

[4] The Deloitte study also predicted an 80 percent decline in the poultry industries. The study assumed that the poultry commodities would be included in the FTA. As this did not come to pass, it will be excluded from this analysis.

[5] For a discussion of the trade wars leading up to the current GATT talks, see Kerr (1989, pp. 49-53).

[6] The OECD actually considered a number of cases with different assumptions regarding the composition of the 10 percent cut. This particular scenario was chosen because the implied equity in reduction might be considered politically tractable (Organization for Economic Cooperation and Development, p. 144).

[7] The Producer Subsidy Equivalent (PSE) and Consumer Subsidy Equivalent (CSE) were used as the means to standardize the value of agricultural support across countries and across programs. The PSE is defined as the payment that would be required to compensate farmers for the loss of income resulting from the removal of a given policy measure. The CSE corresponds to the implicit tax on consumption resulting from a given policy measure and to any subsidies to consumption.

References

Bruce, C. J., and W. A. Kerr. "A Proposed Arbitration Mechanism to Ensure Free Trade in Livestock Products." *Canadian Journal of Agricultural Economics* 34(November 1986):347-360.

Deloitte, Haskins, and Sells Associates. "Canadian Agricultural Trade Issues," Vol. 1: *Free Trade with the U.S.A., Executive Summary and Conclusions.* Regina, Saskatchewan: Prairie Pools, Inc., 1985.

Gillis, K. G., C. D. White, S. M. Ulmer, W. A. Kerr, and A. S. Kwaczek. "The Prospects for Export of Primal Beef Cuts to California." *Canadian Journal of Agricultural Economics* 33(July 1985):171-194.

Hathaway, D. E. *Agriculture and the GATT: Rewriting the Rules.* Washington, D. C.: Institute for International Economics, 1987.

Kerr, W. A. "Accessing the California Market—A Strategy for Free Trade Negotiations." Proceedings of a Conference on Profiting in Difficult Times, Saskatchewan Agriculture and the Saskatchewan Livestock Association, Regina, February 4-6, 1986.

_____. "The Diversification of Prairie Agriculture: Opportunities Arising from Changes in the International Trading Environment." Discussion Paper No. 359. Ottawa, Ontario: Economic Council of Canada, 1989.

Organization for Economic Cooperation and Development (OECD). *National Policies and Agricultural Trade.* Paris, France: OECD, 1987.

Schmitz, A. "Prospects for Change in Livestock Production and Trade." In *World Agricultural Policies and Trade*, edited by G. E. Lee. Saskatoon: University of Saskatchewan, 1984.

Statistics Canada. *Cereals and Oilseeds Review,* Cat. No. 22-007. Ottawa, Ontario: Minister of Supply and Services.

_____. *Exports: Merchandise Trade,* Cat. No. 65-202. Ottawa, Ontario: Minister of Supply and Services.

_____. *Flour and Breakfast Cereal Production Industry,* Cat. No. 32-228. Ottawa, Ontario: Minister of Supply and Services.

_____. *Fruit and Vegetable Production,* Cat. No. 22-003. Ottawa, Ontario: Minister of Supply and Services.

_____. *Grain Trade of Canada,* Cat. No. 22-201. Ottawa, Ontario: Minister of Supply and Services.

_____. *Honey Production* Cat. No. 23-007. Ottawa, Ontario: Minister of Supply and Services.

_____. *Honey Production and Value,* Cat. No. 23-210. Ottawa, Ontario: Minister of Supply and Services.

_____. *Imports: Merchandise Trade,* Cat. No. 65-203. Ottawa, Ontario: Minister of Supply and Services.

_____. *Production of Poultry and Eggs,* Cat. No. 23-202. Ottawa, Ontario: Minister of Supply and Services.

_____. *The Dairy Review,* Cat. No. 23-001. Ottawa, Ontario: Minister of Supply and Services.

_____. *Wool Production and Supply,* Cat. No. 23-205. Ottawa, Ontario: Minister of Supply and Services.

4

Growth and Development of Value-Added Activities

K. K. Klein and L. Chase-Wilde

Introduction

Value-added activities in agriculture can mean anything from manufacture of inputs, which are required for growth of bulk commodities on Canada's farms, to further processing of consumer food items. Food products may be processed for consumers in the immediate area as well as for those in locations far from the area of production. Some additional value to primary grain production can occur at the farm level; for example, the feeding of cereal grains to livestock increases the value of the primary products. However, the major employment and economic development opportunities arise from off-farm processing activities. Slaughtering, packing, and further processing of livestock products in the immediate area increase the meat product value in Western Canada. Processing of grains in the area where they are produced increases the level of economic activity and income in the region. Many other value-added agricultural activities currently provide employment for thousands of Manitoba, Saskatchewan, and Alberta residents: refining of sugar, storage and dehydration of potatoes, dehydration and pelleting of alfalfa, crushing of canola, and canning of vegetables, to name a few.

It is not only the activities at the farm gate and beyond that can increase the value of production in the agricultural industry in Western Canada. If inputs required for the production of agricultural products are manufactured in the region of their use, regional economic activity will be stimulated. For example, many farm implements have been developed, modified, or improved in farmers' machine shops in Western Canada. Several of these have spawned short-line machine companies. Feed mills, as well as agricultural chemical and fertilizer manufacturers, are other examples of agricultural input supply industries that have diversified the economies of the Prairie provinces.

In 1987 the government of Canada established a Western Diversification Office headed by a Minister of the Crown. Its purpose is to provide financial assistance to private firms that are engaged in the business of diversifying the Prairie economy.

The objectives of this chapter are to:

1. Examine growth of value-added activities in selected agricultural industries in the Prairie region of Canada.

2. Identify and critically evaluate constraints that may have impeded growth of these activities.

3. Assess prospects for further development of value-added activities in Prairie agriculture.

Resources do not permit a comprehensive evaluation of all types of agriculturally related industrial activity in Alberta, Saskatchewan, and Manitoba. By examination of selected value-added industries, it is hoped that major environmental, economic, social, and political constraints to development of these agricultural-industrial activities will be identified. Specific industries selected for analyses in this study are (1) grain processing, (2) red meats processing, (3) short-line machine companies, and (4) biotechnology.

Industry Profiles

Wheat and Wheat Products

Processing facilities, such as flour and feed mills, initially developed as communal mills for farmers to process their own production. As western agriculture progressed, mills served local markets. Railroads made viable the shipments of grains and grain products, but transportation was costly. It was rational to process grains locally and ship the lighter weight, higher value products.

In 1897 the federal government and the Canadian Pacific Railroad signed the Crow's Nest Pass Agreement which reduced freight rates on western-bound manufactured products from Central Canada and on eastern-bound grains and flour from the Prairie region. In 1926 these rates were extended to the Canadian National Railway; thus, shipments made from all western railway points qualified for the "Crow" rates. Soon, thereafter, shipments made to Pacific and Churchill ports were included in the Crow rate structure, while manufactured goods were removed.

This spur to exports came at a time when world demand for hard spring wheat and flour was growing. Concurrently, many larger Canadian mills were constructed between 1880 and 1915. Some restructuring of the industry occurred during the 1920s and 1930s, particularly when the adoption of electricity allowed relocation of newer technology mills away from sources of water power [Department of Regional Industrial Expansion (DRIE) 1984, pp. 1 and 2].

The structure of the grain-processing industry changed as transportation technology changed and shipping costs declined. Hopper cars, which were easier to load and unload, replaced box cars. Flour, which is usually bagged for shipment, became comparatively more costly to transport. The growth of an eastern livestock industry developed a market for millfeeds. Thus, milling capacity began to shift from Western to Eastern Canada.

In the mid-1980s, 47 flour mills existed in Canada, 15 of which were located in Western Canada. Multiplant companies own six of these western mills; virtually all of the larger mills belong to companies with diversified interests. Many of the so-called independent mills are subsidiaries of larger firms in the food industry, while other major national and international firms, with a diversified line of products, have interests in Canadian flour mills. Thus, the Canadian milling industry is both horizontally and vertically integrated. Much of the corporate control is in Eastern Canada, often with international ties (DRIE 1984, pp. 3 and 4).

Geographical dispersion of wheat milled in Canada is shown in Table 4.1. Although virtually all red spring wheat and durum is produced in Western Canada, most of it is milled in the East. Moreover, eastern wheat does not move west for milling.

Approximately 27 percent of the milling capacity is located west of Ontario, a decline from 31 percent in 1975. Over the same period, an increase of 25 percent in the East has meant that total Canadian capacity has increased. Utilization of milling capacity rose during the 1970s and fell between 1979 and 1983. Since then, the trend has been slightly upward, with current utilization of about 72 percent capacity across Canada. Western utilization has been consistently lower than that in the East. However, a rationalization in some western mills has somewhat lessened the disparity (Statistics Canada, 22-502). Ogilvie Mills, Ltd., announced the closing of their Winnipeg milling operation at the end of January, 1989. Due to overcapacity in the industry, there has been only moderate

TABLE 4.1

Wheat Milling in Canada, 1975-76, 1980-81, and 1986-87[a]

	1975-76			1980-81			1986-87		
	East	West	Total	East	West	Total	East	West	Total
				thousand tonnes					
Wheat milled									
Canadian western red spring	1,325	748	2,073	1,343	712	2,055	1,277	604	1,891
Canadian western red winter[b]	42	55	47	c		37			20
Canadian western soft white[d]	95	27	122	90	30	120		e	94
Durum	95	9	104	110	11	122	141	20	160
Ontario winter	189	0	189	170	0	170	240		240
TOTAL	1,762	789	2,551	1,750	756	2,506	1,791	675	2,466
Flour produced									
All flour except durum	1,201	562	1,763	1,210	547	1,758	1,247	496	1,743
Durum semolina	70	7	77	84	8	92	107	15	122
TOTAL	1,271	569	1,840	1,294	555	1,850	1,337	498	1,835

[a]Figures may not sum due to rounding and conversion.

[b]Formerly Alberta red winter.

[c]Blanks indicate no data available.

[d]Formerly Alberta soft white spring.

[e]Proportions for the previous two-crop years were roughly: 68 percent milled in the East and 32 percent milled in the West.

Source: Statistics Canada, *Grain Trade of Canada*, Cat. 22-201.

reinvestment, most for modernization or improvement of employee safety (DRIE 1984, p. 10).

The problem of overcapacity has resulted from declining demand, most of which has been export demand. Table 4.2 shows the substantial decline in Canadian exports of wheat flour over a span of 10 years from 926 thousand tonnes in 1975-76 to 321 thousand tonnes in 1985-86. Canada's share of the international market exceeded 30 percent in 1960; by the mid-1980s, it was down to about 5 percent. Policies of other countries, our own domestic policies, and events taking place in deficit regions of the world all play a part in this decline. More will be discussed concerning the constraining impacts of such policies later in this report.

Declining international demand can also be attributed to several nonpolicy factors, largely representing fundamental long-term shifts in demand (Mitchell and Ross). There has been substitution of flour imports with domestic milling capacity, especially in less-developed countries; three nonwheat producing countries (Indonesia, Nigeria, and Sri Lanka) have three of the largest mills in the world. This shift has been coupled with growing self-sufficiency of former major importers, such as India. Changes in milling and baking technologies, along with changes in taste, have helped to curb growth in demand for the typically hard Canadian wheats. Gluten levels may now be kept lower for baking the same quality of bread. Further, bread of a type different from that typically found in North America is more prevalent in newer markets. Lower quality, less-expensive wheats serve these needs as well or better.

TABLE 4.2

Canadian Exports of Wheat Flour

Year	North America	South America	Europe	Africa	Mid-East	Asia	U.S.S.R.	Other	Total
				thousand tonnes					
1975-76	44.4	a	4.8	19.2	21.8	389.5	429.0	18.0	926.7
1980-81	217.0	6.3	11.0	44.8	15.4	14.7	243.6		552.8
1985-86	106.8	9.2		80.8	27.4	44.4		52.6	321.1

[a]Blanks indicate negligible.

Source: U. S. Department of Agriculture, Foreign Agricultural Service.

Flour consumed in Canadian homes has remained relatively stable. Increases in population have been offset largely by decreases in per capita consumption of wheat flour. However, demand for higher value products, such as biscuits, has been on the rise. Reduced demand for export flour has been taken up in part by increased demand for flour products consumed away from home. Approximately two-thirds of Canadian flour is consumed domestically (Veeman and Veeman, pp. 70 and 71).

Table 4.3 provides estimates of value-added activity (including employment) of wheat processing in the Prairie provinces. The bakery goods produced in-store at major grocery outlets are not included in the 1985 data. To the extent that these are not counted, and particularly to the extent that this has been a growth area for bakery products, Table 4.3 underestimates the contribution of the flour industry to total value-added activities. The value-added estimates are also low because of estimation methods necessitated by confidentiality requirements of Statistics Canada.

Prairie Malt Processing

The first malting facilities in Western Canada were constructed in the early 1900s. Canada Malting Company, Ltd., built the first plant in Winnipeg in 1907 (now closed) and the second in Calgary in 1912. The Dominion Malting Company began production in Winnipeg in 1928. The next and latest facility came 50 years later with the opening of Prairie Malt, Ltd., in Biggar, Saskatchewan, in 1978. These three existing plants represent almost 60 percent of Canadian capacity.

Canada Malting Company, Ltd., the largest firm in Canada, controls 71 percent of total industry capacity. It is vertically integrated with Labatt Brewing Co. and Molson Breweries of Canada, Ltd., each owning 20 percent of its shares. Due to the regional dispersion of its plants, Canada Malting Company, Ltd., is well positioned to service the domestic industry.

Dominion Malting Company in Winnipeg has been modernized and expanded over the years, with the latest construction occurring in 1974. At one time, the firm was affiliated with Carling O'Keefe, and it has maintained a close relationship with this brewer. Today Dominion Malting Company is owned by Sumitomo, Ltd. (a Japanese trading company), and by Western Dominion Investment Corporation (DRIE 1987, pp. 7 and 8).

Prairie Malt, Ltd., was constructed originally, with some private investment, as a development project of the Saskatchewan government. Today the provincial government completely owns the company. Most

TABLE 4.3

Estimates of Additional Value in Wheat Processing The Prairies, 1975, 1980, and 1985[a]

Industry	Number of: Plants			Number of: Production employees			Value-added manufacturing ($ million)		
	1975	1980	1985	1975	1980	1985	1975	1980	1985
Products[b]	17	15	12	712	697	569	23.661	39.522	51.348
Cereal grain flour[c]	d		4			475			40.042
Flour mixes/ cereal foods[e]			4			94			19.306
Other products									
Biscuits[f]	6	3	4	466	212	295	9.993	5.624	10.801
Bread/bakery products[g]	330	285	74	2,238	2,465	1,578	52.518	90.593	89.344

Industry	Total number of employees			Total value added ($ million)		
	1975	1980	1985	1975	1980	1985
Products[b]						
Cereal grain flour[c]	d		719			40.102
Flour mixes/ cereal foods[e]			180			11.200
Other products						
Biscuits[f]	831	416	553	10.137	6.078	16.245
Bread/bakery products[g]	3,335	3,451	2,309	54.568	93.007	92.402

[a]Estimates are based on Statistics Canada, *Biscuit Industry*, Cat. 32-202; *Bread and Other Bakery Products Industry*, Cat. 32-203; *Miscellaneous Food Processors*, Cat. 32-224; and *Flour and Breakfast Cereal Products Industry*, Cat. 32-228. Confidentiality requirements necessitate reconstruction of figures for the Prairies.

[b]S.I.C. 105; flour and breakfast cereals.

[c]S.I.C. 1051.

[d]Blanks indicate no data available.

[e]S.I.C. 1052.

[f]S.I.C. 1071.

[g]S.I.C. 1072. A reclassification of this category occurred between 1980 and 1985. In 1985, firms combining bakery and retail functions were dropped from S.I.C. 1072.

Source: Statistics Canada, *Manufacturing Industries of Canada: National and Provincial Areas*, Cat. 31-203.

production of Prairie Malt, Ltd., is destined for the export market. The plant represents 13 percent of Canadian capacity (DRIE 1987, p. 8).

There are three basic types of malt. In Canada 95 percent of malting barley processed becomes brewers' malt and is used in beer production. This compares with an average 85 percent for worldwide production. Distillers' malt represents 4 percent of malt production in Canada. The remaining 1 percent of malt production is a specialty malt used in specialty beers and in the food products industry.

Since beer production is the major use of malt, the demand for malt is directly related to the demand for beer. In Canada, per capita and total consumption of beer has been declining, associated with a general decline in alcohol consumption. The decline in beer consumption has been attributed to a variety of factors, including higher levels of product prices and taxes, social and legal pressures, changing lifestyles and tastes, and an aging population. These factors have also contributed to increased consumption of beer with a lighter alcohol content which requires less malt to produce. Slow population growth and decreased per capita beer consumption will result in a stable or declining domestic market in the future unless Canadian brewers can expand into export markets. The proposed merger of Molson and Carling brewing companies is an attempt to become more internationally competitive and to increase their share of the U. S. beer market.

The fact that Western Canada has almost 60 percent of the country's malting capacity and less than 25 percent of brewing capacity seems like an anomaly until the export market is considered. As seen in Table 4.4, about 35 percent of Canadian malt output is exported. Japan and the United States, Canada's largest malt customers, traditionally accounted for approximately 65 percent of export sales. By 1985, these markets produced about 85 percent of malt sales. The United States is both an exporter and an importer of malt with most imports going into the deficit western region (DRIE 1987, p. 14). However, as reported by Canada's External Affairs Department, "considerable potential is foreseen in South America, Africa, South-East Asia, and the Caribbean." In developing countries beer consumption is just beginning to gain a foothold. These countries show the most potential for market growth, though this is dependent on increases in consumer income and the availability of foreign exchange.

The international market for malt is highly volatile. For some markets such as the United States and the United Kingdom, malt imports are a residual source of supply, to a large extent dependent on the quantity of domestic production. To meet competition, price changes may be erratic.

TABLE 4.4

Supply and Disposition of Malt, Canada, 1976, 1980, and 1984

Supply and disposition	1976	1980	1984
		thousand tonnes	
Consumed			
Brewing	293	314	314
Distilling	19	15	7
Food industry	4	4	4
SUBTOTAL	316	333	325
Exported	156	236	179
TOTAL	472	569	504

Source: Canada Grains Council, 1988, p. 61.

Subsidies by other exporters, particularly the European Community (EC), have been a major competitive factor. Such problems are exacerbated by recent increased trade tensions between Europe and the United States.

As far as Canadian exports are concerned, Canadian maltsters, since their product is differentiated, had been able to keep a number of their markets until 1983. However, between 1983 and 1985, sales to developing countries declined significantly. This was a period of aggressive marketing strategies (export subsidies) by Canadian competitors, particularly the EC and the United States. Over this period, there were shortages of malting barley in Canada due to severe droughts in the southern Prairies during 1984 and 1985, higher transportation costs due to adjustments to statutory freight rate structure for Prairie grains, and a relatively high-valued Canadian dollar (DRIE 1987, pp. 14 and 15).

The demand for malt is more inelastic in the short term than it is in the long term. Over time, production processes can be adjusted or consumer tastes changed to accept a slightly different brew. In the short term these options are not available. However, if malt prices become too high, brewers can and will use substitutes, including synthetic enzymes. The substitution of synthetic enzymes, lessening the need for larger quantities of malt, has already occurred.

Value-added activities in the malting industry are difficult to estimate since confidentiality requirements of Statistics Canada prevent publication of industry figures. Table 4.5 provides estimates for both the brewing industry and the "Other food products" category. The malting industry in Western Canada is likely to make up the vast majority of this latter group, perhaps $30-$35 million of additional value in 1985. The brewing industry in 1985 produced about $121 million in additional value for processing in the Prairies. Employment figures from the "Other food products" category likely overstate employment in the malting industry since malting requires relatively little labor. Indirect employment occurs particularly in the grains. (Dollar figures in this chapter are Canadian dollars.)

The Pasta Industry

Pasta here refers to spaghetti, noodles, and the various types of macaroni products that utilize the by-products of durum wheat in their manufacture. North American consumption per capita of these products increased from approximately 1.8 kilograms in 1970 to an estimated 3.8 kilograms in 1986. Pasta production in Canada has increased over the past decade at a rate similar to the increase in demand.

Western Canadian production made up about 19 percent of total Canadian output in 1985 (Table 4.6). This represents 23 million kilograms of pasta, compared with total Canadian production of 122 million kilograms for the same period. This situation occurs in spite of the fact that virtually 100 percent of Canada's supply of amber durum wheat, the main ingredient in pasta, comes from the three Prairie provinces. Saskatchewan alone produces approximately 80 percent of total amber durum wheat.

The Canadian pasta industry has been protected to some degree from outside competition by market-share regulations embedded in federal legislation. However, regulations on related markets have also had an effect on the Canadian pasta industry. Until 1988, the Canadian government imposed a two-price system on the sale of Canadian wheat: one price for domestic consumers and a competitive price on wheat for export. Domestic consumers of wheat, including pasta manufacturers, were required to buy their major input (wheat) from the Canadian Wheat Board (CWB) which administered the two-price policy. In some years, the domestic price was below the export price; in those years, Canadian pasta manufacturers and

TABLE 4.5

Estimates of Value Added in Malt Barley Processing[a]
The Prairies, 1975, 1980, and 1985

	Number of:						Value-added manufacturing		
	Plants			Production employees					
Processing	1975	1980	1985	1975	1980	1985	1975	1980	1985
								million dollars	
Other food products[b]	c	36			745[d]				42.992
Breweries[e]	16	13	12	1,288	950	1,276	58.779	82.268	157.857

	Total number of employees			Total value added		
	1975	1980	1985	1975	1980	1985
					million dollars	
Other food products[b]	c	1,092[d]				47.151
Breweries[e]	1,734	1,408	1,736	58.599	82.818	121.363

[a]Estimates are based on Statistics Canada, *Alcoholic Beverage Industry*, Cat. 32-232, which was discontinued in 1985 and replaced by Statistics Canada, *Beverage and Tobacco Products Industries*, Cat. 32-251.

[b]S.I.C. 1098. Statistics for the malting industry are grouped with other food products to maintain confidentiality. Although malt processing makes up a major portion of this category, figures shown overstate the degree of activity in the malting industry.

[c]Blanks indicate no data available.

[d]Overestimates for malting; a better estimate of 300 persons from DRIE, 1987.

[e]S.I.C. 1131.

Source: Statistics Canada, *Manufacturing Industries of Canada*, Cat. 3-203.

TABLE 4.6

Pasta Production, Consumption, and Trade Canada, 1985 and 1986[a]

Year	Production[b]	Imports		Exports		Domestic consumption[b]
	thousand kilograms	thousand kilograms	thousand dollars	thousand kilograms	thousand dollars	thousand kilograms
Saskatchewan						
1985	c	166.5	167	0.0	0	
1986		3.2	6	10.9	26	
Alberta						
1985	7,300	220.1	253	23.7	27	
1986		126.6	166	36.6	53	
Western Canada [d]						
1985	23,000	4,556.3	5,337	142.6	232	28,300
1986		2,414.3	2,722	123.5	179	
Canada						
1985	122,000	15,787.1	15,881	12,596.5	15,946	125,190
1986		8,437.4	8,803	7,183.5	8,648	

[a]First half of 1986.

[b]Estimated.

[c]Blanks indicate no data available.

[d]Includes Manitoba, Saskatchewan, Alberta, and British Columbia.

Source: Peat, Marwick, and Partners.

millers benefited. In some recent years, however, the domestic price has been above the export price, often substantially above the export price. This has affected recent competitiveness in the pasta industry in a negative way.

Export markets have traditionally been insignificant for Western Canada. On a national basis, however, an increasing proportion of Canadian pasta is being sold on the export market. Exports have continually exceeded imports for a number of years. For example, in 1985 and again in 1986, Canada exported pasta products to 24 different countries. Close to 90 percent of all exports are to the United States, a market almost 10 times larger than the Canadian market. In June, 1985, the United States increased its duties on pasta imports, largely in retaliation for the EC preferential tariff treatment of Mediterranean oranges and lemons. Such action is in accord with U. S. trade laws and yet has the potential to affect Canadian product movement. These tariffs will be removed as the U. S.-Canada Free Trade Agreement takes effect. The existence of the U. S.-Canada Free Trade Agreement may add stability to Canadian exporting industries. However, with American capacity in semolina production increasing substantially in the very near future, it is unlikely to have much positive effect on Canadian pasta manufacturers.

Estimates of added value of production for the existing pasta industry in the Prairie provinces can be made on the basis of average gross returns for the Canadian industry. Using a Western Canadian production level of 23 million kilograms, 1985 added value of production for the Prairies is estimated at $5.6 million (that is, $0.24 per kilogram gross profit multiplied by 23 million kilogram output).

Three major pasta makers manufacture almost 60 percent of annual production in Canada. Four other companies produce most of the remaining 40 percent. Table 4.7 lists the major brands and manufacturers, their plant locations and approximate share of annual output. Approximately 85 percent of Canada's total pasta production occurred east of the Prairies. Only two plants were located in Western Canada—one in Winnipeg and the other in Lethbridge. Several small-scale manufacturers were scattered throughout Western Canada, but their combined output was no more than 3 percent of the total. In 1987 Primo Foods, Ltd., acquired Facchin Foods, Ltd., Edmonton, and together their capacity has increased considerably.

The Canadian pasta industry is highly concentrated. Although there are relatively few manufacturers in Canada, the majority of these have

TABLE 4.7

Pasta Industry Profile, Canada, 1985

Brand	Manu-facturer	Parent company	Company location	Plant location	Production
					percent
Catelli	Catelli Ltd.	Borden, Inc. U.S.A.	a	Montreal Toronto Lethbridge	22
Lancia	Bravo Foods	General Mills	Rexdale, Ontario	Toronto	20
Primo	Primo Foods, Ltd.	Pet Foods U.S.A.		Weston, Ontario	15
Creamette	Creamette Canada, Ltd.	Borden, Inc. U.S.A.		Winnipeg	13
Kraft	Kraft, Ltd.		Montreal, Quebec	Montreal	10
Romi	Ault Foods, Ltd.		Weston, Ontario	Weston	9
Napoli	Pasta Napoli, Inc.		Mississauga, Ontario	Mississauga	8
TOTAL					97

[a]Blanks indicate not applicable.

Source: Peat, Marwick, and Partners; updated by authors.

well-established and integrated operations. As a result, they have the capability to influence markets and create barriers for new entrants. The main domestic pasta markets are retail, food services (institutional and restaurant), and specialty outlets. Retail markets absorb 65 percent of Canadian supply; food services, 30 percent; and specialty markets, 5 percent. Approximately 40 percent of pasta production is marketed through wholesalers.

Beef Industry

Western Canada's share of total cattle and calf production has been declining since 1975. In 1976 Western Canada produced over 60 percent of total cattle and calves produced in Canada. By 1986, this portion had dropped to less than 57 percent. The number of heifers for slaughter and the number of steers followed much the same pattern. There are indications that this trend may be reversed although data are not yet available to confirm a new trend (*Western Producer*).

Cattle slaughter in Western Canada has followed much the same pattern as cattle production (Table 4.8). Western Canada's share of cattle slaughter has declined from 62 percent in 1976 to 55 percent in 1986.

In terms of the regional distribution of slaughter (supply) and percent of human population (demand), the beef deficit regions are British Columbia, Ontario, Quebec, and Atlantic provinces while the Prairie provinces are a surplus-producing region. This pattern has remained relatively stable with Quebec's share of slaughter rising somewhat in the last 10 years.

A list of beef-slaughtering plants, as well as their functions, operating in Western Canada is provided in Table 4.9. Most plants in the Prairie provinces and Ontario have large capacities and are seeking to lower unit costs, while those in Quebec are smaller and much more oriented toward local markets. Cargill's large beef processing plant in High River, Alberta, which opened in 1989, may lead to further structural changes and lower unit costs in beef processing plants on the Prairies.

Pork Industry

The Canadian hog/pork industry is nationwide and is a reputable pork exporter. A minimal quantity of hogs and pork are imported (Agriculture Canada 1987, Table 8, p. 12). This industry is predominantly market driven, with different levels of government facilitating market functions such as selling and provision of certain inputs.

TABLE 4.8

Federally Inspected Cattle Slaughter[a]
1976-1986

Year	Alberta	Saskatchewan	Manitoba	Canada	Western[b] Canada as a percentage of Canada	Total western Canada
	thousand head				percent	thousand head
1976	1,537.6	199.1	537.7	3,676.3	62	2,274.4
1977	1,590.2	208.9	515.0	3,761.4	63	2,386.1
1978	1,381.3	223.4	424.1	3,439.8	59	2,028.8
1979	1,256.6	184.4	314.4	2,954.3	59	1,755.4
1980	1,251.2	185.5	302.6	3,059.5	57	1,739.3
1981	1,307.8	214.8	311.2	3,292.1	56	1,833.8
1982	1,338.0	251.1	346.8	3,398.6	57	1,935.9
1983	1,324.7	242.7	340.0	3,340.7	57	1,907.4
1984	1,276.3	262.4	332.9	3,214.4	58	1,871.6
1985	1,285.2	261.2	369.6	3,273.3	59	1,916.0
1986	1,142.2	262.9	377.4	3,234.3	55	1,782.5

[a]Data since 1981 include slaughter in provincially inspected plants.

[b]Excludes British Columbia.

Source: Agriculture Canada, Market Information Service.

TABLE 4.9

Location of Slaughtering Plants
The Prairies, Canada, July, 1987

Establ. number	Name of firm	Location	Function code[a]
	Alberta		
1A	Burns Meats, Ltd.	Lethbridge	1A, 1B, 3X, 4, 7, 8, 9
7L	Canada Packers, Inc.	Lethbridge	1A, 1B, 4, 8
7Z	Canada Packers, Inc.	Red Deer	1A, 1B, 3X, 4, 7, 9
18B	Gainers, Inc.	Edmonton	1A, 1B, 1D, 2X, 3X, 4, 5, 6X, 7, 9
21	Lacombe Meat Research	Lacombe	1A, 1B, 1D, 3X, 7, 9
38	Lakeside Packers, Ltd.	Calgary	1A, 1B
152	Dvorkin Meat	Brooks	1A, 1B, 7, 9
231	Capital Packers, Ltd.	Edmonton	1A, 1B, 1C, 1D, 3X, 7, 9
401	X-L Beef	Calgary	1A, 1B
	Saskatchewan		
69	Intercontinental Packers, Ltd.	Saskatoon	1A, 1B, 1D, 2X, 3X, 4, 6X, 7, 9
219	Town & Country Meats (1986), Ltd.	Estevan	1A, 1B, 1C, 1D, 1E, 3X, 6X, 7
394	Moose Jaw Packers (1974), Ltd.	Moose Jaw	1A, 1B, 1D, 3X, 7

(Continued on next page.)

TABLE 4.9—continued.

Establ. number	Name of firm	Location	Function code[a]
	Manitoba		
7P	Canada Packers, Ltd.	Winnipeg	1A, 1B, 4, 7, 8
1	Burns Meats, Ltd.	Winnipeg	1A, 1B, 1D, 3X, 4, 5, 6X, 7, 8, 9
1F	Burns Meats, Ltd.	Brandon	1A, 1B, 3X, 4, 5, 6X, 7
33	East-West Packers, Ltd.	Winnipeg	1A, 1B, 1D, 3X, 4, 7, 8
41	Best Brand Meats, Ltd.	Winnipeg	1A, 1B, 1D, 3X, 4, 6X, 7, 8
58	Winkler Wholesale Meats, Ltd.	Winkler	1A, 1B, 1C, 1D, 3X, 4, 6X, 7, 9

KEY TO FUNCTION CODES

1A = Slaughter beef

1B = Slaughter calves

1C = Slaughter lambs, sheep, and goats

1D = Slaughter swine

1E = Slaughter horses

2X = Canning red meat

3X = Boning and cutting of red meat

4 = Edible rendering

5 = Casing preparation

6X = Other processing of red meat

7 = Packaging and storage

8 = Inedible rendering

9 = Establishments classified under 1, 2, 3, 5, 6, and 7 have facilities for the reinspection of imported meat products.

Source: Agriculture Canada, Livestock Development Division.

Hog production in Canada has increased unevenly over time. Table 4.10 shows Canadian production levels and regional distribution of this output for census years since 1971. In 1985 Eastern Canada had 66.2 percent of hogs while the Prairies had less than half that amount, whereas in 1971 the two regions were approximately equal in hog output. Within the Prairie region, the majority of hogs have been produced in Alberta, with Manitoba production a close second.

The majority of hogs produced in Western Canada are slaughtered in their province of origin (Agriculture Canada, Agricultural Development Branch 1987b, p. 25). Limited interprovincial movement occurs. More significantly, prior to 1985, approximately 10 percent of Prairie slaughter hogs were exported to the United States (Agriculture Canada, Agricultural Development Branch 1987b, p. 27). Exports of Canadian hogs and pork to the United States have been under pressure since the mid-1980s. A U. S. countervailing duty in mid-1985 disrupted flows of hogs and pork. The action has continued since July, 1985, on live hogs only, although the amount of this countervailing duty was reduced by half in 1988. In early 1989, action against exports of Canadian processed pork has been undertaken in spite of U. S.-Canada Free Trade Agreement.

TABLE 4.10

Production of Hogs and Their Regional Distribution Canada, 1971, 1976, 1981, and 1985

Year	Total production	Eastern Canada	Prairies				British Columbia
			Total	Manitoba	Saskatchewan	Alberta	
	thousand head	percent					
1971	8,106.9	49.3	49.7	13.2	14.1	22.4	1.0
1976	5,768.1	64.6	34.5	10.8	8.5	15.2	1.0
1981	9,875.0	70.6	26.8	8.9	5.8	12.1	2.6
1985	10,784.5	66.2	31.2	11.1	6.5	13.6	2.5

Source: Agriculture Canada, 1987, Table 1, p. 6.

The majority of hogs are slaughtered in Quebec; it is estimated that between 40 percent to 60 percent (DRIE 1987, p. 29) of Canadian hog slaughter occurs in Quebec. Ontario is the second largest pork-slaughtering province, and Alberta third with approximately 12 percent of Canadian hog slaughter (Table 4.9 shows the location of pork slaughtering facilities in the Prairie provinces).

Estimates of added value from all meat and meat product industry activity (hogs, beef, sheep, and lambs) in the Prairie provinces were in the neighborhood of $340 million in the mid-1980s (DRIE 1987, p. 30). This estimate is for slaughter only. Alberta Agriculture (1988) has estimated added value in the pork industry in Alberta for the first quarter 1988 at $65 million. On an annual basis, this would result in an estimated added value for pork in Alberta of $260 million. This estimate does not include added value through curing and further processing of the hog carcasses.

If the Alberta Agriculture methodology were applied to all Prairie provinces using 1985 slaughter numbers and 1988 primal cut prices, added value in the Prairie pork industry would be an estimated $678 million. Of course, many assumptions inherent in the Alberta Agriculture methodology (such as average carcass weight, proportion going to fresh and processed products, and proportion of various primal cuts) may not be appropriate for other provinces.

Short-Line Machinery

The small and specialized equipment industry commonly known as "short line" is well established in Western Canada. Short-line machinery generally is produced by smaller companies which manufacture most of the machinery and equipment that are required for the special conditions facing Prairie farmers. Much of the short-line equipment improves, modifies, or complements large imported machines.

The short-line machinery industry in Western Canada began to emerge after the 1930s. The majority of the innovations in this industry were responses by farmers to problems encountered with imported machines. The depression of the 1930s motivated individuals to alter machine designs in order to conserve soil moisture, minimize soil erosion, and reduce costs. The Prairie Implement Manufacturers Association (PIMA) estimates that 80 percent of all farm machinery innovations in Western Canada can be attributed to farmers.

In 1989 there were over 200 manufacturers of farm machinery and equipment in the Prairie provinces. Of these, over 97 percent produced strictly short-line equipment. This represents 38 percent of the total number of short-line manufacturers in Canada (Industry, Science, and Technology Canada). The total farm machinery industry annually produced about $1 billion worth of machines and equipment in 1988. Short-line machines and equipment made up close to 70 percent of the total or about $684 million in 1988. The industry employed over 7,800 in 1988. This figure is down from a peak of approximately 9,500 in 1980. Only about 10 percent of firms employ more than 100 people.

On the national level, Canada imports close to 90 percent of all farm machine requirements. Of these imports, 90 percent come from the United States. The Prairie provinces import less than 90 percent of their equipment needs due mainly to the quantity of implements manufactured domestically. Such implements are designed to meet the requirements of Prairie agriculture.

About 35 percent of the total production of short-line equipment in the West is exported. Slightly more than 40 percent of total exports go to the north central and western United States where agricultural production characteristics are similar to those on the Canadian Prairies.

Constraints to Prairie Agricultural Processing

Four broad categories of constraints exist to enhance value-added activity in Western Canada: (1) physical commodity and environment, (2) market structure, (3) domestic policy, and (4) international policy. Each type of constraint pertains to a particular level of decision making. For example, remedies for domestic policy constraints must emanate from the federal or provincial levels of government. International policy constraints lead to a focus on bilateral or multilateral negotiations. Market structural constraints suggest a need for changes to improve the behavior of firms operating in the industry.

Although four types of constraints are used for purposes of exposition, the categories are not mutually exclusive. For example, characteristics of demand—either international or domestic—will be discussed as part of market structure. It is obvious, however, that demand expressed at the market level will often be affected by policies—either international or domestic— which regulate the industry in some way.

Physical Commodity and Environmental Constraints

Those products with values that are relatively higher than their associated transportation costs will tend to be produced near the source of supply. Perishables difficult to transport or those products whose processing does little to reduce weight and/or volume are usually produced close to consumption centers. This principle has been used to explain the location of grain-based manufacturing industries: "the dominance of eastern Canada presumably reflects primarily the location of demand and the relative costs of shipping raw and finished products" (West, Lacasse, and Weedle, p. viii). The perception of the Prairies as a source of raw materials for Eastern and Central Canada corresponds closely to the first viewpoint stated above: Locational advantage is determined within a national context. It concurs with a recent assessment of Canada's historical performance in the area of value-added activities: "the Canadian food processing industry has traditionally concentrated on meeting domestic market requirements" (Canada Grains Council, 1985c, p. 252) and has been applied to livestock products as well as to grains. Policies to stimulate value-added activities on the Prairies would be more efficient if they focused on products which adhere to locational advantages near the source of supply rather than if they attempted to overcome the pressures of locating secondary industry near the source of demand.

Perhaps it is not surprising that the vast majority of Canadian grains for domestic use are processed in Eastern Canada. The processing of grains removes little of their weight. Estimates for products including wheat flour, breakfast cereals, biscuits, and malt show that production in Eastern Canada is in the range of 75 to 80 percent of the national total in 1980. The preceding profiles noted an upward trend in the location of these activities outside the Prairie region during the 1970s. Value-added activities in the livestock sector, however, should be more efficient if located near the source of supply (e.g., U. S. location of boxed beef plants near large feedlots). Even with some types of grain products, value-added industrial activity near the source of supply may be rational. For example, high technology extraction of starches and glutens and their further processing in Western Canada would be consistent with the locational principle of transporting lower quantity, higher value products.

The other side of the locational coin is transportation cost. Domestic policies have played a dominant role in establishing the level of transportation cost. Some policies will be discussed in more detail later under domestic policy constraints, including statutory and compensatory freight rates, and the two-price wheat policy. Domestic policies have had the overall effect of increasing the opportunity cost of adding further value to grains on the Prairies and reducing input costs to eastern-based grain processors. The location of livestock production and processing in Canada has also been affected indirectly by domestic transportation policies and by other policies such as feed freight assistance.

The profile on the pasta industry points to a less obvious impact of locational disadvantage. Processing of durum into semolina is a prior step to production of pasta. Despite the fact that virtually all of Canadian durum is produced in the Prairie provinces, only about 12 percent is milled in Western Canada. The vast majority of semolina processing is done in Eastern Canada. There is current capacity in Prairie-based plants for semolina production to increase. If this capacity is not utilized, western expansion of pasta manufacturing will be limited by the availability of semolina from outside sources.

This relationship between semolina and pasta exists also for other products. Lack of capacity in processed product inputs can limit the development of further processing opportunities. Growth of the Prairie meat processing industry in the Prairie provinces is constrained since most slaughter capacity is geared to the basic steps of killing, chilling, and shipping. The malt industry on the Prairies has the opposite problem: excess capacity. However, the domestic brewing industry is inflexible in location in large part due to domestic regulation of liquor sales. The important point here is that the shipment of primary agricultural products from the Prairies creates a "catch 22" situation: If secondary processing is limited, so is tertiary processing.

Cost of transportation is related to the level of technology in handling, packaging, shipping, and storing. As technologies improve, locational disadvantages due to transportation costs lessen, and the geographic market increases. Prairie products which are more easily differentiable—such as higher value-added bakery products or beer—are in a position to gain from improvements in technology. This is a promising area for research, especially if that research is undertaken in the Prairie region.

Malt serves as a good example of the importance of considering an international as well as a domestic viewpoint. Western malt production has increased during the 1980s (DRIE 1984 and 1987). With a product such as malt, which is already moved in bulk, it is less likely that significant changes in transportation costs will affect its locational disadvantage. However, Prairie plants are well located to service the major export markets of the western United States and countries in the Pacific Rim. Changes in the industry which have allowed for the separation of the malting process from brewing have opened up advantages for plants located either near the growing region or near an export port. Another factor favoring western locations is the availability of relatively inexpensive land and energy compared to the cost of those resources in Central Canada.

The western beef and hog industries are located far from major consumption areas within Canada but not necessarily in North America. Traditionally, most of the regional exports of beef and beef products have gone east to Montreal and Toronto markets. Viewing Canadian pork and beef products as part of the larger North American market shows how the Prairie provinces and, in particular, Alberta are well situated to serve major U. S. and Pacific Rim markets. The relative costs of producing hogs vis-à-vis Quebec, Ontario, and the U. S. Midwest (Dawson, Dau, and Associates, Ltd., Table 6-10) as well as the opportunity for backhauls from the western United States, particularly California, lend themselves to locational and comparative advantage.

A strong factor in the short-line industry in the Prairie provinces is the ability of manufacturers to design specialized machinery for problems that are unique to farming practices and agricultural characteristics of Prairie agriculture. This characteristic may become even more significant as primary agriculture moves to more specialized technologies. Advances in biotechnology, no-till cropping techniques, and an increased emphasis on soil and water conservation will require specialized equipment. The presence of local manufacturers is particularly beneficial in Western Canada where farming conditions vary significantly from district to district.

Of equal importance is the increasing need for machinery that maximizes efficiency in agricultural production in the presence of high input cost and low product price markets. More precise monitoring of fertilizer, chemicals, and seed applications will spur the development and the demand for appropriate technologies.

Market Structure Constraints

Market structure here refers to characteristics of demand and supply which determine the size of the market and degree of competitiveness in an industry. The facilitative functions of marketing and financing are included in this analysis with the exchange functions of buying and selling.

Demand relative to the size of the population is tempered by a number of factors, including age, composition, and tastes of consumers. These factors combine to set limits for output. Similarly, supply or access to market is dependent on such characteristics as the number of sellers, their level of concentration, vertical integration, ownership, and level of technology. This category of constraints includes both domestic and international concerns.

All of the industry groups covered in this study suffer to some extent from weaknesses in demand: domestic, international, or both. Grains and grain products in general suffer from low price elasticities of demand and low or negative income elasticities. This is particularly true for flour and the many flour products consumed in the home.[1] There has been growth in demand in certain market niches, such as health and ethnic foods; however, the added costs of filling such market niches detract from capturing significant gains (*Milling and Baking News*, pp. 1, 13, 16, and 17).

Increasingly, Prairie hard wheats may be at a competitive disadvantage in world markets. Growth markets, such as lesser developed countries and those with centrally planned economies, have greatly expanded their imports of lower quality wheat; this trend is likely to continue (Carter, McCalla, and Schmitz). Factors which encourage technologies that displace use of hard wheats may accelerate this development. Such long-term changes in demand may make it more attractive to pursue potentials for lower quality wheats such as the medium quality varieties that are agronomically suited to Western Canada. On the other hand, a major switch in production from traditional high quality to lower quality wheat may result in a loss in markets that have traditionally been open to Canada. It may be possible to retain sufficient production of hard wheats to supply Canada's traditional customers as well as to develop new markets for lower quality wheats. The challenge will be to develop regulations for the Canada Grain Commission that can allow adequate production of both qualities of wheat in the Prairies.

The demand for malt is closely related to the demand for beer. Beer producers suffer from slow or no growth in domestic demand. This characteristic stems not only from income and price factors but also from changes in public acceptability of alcohol consumption in general. On the other hand, international growth in beer consumption is likely to continue, albeit subject to income growth in developing countries. Depending on foreign brewing capacity, markets for malt may open up some opportunities. Even on the domestic side, the demand for product from "cottage" or specialized brewing firms may provide opportunities for local maltsters. The creation of these market niches require changes to provincial regulations which may be encouraged at a federal level.

Pasta producers are not suffering from declining demand. The pasta industry has experienced a steady growth rate in both Canada and the United States over the past decade. There is a fad element in this demand; yet, price-conscious consumers have begun to regard pasta products as being a substitute for other protein sources.

Indeed, experience from the pasta and to some extent the cottage brewery industries may herald what is possible for development of new products. In particular, demand is rising for such untraditional products as wheat germ oil, defatted wheat germ, pharmaceutical products, wheat gluten (for food and other industries), vitamins, amino acids, xanthum gums, monosodium glutamate, defatted bran, and new snack foods. Such products offer new market niches. Production capacity for such products established in Western Canada in the late 1980s puts this region on a more equal footing with more established firms in Central and Eastern Canada.

Where this potential exists, availability of financing may become the effective constraint. There are definite risks to launching a new and untried product. Regional incentives, such as those provided by the Western Diversification Office, will help to overcome the effects of existing barriers and to assist local "infant industries" to achieve their comparative advantage.

Demand characteristics for livestock products differ somewhat from those of grain products. Table 4.11 shows changes in per capita consumption of meats in Canada and the United States between 1975 and 1984. Beef has lost ground to both pork and poultry. Perhaps this has been due to the image that beef is both too fatty and too expensive relative to poultry. The shift away from consumption of red meats has been somewhat mitigated for pork products by recent marketing efforts to class pork as a healthy alternative "white" meat (Alberta Agriculture 1986). Hogs in Canada are marketed on an index system that "downgrades" fat carcasses. This creates a production incentive for leaner meat. This is a characteristic desired by consumers and, therefore, packers. Because the U. S. producers sell hogs on a liveweight basis, Canadian products tend to be favored.

From an international demand perspective, the sheer size of the U. S. market makes it more attractive than the Canadian one. From 1985 to 2000, population in Canada is estimated to rise by 2 million people—from 25 million to 27.7 million. In the United States, population is projected to increase by 31 million—from 237 million to 268 million by the year 2000.

In the short-line machinery industry, the matching of supply and demand is a problem. Success in this industry has hinged on market access

TABLE 4.11

Per Capita Consumption of Beef, Pork, and Poultry Meats Canada and the United States, 1975 and 1984

Country	Consumption of meats	
	1975	1984
	kilograms per capita	
<u>Canada</u>		
Beef	48.4	38.3
Pork	24.1	27.9
Poultry	<u>19.0</u>	<u>23.0</u>
TOTAL	91.5	89.2
<u>United States</u>		
Beef	54.0	48.0
Pork	25.2	29.8
Poultry	<u>22.3</u>	<u>30.5</u>
TOTAL	101.5	108.3

Source: Alberta Agriculture, 1986, Table E-1, p. 101.

and the ability to compete with full line and established foreign manufacturers. Successes and failures are also tied to the health of the grain economy. It is essential to gain access to efficient distributors and/or dealer networks. Large established firms are able to hire the necessary sales staff and offer financial incentives to dealers. Such incentives often include financing arrangements and consignment sales. The regional Prairie implement manufacturing firms have historically had a problem in marketing their products. Marketing problems stem mainly from a lack of capital and management abilities. This situation is especially applicable to the majority of smaller firms, whose markets tend to be local. Smaller firms do not have the financial resources to have large quantities of inventory on consignment with dealers. Similarly, many dealers cannot afford, or are not willing to purchase, quantities of equipment from a manufacturer. For these reasons, domestic market penetration has been a problem for western-based firms. The industry sees a need for equity investment. Many smaller firms are technically adept but lack expertise in business management. Recent efforts have been initiated by the PIMA to address this management issue.

It is important for short-line machine manufacturers to gain a foothold in the domestic market because, although this market may not be growing, it provides some stability in times of contraction and/or instability in the international market. Foreign buyers may require or desire a consistent source of supply. Unless Prairie manufacturers and their input suppliers are willing or able to maintain supply above domestic requirements, exports to fill outside markets will be impeded.

In the malting and milling industries, new entrants are discouraged by current highly concentrated and vertically integrated market structures. Canada Malting Co., with 71 percent of Canadian market share and vertical integration with brewing firms, has an advantage in servicing the domestic market. Moreover, this type of market structure makes it possible for integrated firms to transfer market risk to independents (Ulrich, pp. 27-32) and/or spread its risk over a wider range of market activities. In milling, the three-firm concentration ratio is over 70 percent, and these firms are vertically integrated with well-known brand names. It is a similar situation with pasta; the current Canadian pasta industry resembles an oligopoly with domestic and European producers having well-entrenched domestic market shares. In these industries, a new entrant would have to gain domestic market share at the expense of existing firms; yet, the existing economic infrastructure makes this unlikely. Consequently, new production from the Prairies would have to rely heavily on export markets.

Effects of Domestic Policies

Government policy pervades all aspects of society, of which agriculture is a part. In this section only a few types of policies will be analyzed—those that have been recognized as having direct bearing on Prairie agriculture and value-added activities. The two main areas of concentration are transportation policy and pricing policies, including regional subsidies. Other policies considered particularly relevant to growth of value-added activities on the Prairies include patents, ownership rights, rules of conduct, and research. Overriding macroeconomic policies are discussed briefly as an introduction to this domestic policy section.

Effects of macroeconomic policies serve as a reminder of the interdependence of the policy environment. The combination of monetary and fiscal policies followed by the federal government and the Bank of Canada throughout the 1980s, has raised real interest rates and the exchange value of the Canadian dollar. These outcomes have had the effect of inhibiting investment in plants and equipment as well as depressing expansion of exports to the United States. Where new capital is required—e.g., for developing infrastructure for biotechnological research or for replacing worn-out meat-packing facilities—such constraints may be felt more acutely. Policies can be directed to offset these constraints; for example, federal and/or provincial industrial expansion grants in some sectors have been used to spur investment. The danger here lies in policy proliferation, often at different levels of government, such that actual effects are difficult to discern. Exchange rates are affected by monetary and fiscal policy as well as by interest rates and currency level objectives. There is a large amount of literature analyzing the effects of exchange rate changes on agriculture (Chase-Wilde *et al.*). Care must be taken when assessing the long-term viability of capacity development or expansion when trade is necessary for either disposal of products or purchase of inputs.

Transportation policy in the Prairie region has been dominated by the statutory freight rates for moving grain [formerly Crow's Nest Pass rates, now Western Grain Transportation Act (WGTA) rates]. The major economic incentive of the statutory freight rates has been to move grain off the Prairies to export points. Other transportation policies such as the Feed Freight Assistance Act have exacerbated this trend. The current freight rate policy, enforced through the WGTA, requires the government to make an annual payment of $658 million to the railways. An additional annual subsidy required, because of the costing formula adopted, has raised the total annual subsidy for transporting Prairie-produced grain to $720 million.

This policy has inhibited the further processing and feeding of grain on the Prairies (Canada Grains Council 1985a, pp. 29-31). However, this policy has not affected the production of malt since the subsidy has applied to both malting barley and malt.

In addition to dampening incentives for livestock production on the Prairies, this transportation policy has other features which distort regional advantage. Foremost, it has benefited Eastern Canadian feed grain users since not all grain moving east under subsidized rates is exported (Canada Grains Council 1985a, p. 30). Klein, Webber, and Graham have estimated that, should the Crow benefit be paid to western producers rather than to railways, the long-run effects would be a net injection into western agriculture of $100 million (crop producers lose by $240 million; livestock producers gain up to $340 million). The impact on eastern agriculture is the reverse and much smaller (crop producers gain $6 million; livestock producers lose $8 million). Secondly, as the subsidy is felt most where shipment distances are the longest, the constraining impacts on processing have been felt more in Alberta than, for example, in Manitoba. Since Crow policies have been in operation for almost a century, it can be expected that many distorting effects have been built into the structure of Canadian agriculture. Many dislocations of existing operations would likely be the result from returning the freight rate structure to one that is neutral regarding comparative advantage.

Reduced incentives for feeding grain to livestock on the Prairies has had a negative impact on packers. Not only has their major input been reduced in supply, security of supply has also been reduced. With the Alberta Crow Offset Program in place, some of the distortions in livestock supply caused by WGTA rates may have been neutralized in that province, thus adding to supply and stability of supply of finished beef and pork to Alberta packers.

Pricing of products has a direct impact on economic incentives. The CWB has a major role in the pricing of grains; liquor control boards are important in the pricing of malt. For example, the two-price policy for wheat which, while in effect maintained a distinction between export and domestic prices, impacted producers differently. Those areas producing largely for domestic uses (Ontario) were gaining proportionally more than those areas producing largely for export (Prairies) over the period (Saskatchewan Wheat Pool, p. 185). The changes of this policy, announced in mid-1988, should help to redress some of these distortions. However, the location of capital currently in the industry will continue to have a distorting effect for some time to come.

Regulations require that Canadian processors of wheat, malting barley, or semolina (milled durum flour) purchase their requirements only from the CWB. The purchase may be made only from producers who have permit books and a legitimate quota for delivery. The only exception is a situation where the processing plant utilizes its own grain production. In this case, the processor is not subject to quota regulations, but it cannot market the processed product outside of the province of production. The processor is required to charge the domestic price for its products (a price typically higher than the world price). In the case of malting barley, the processor must pay the price guaranteed to the producer (i.e., the initial price plus interim and final payments). Exceptions on input pricing may occur if deemed desirable by the CWB, for example, in facilitating an export sale of the processed product. These regulations have altered prices and domestic use in the past. The involvement of the CWB in allocating available supplies has the potential to put domestic use in direct conflict with export use. The CWB has the option of exporting a high or low quality product while providing domestic users with products of the opposite quality. An example of this occurred in 1985 when shortages of #1 Canadian western red spring wheat for export led the CWB to place an embargo on its domestic use.

Provincial pricing regulations for liquor influence, artificially and negatively, affect the demand for beer and increase regional costs of supply by restricting the movement of beer across provincial boundaries. Because demand for malting barley is closely related to that for beer, this adversely affects the potential for malt processing on the Prairies. Economies of scale are thus difficult to realize. Both effects constrain the profitability and potential of the brewing industry in Canada and, therefore, constrain the profitability and potential of the malting industry. These effects are most constraining in areas with a relatively small population, such as in the Prairie provinces.

It is imperative that restrictive provincial regulations on beer supply and demand be changed in order to allow and encourage consolidation and rationalization of the brewing industry in Canada. If these changes were to occur, it is anticipated that costs and prices of domestic beer would decrease (the degree related to competition in the industry). The demand for domestic malt would also increase. Depending on the flexibility of capital invested in beer production and considering historical precedent, some of this increased demand for malt would be produced on the Prairies.

A recent General Agreement on Tariffs and Trade (GATT) panel ruled against pricing by Canadian liquor control boards which discriminates between foreign and domestic product. As changes are brought about in accord with this ruling, consolidation and rationalization of the brewing industry will likely occur. The recent proposed merger between Molson's and Carling's brewing companies may be the first tentative steps to eventual rationalization and better resource allocation in the Canadian brewing industry.

Provincial agricultural subsidies have been used extensively in Canada (e.g., Harling and Thompson; Menzie; Clements and Carter; and Intercambio, Ltd.). These have distorted industrial structure within Canada and changed Canadian competitiveness vis-à-vis foreign producers. This is not to imply that subsidies have been the only factor in structural change among regions; however, their influence surely has been felt.

Suffice it to say that, until competition among regions in Canada is based more on market factors and less on provincial and/or federal expenditures, the Prairie region will be hard pressed to compete in areas that are significant to agricultural processing. This is not to say that all subsidies should be removed. What is required is that efforts be made to standardize the level of these subsidies so as to not distort comparative advantages of production in the different regions. Tripartite stabilization programs, where producers must pay one-third of the total cost, have been introduced for some farm products in the latter half of the 1980s. These programs will reduce the interregional distortions in production incentives.

Government licensing procedures on use or marketing can have effects that are similar to ownership regulations. One example of licensing constraints is the restriction on types of wheat that can be handled by the Prairie grain marketing system. Even though the use of high-yielding wheat has been on the rise, especially in the relatively fast growing developing countries (Carter, McCalla, and Schmitz), Canadian institutions have been reluctant to change their focus from traditional wheat varieties. In part, this reluctance stems from a grading system geared to visual inspection (Loyns and Carter, p. 84). These constraints to change continue to exist in the inspection, grading, and certification activities carried out by government agencies. It is recognized that trade is facilitated by the existence of such regulations. However, product diversification on the Prairies is being stifled by those regulations.

Patents and related protective regulations provide both constraints and incentives to innovation. Patents are intended to induce investors to make their ideas known to the public. In return, the inventor is given exclusive ownership and control of the invention for 17 years. However, when the patent is issued, the information pertaining to the patent is disclosed to the public. This is an area most relevant to machinery industries and biologically related agricultural industries.

Plant breeders' rights serve the same function as patents; they are expected to provide protection for inventors and serve as a means by which royalties can be collected. The first Plant Breeders' Rights Bill, Bill C-32, was introduced in the House of Commons in May, 1980. Objectives of the Bill were to: "1) grant legal recognition, similar to a patent, that a new variety is the property of the breeder; 2) provide a legal means by which royalties may be collected on varieties in Canada; 3) stimulate further plant breeding in Canada; 4) facilitate royalty collection on Canadian varieties used abroad; and 5) facilitate the importation and distribution of protected foreign varieties for use in Canada." These objectives were meant to protect the breeders' rights for a period of up to 18 years. Although the Bill was first introduced in 1980, it has not yet been passed.

Without legislation to protect inventors' and plant breeders' rights, the biotechnological industry will not expand at a natural rate. Caution does not lend itself to innovation. While rights remain unprotected by law, the industry will also be stifled by the inability to attract research dollars into Canada. Cangene, a Canadian based company, has established a U. S. subsidiary in San Francisco called Cangene U. S., Inc. It will benefit from the U. S. patent laws which are more protective than Canadian patent laws. Other biotechnological companies in Canada have taken out patents in the United States and will continue to do so in order to remain competitive in the industry.

In the short-line machinery industry on the Prairies, Canadian patent laws are perceived to be an inhibiting factor in industry expansion. Many innovative designs for machinery and machinery improvements have not been incorporated into the industry because of the costs and difficulties of patenting.

Research is a key to change. It can be done in the public and private sectors. Overall, Canadian levels of research activity must be considered in addition to regional research when considering constraints to development of value-added activities in Prairie agriculture. This acknowledgment points to one of the difficulties in maintaining a well-funded research program: The gains to research are often difficult to capture. Many observers have noted that public support of agricultural research is inadequate (Oehmke,

p. 53). This comment has also been made about overall research activity in Canada (Tung and Strain, p. 42).

The benefits of public research that accrue to the Prairie region depend in part on the industry groups that are emphasized and in part on the way in which research is conducted. The adaptability of the products of research to Prairie conditions is important. For example, public research in beef, hogs, grain, and oilseed varieties has had a positive influence on agricultural productivity (Tung and Strain, p. 37) and these are commodities produced in Western Canada. Intercambio, Ltd., attributes much of the increased productivity in beef production on the establishment of new grading systems and new exotic breeds of cattle imported from Europe. These were supported by extensive animal productivity research. A similar situation exists for hogs where the Canadian hog index system and research establishments at the federal and provincial levels have contributed to the international competitiveness of Canadian pork.

It should be recognized that research on such services as packaging, shipping, and storing is vital to Prairie competitiveness in many potential value-added activities. New and not-so-new grain products, such as pasta, bakery goods, and beer, would have a larger market and, potentially, greater access if such services could be improved. Some significant opportunities have already been identified in further processing of grains (Evans), but realization of these opportunities requires support of public agencies in patents or in applied research. If this research, itself, were carried out on the Prairies, added value would be created in the form of employment and industrial activity.

Effects of the International Environment

International policies by their nature affect the whole of Canada; however, regional impacts can be experienced because of factors such as different crop mixes, degree of export orientation, ability to withstand uncertainty, and targeted disputes. In general, tariffs are no longer a major problem in international trade. As the general level of tariffs has fallen through the GATT, other more disguised impediments have risen to the fore. Two major forums are being used by Canada to alleviate trade problems: multilateral negotiations through the GATT and bilateral negotiations through the U. S.-Canada Free Trade Agreement.

Although tariffs typically are not a significant cause of trade problems, their use varies with the level of value-added activity. Tariff schedules tend to move up as value is added to a product. Countries have a propensity to protect domestic value-added industries more than they do primary industries. This tendency is evident from looking at almost any list of tariff rates (Canada Grains Council 1985b). Since this tendency is not likely to change, effective development of Prairie value-added industries should include a conscious tariff strategy to complement other efforts.

Aggressive marketing practices, including inappropriate use of subsidies, are major impediments to Canada's export opportunities. The United States and the EC are key sources of these problems. Practices include the U. S. Public Law 430 foreign-aid program, the U. S. export enhancement program, and the European system of export restitutions. It is unlikely that these policies will abate in the near future. They have been well documented elsewhere;[2] thus, descriptions will not be repeated here. The focus of these policies typically is on grains, although livestock and meat products may also be affected. For example, the Canadian beef market was disrupted recently by EC exports which were induced largely by a change in European dairy policy in 1985.

Canada may attempt to counteract the effects of these policies by offering its own credit arrangements and other export-enhancing programs. These could be targeted to specific products produced on the Prairies and even to specific sales. Such market tactics have been used effectively against Canada; for example, recent U. S. attempts to economically injure the Europeans indirectly injured the Canadian malt market in Japan. However, Canada also has chosen this route to protection. Some malt sales, for example, have been encouraged by adjusting the malting barley price in light of foreign competition.[3] Targeted subsidies can help to keep the cost of retaliation to a minimum but, as seen above, they increase the cost of exporting.

Besides export subsidies, nontariff barriers of all sorts constrain the export of Prairie products. The Organization for Economic Cooperation and Development has identified and grouped numerous "domestic" policies which directly or indirectly affect international trade. Negotiations at the multilateral GATT level will attempt to address some constraints associated with nontariff barriers but change will be slow in the making and slow in the implementation.

Some policies such as U. S. quotas on beef, which result from the triggering of the U. S. Meat Import Law, must be reviewed in a slightly different light. These are publicly announced, institutionalized barriers designed ostensibly to protect domestic industry from sudden market disruption. Often the problem is not with the intent of the policy; rather, it is with the way the policy is used or misused. Health and safety regulations exist to protect product quality and thus domestic consumer or producer safety. If such regulations are abused by prolonged inspections and/or rejections, their effects tend to be destabilizing. They can add risk and uncertainty to the economic environment and can cause mistrust between trading partners. The chloramphenicol ban by several northern U. S. states that disrupted Canadian hog exports in the mid-1980s is a case in point.

The imposition of a temporary countervailing duty on Canadian hogs and pork is another example. Upon investigation, even though the final outcome led to the removal of the duty on pork products (thus being beneficial to more Prairie value-added processing activities), the disruption identified a potential source of market uncertainty out of the control of exporters.

Many nontariff barriers are disguised; they often result from producer influence.[4] Disguised nontariff barriers may be justified by producers on the grounds of ensuring the protection of consumer health. Packaging requirements provide other examples of disguised nontariff barriers. Canadian beef shipped to the United States is not allowed to have the Canadian grade stamp rolled down its side. This stipulation acts as a nontariff barrier in that exported beef must be freshly slaughtered rather than simply shipped from storage. Under a fluctuating market for beef, this may impose delays on producers exporting beef as they wait for optimal market conditions. Producers may incur losses from foregone opportunities. These types of nontariff barriers have discouraged the export of beef to the United States.

In the past, officials in the United States have made periodic use of another disguised nontariff barrier where import restrictions are suddenly changed and not made known to the exporter. As a result, beef exports are sometimes rejected when packaging requirements or transportation facilities do not comply with import regulations. This deters future exports, depending on how often export shipments are rejected. This continuation of disguised nontariff barriers imposes a greater threat of beef exports to the United States than do formal trade barriers already in effect.

As opportunities in the American far western states increase for the export of Canadian beef, nontariff trade barriers have increasingly become a problem. Surplus beef in Western Canada is generally shipped east to Central and Eastern Canada rather than south to the United States

It is estimated that the U. S.-Canada Free Trade Agreement is likely to have positive effects for those industries under consideration (Deloitte,

Haskins, and Sells). If other constraints can be overcome, as well as the development of more harmonious management-labor relations in Prairie meat packing, both the size of the U. S. market and the preference for leaner meats should enhance opportunities for beef and pork products coming from Western Canada. Further, the Agreement should help to define legitimate subsidies, thus reducing the risk of policy retaliation; harmonize technical standards; and make explicit health and safety regulations.

There may be fewer opportunities for gain in grain products. Under the Agreement, import permits on certain grains are to be removed when the subsidy levels in each country reach similar levels. Subsidy levels for oats are now relatively similar and thus import permits on this grain should be removed shortly after the Agreement takes effect. However, subsidy levels for barley and wheat in the late 1980s have been quite different in the two countries. It is likely to be some time before subsidies on production of these grains are equalized in the two countries. The development of a North American market for malt and beer may provide an opportunity or a threat, depending on the responsiveness of government legislators and of the industries to changes in market conditions.

Opportunities for Growth

In spite of the numerous constraints to diversification listed in the previous section, the Prairie provinces have many advantages for increasing value-added activities in agriculture. It is a large geographical area that has favorable soil and climatic conditions for growing a wide range of plant and animal products. Much agricultural research has been undertaken which removes some of the natural barriers to agricultural production. In the international area, federal and provincial governments have dedicated teams of officials involved in negotiations, trade promotion, and other activities in an effort to release some of the international barriers to trade. Despite even the agricultural trade war between Canada and the United States and the EC, Canadian agricultural exports continue to increase. Notable progress has been made in alleviating domestic policy constraints with a number of major shifts in economic and agricultural policies in recent years.

Grains Industry

The Prairie provinces are large producers of wheat (including durum), barley, oilseeds, and specialty crops. There are potential cost savings on transportation and handling of the primary input to western processing facilities; however, these must be compared to the value of the products being produced. Potential savings on raw product shipments off the Prairies could offset at least part of the additional costs of transporting finished products like flour and pasta to the major areas of consumption in Central and Eastern Canada. For more highly value-added products from grains, it would appear possible to expect and/or encourage processing to take place locally in the Canadian west. These products offer new market niches, putting additional capacity of Western Canadian firms in an advantageous position relative to more established, but older firms in other parts of the country.

Markets continue to grow for grain products, both domestically and internationally. The pasta industry has experienced a steady rate of growth in both Canada and the United States over the past decade. Price-conscious consumers have begun to regard pasta products as being a substitute for other protein sources such as red meats. This shift in demand may open up long-term comparative advantages for Prairie processing.

World growth in beer consumption is expected to continue, subject to growth of income and access to foreign exchange in the developing world. This growth in beer demand is likely to be translated into increases in demand for Prairie-produced barley malt, given the existence or creation of brewing production capacity in Western Canada.

Improvements in transportation, handling, packaging, and preservation technologies have expanded and will continue to expand the geographic market for grain products manufactured in the Prairie region. These technological improvements offset to some extent the distance constraint faced by Western Canadian processors. Any such benefits will be augmented if markets in the western United States become more open to Canadian products in an era of fewer trade restrictions. The realization of these opportunities is dependent, however, on the ability of Western Canadian processors to be competitive.

A number of changes in regulation of the grains industry can generally be described as positive for expansion of whole grain processing in the Prairie region. The most recent change in economic regulation of the industry should be very advantageous to the industry. As of August 1, 1988, a two-price system for wheat in this country no longer exists. Processors will be able to purchase their raw materials at North American prices. For the last several years, they had to pay a fixed price for wheat used in the domestic market. This was often considerably higher than wheat sold for export or that used for livestock feed. Processors have lobbied hard for this change and are expected to increase their output in line with their increased competitiveness relative to their U. S. competitors.

The Western Grain Stabilization Program provides a degree of protection for Prairie grain producers from fluctuations in net income. Recent changes announced by the government of Canada (broader commodity coverage and higher levies by both farmers and government as well as incentives for a higher rate of farmer participation) should make this Program financially sound. The enhanced financial stability of producers of the raw materials is a positive development for industries dependent upon primary products as inputs.

The government of Canada passed the WGTA in November, 1983. This Act replaced the formerly fixed statutory freight rates on transportation of grains and oilseeds to export terminals with rates that were meant to reflect changing costs of grain transportation. The Act allows for a gradual increase in the proportion of total transportation costs paid by farmers, while the government of Canada pays a "Crow benefit" subsidy that over time would be an increasingly smaller proportion of the total cost of grain transportation. These changes were intended to induce the grains processing industry to expand capacity in the Prairie region since the cost of the raw material would be lower. However, the government of Canada has passed a series of amendments to the Act during the past three years which have shielded producers from paying the level of freight rates originally prescribed in the Act. Moreover, the government of Canada has continued to pay the Crow benefit subsidy to the railroads rather than directly to the producers. This choice has had the effect of keeping farm-gate grain prices at a higher level on the Prairies than if the subsidy was paid directly to the producers. An experiment in the Province of Alberta to offset this distribution is discussed below in the meat-processing section. The federal government remains committed to implementation of the WGTA as adopted in 1983 despite the recent amendments. Thus, there is reason to believe that when the present financial difficulties in primary agriculture are eased, primary producers will pay a higher proportion of the total grain transportation costs. This development should ease some constraints to Prairie agricultural processing. In addition, Prairie processing would be encouraged if efforts of the Alberta government and some producer organizations eventually alter the way the Crow benefit is paid where at least some producers on the Prairies would receive the subsidy directly. Any reductions in price distortions due to freight rate subsidies on raw grains and oilseeds will benefit the grain processing industry in the Prairies.

The recent licensing of 3-M varieties of wheat is another positive factor for the grain processing industry. These varieties are of lower quality than are the standard varieties of hard wheat that have traditionally been grown in the Prairie region. However, they are higher yielding and lower priced than are the traditional varieties. Availability of these varieties may help to expand and/or diversify Canada's export opportunities for wheat flour and for Prairie products that are made from this lower quality wheat. This outcome would be a net plus to the Prairie region as long as existing markets for hard wheats and wheat flour are not lost in the process. More knowledge is required of the production and marketing trade-offs in this area.

Grain processing in the Prairie region should be assisted by reduction of trade barriers with the United States. The Prairie region is geographically closer to highly populated regions of the United States than it is to Central and Eastern Canada. Reductions of U. S. import constraints may provide expanded opportunities for the grain processing industry in the Prairie provinces.

Red Meats

The U. S.-Canada Free Trade Agreement offers the greatest prospect for increased capacity and output for processing of red meats in the Prairie provinces of Western Canada. The Agreement will provide for more secure access to U. S. markets of Canadian produced beef and beef products. The size of this market makes it an attractive one. Canadian producers will be exempt from provisions of the U. S. Meat Import Act. The two countries have agreed to harmonize their respective technical and health regulations as well as to remove other nontariff barriers to trade. The Agreement also has the potential of reducing insecurities inherent in the application of the U. S. trade law. Canadian and U. S. trade laws will still be in effect; but the proposed bilateral panel for dispute settlement, as well as provisions requiring notification of changes to laws, will help to make the environment within which trade takes place less risky.

Alberta, in particular, has a locational advantage for supplying increased quantities of red meats to the highly populated areas of southern California and the Pacific Northwest. Moreover, with deliveries of Western Canadian meat products to western United States, particularly to California, backhauls of fruits and vegetables make transportation costs relatively less expensive. With more secure access to these markets, production and processing of red meats in the Prairie provinces can be expected to grow.

The meat processing industry in Western Canada has already begun to adjust in preparation for the expected increase in export opportunities. Cargill Grain, Ltd., has recently announced the construction of a major new packing plant at High River, Alberta. A major new hog facility in southern Alberta, as well as expansion of an existing plant in Saskatchewan, have also been announced. The beef packing plant at High River, Alberta, will be the largest packing plant in Western Canada and should be competitive with major packing plants in the United States. Other packing plants in the Prairie provinces have also announced plans to become more competitive with plants in the United States. Some have modernized their plants and equipment. All have forced downward revisions in their compensation to employees.

An added incentive for modernization and expansion of packing plants in the Prairie provinces has been available in the form of federal and provincial government assistance. The Nutritive Processing Agreement in Alberta and subsidiary agreements under the Economic and Regional Development Agreement in Manitoba and Saskatchewan have provided grants, loans, and loan guarantees to enable meat-processing facilities in Western Canada to become larger and to use more modern equipment. These subsidies counteract to some degree the negative effects of federal monetary policies.

The red meats industry in Canada produces leaner carcasses than that in the United States. This provides a competitive edge for Canadian producers over their U. S. counterparts, especially among health-conscious consumers who are attempting to decrease their intake of saturated fats. Grading systems in Canada, unlike that in the United States, penalize overfat pork and beef carcasses. Choice grade of beef in the United States is equivalent to the A3 grade in Canada. Canada's grades A1 and A2 are available only for leaner carcasses. The same is true for hog carcasses. Canadian hog carcasses are marketed on an index system that penalizes fatter carcasses. In the United States, most hogs are sold on a liveweight basis that does not so clearly distinguish fatter and leaner carcasses. The Canadian grading systems for red meats, therefore, puts Canadian firms in an advantageous position for expanding exports to the United States.

As noted earlier, one of the major constraints to expansion of the red meats industry in Western Canada is the method of paying the transportation subsidy for movement of Prairie-produced grains off the Prairies. The serious consequences of this policy have been at least temporarily alleviated in Alberta by the Alberta Crow Offset Program. This program began on

September 1, 1985, and is planned to be continued until at least 1990. Its purpose is to redress the distortions in the Alberta feed grain market created by the method of payment of the Crow benefit under the auspices of the WGTA. It is designed to maintain Alberta's comparative advantage in livestock production and processing. The program in 1989 pays $13 per tonne for grain fed to animals in Alberta. This permits feeders to purchase feed grains at a lower net price than if they were competing directly with CWB purchases of feed grains. The government of Alberta has requested the federal government to pay the subsidy directly to farmers rather than to the railroads. This would reduce the farm level price of grain enabling livestock feeders to purchase feed grains at a lower price.

Although it is too early to evaluate fully the impact of the Alberta Crow Offset Program on the marketplace, it would appear that it has made the feeding of cattle and hogs more profitable in Alberta. However, the program produces its own distortions by changing the relative prices for a major input (feed) in that province as compared to prices in Saskatchewan and Manitoba.

The proliferation of provincial subsidy programs that have sought to alter comparative advantages of production is another issue that has constrained growth of the red meats industry in the Prairie provinces. These provincial policies also invite retaliation by Canada's trading partners. A hopeful sign in resolving this problem has been the development of tripartite stabilization plans in Canada. These programs are based on equal-sized contributions by producers, provincial governments, and the federal government. Premiums and support levels would be available only for the proportion of the commodity that is consumed domestically. Provincial producers can join only if their province does not "top load" provincial subsidy programs onto the tripartite programs. The tripartite programs have only a modest subsidy component and thus offer producers a more stable output price regime with little distortion in comparative advantage. This type of program is a great advantage to an industry that is cyclical in nature and requires a long-term investment commitment. Unfortunately, only two provinces in Western Canada (Alberta and Saskatchewan) joined the beef tripartite stabilization program at its outset in 1988, although all are participants in the hog plan. Subsequent to the beef and hog tripartite plans, tripartite stabilization programs for other products are being negotiated.

With approximately 14 percent of the Canadian meat and meat products industry being exported (DRIE 1987, p. 35), and a higher proportion from the Prairies, the reaction of other countries to Canadian agricultural policy should be taken into account. It is important that domestic policy be viewed as nontrade distorting tripartite programs which offer stability with little or no support and are more likely to be acceptable than are subsidy programs. Discussions at the current round of GATT negotiations may provide an opportunity for Canada to rationalize existing agricultural policy.

The pork processing industry has been assisted during 1986 to 1989 by the imposition by the United States of a countervailing duty of 4.4 cents per pound on live hogs (the duty was reduced by half in late 1988). This duty reduced the price of live hogs in Canada because the live product became relatively more expensive to export. This price reduction provided a temporary incentive for additional slaughtering and processing of hogs in Canada since no duty was applied to exports of hog carcasses or processed pork. This type of short-run benefit cannot and need not be sustained to maintain or increase the export of pork products to the United States. Other advantages, including quality of product and proximity to market, exist for Western Canada.

Canada makes a major investment in research for the red meats sector. Most research is conducted by Agriculture Canada and faculties of agriculture at universities. A limited amount of research is conducted by provincial governments and by private firms. Productivity in the red meats sector has increased rapidly over the past 20 years, due mainly to research and importation of different breeds. Changes in the grading system also have contributed to increase in productivity. The need for such research is an ongoing one if Prairie products are to continue to be competitive and especially if this competitive edge is to be enhanced.

Machines

A strong factor in the growth of the machinery industry in the Prairie provinces is the ability of the manufacturers to design specialized machinery which is unique to farming practices and agricultural characteristics of Prairie agriculture. This industry characteristic is likely to become even more significant as agriculture becomes more specialized and as other agricultural technologies emerge. Advances in biotechnology, no-till cropping techniques, and an increased emphasis on soil and water conservation will necessitate the availability of specialized equipment. The presence of local manufacturers is particularly beneficial in Western Canada where farming conditions vary significantly from district to district.

Approximately 75 percent of the articles and materials which are imported for the manufacture of machinery originate in the United States.

Currently, most of these items enter duty free. Exceptions include certain items which can also be used in other industries, primarily operational and handling systems components, destined for installation in structural facilities. The relatively free access to inputs from the United States has enabled Prairie manufacturers to be more competitive, especially in the domestic market.

The export market environment is changing somewhat in scope to the advantage of short-line manufacturers. For example, countries such as the People's Republic of China are increasingly interested in adopting dryland farming techniques common to North America. This could expand export opportunities for Prairie manufacturers of short-line machines.

Biotechnology

Biotechnology is the use of high technology to effect major increases in efficiencies of producing plants and animals. The advanced techniques include the use of recombinant DNA to prevent diseases, cell fusion, embryo manipulation, use of bacteria to develop vaccines, production of microorganisms that aid animal digestion of roughages, and production of growth hormones. Biotechnology represents a major departure from conventional methods used in agricultural research and promises much quicker progress in lowering average costs of agricultural production.

While biotechnology offers the promise of enormous increases in agricultural efficiency, perhaps the greatest obstacle to the adoption of biotechnology will come from a worried public that is afraid to accept the new technologies. Ethical issues are of special importance since biotechnologies developed for animal production may also be used on humans. The public must be educated about biotechnology and its benefits. The pressures that often arise from morality-related issues are likely to be reduced if the general public is made aware of the benefits that are possible from biotechnological applications in agriculture.

The education of users of the new technologies is at least as important as general public acceptance. The appeal to farmers will likely increase once they understand the biotechnological processes. Proper education of users regarding techniques and applications will reduce cost and risk to the farmer when adopting the new methods.

Canada must increase its research into biotechnology in order to remain competitive internationally. The private sector is heavily involved in this research along with public research stations and universities. Coordination of research among researchers in the private and public sectors, as well as the distribution of benefits and costs associated with research in biotechnology, are challenging issues facing research administrators today.

Although profits for private companies have been scarce in the past, the number of private biotechnological firms in Canada is growing. Some of the private firms established in Canada are subsidiaries of U. S. companies. A California firm is establishing a subsidiary in Edmonton, Alberta, called Biosis Alberta, Ltd. The Alberta government has made a $20 million loan to aid in the firm's establishment. It is expected to be manufacturing biological pesticides by 1990. Canadian firms that have a focus on development of agricultural biotechnology include Microbe, Inc., of London, Ontario; Bio-Technica International of Canada, Inc., in Calgary, Alberta; Allelix, Inc., of Mississauga, Ontario; and divisions of major companies such as Esso Chemical to name a few.

Chembiomed in Edmonton, Alberta, is 100 percent owned by the University of Alberta. During 1985-86, it had the highest sales per employee of all biotechnological companies in North America. This company focuses on pharmaceuticals as opposed to agricultural products. The successful approach to biotechnological research at the University of Alberta may be a model for other Canadian universities.

Several questions arise when examining biotechnological research issues in Canada. How much should private firms, government agencies, and universities contribute to the biotechnological research? How should funds be allocated to this type of research? Should the government of Canada provide massive loans to all firms such as the loan made by the government of Alberta to Biosis Alberta, Ltd.? If biotechnological research in Canada does not become profitable for private firms, should the public do it? Should only private companies be allowed to benefit from research? How should royalties and profits from the research be distributed?

It seems that Canada, as a whole, would attain greater benefits if the private and public sectors worked together on biotechnological research activities. By sharing expertise, discoveries, and facilities, Canada may be able to remain competitive with the United States and Western Europe where much of the biotechnological research is currently being conducted. However, the interests of researchers in private biotechnological companies and those in universities may not be compatible. University researchers are often less concerned about patents and protecting their discovery than are researchers in private firms. This makes information exchange less likely to occur.

A major constraint to development of a biotechnological-producing industry in the Prairie provinces (as elsewhere) is the control of intellectual property. At the end of the 1980s, few rules exist in the assignment of intellectual property rights. This is a major issue in The Uruguay Round of negotiations at the GATT. Until the issue of patent rights on intellectual property is resolved internationally, private firms will have reduced incentives to invest heavily in biotechnological research.

The Canadian Federation of Agriculture has recommended there be a national biotechnological advisory committee that is to include appropriate representation from the federal government, provincial governments, universities, private research companies, and producers. Such an advisory council would be specifically charged with establishing guidelines for agricultural biotechnological research in Canada, including legislative and regulative changes where required. The Council would also advise on the desired mix of public and private funding for biotechnological agricultural research in Canada (Agricultural Institute of Canada). However, increased research will not benefit the economy unless it is understood and applied by the industry.

In a report prepared by Agriculture Canada (June 1987a), it was suggested that Alberta is becoming a leader in animal biologics, especially in embryo transplants, and has begun to export these technologies. As biotechnology is made more available to agricultural producers, biotechnological research can be expected to grow. Competitiveness should produce an efficient and aggressive research industry that may provide for significant diversification in the Prairie provinces.

Concluding Comments

This chapter has focused on value-added agricultural activities on the Canadian Prairies. Canada and the United States are at the bottom of the industrialized nations in terms of the degree of value-added exports. Data available for a more detailed study of this important sector are piecemeal and in short supply. Information is often not available and, if available, not in a form that is comparable across industries or provinces. This lack of data makes it difficult for public evaluation and assessment of value-added agricultural activities.

The economic environment at the end of the 1980s in which businesses involved in value-added activities are making their decisions is conducive for the creation and enhancement of Prairie-based, value-added agricultural activities. However, a number of improvements to the economical environment would provide additional incentives. Several suggestions for changes are summarized below.

1. If locational advantage is to be realized for value-added activities, transportation subsidies which aid in the removal of raw products from the Prairie region should be discouraged.
2. Assistance may be needed in developing or understanding new market situations (new market and/or new product) (e.g., effects of beef trade liquidization in Japan).
3. Packaging and preserving technologies should be encouraged. These activities would ideally be located on the Prairies.
4. Rapid implementation of the U. S.-Canada Free Trade Agreement should be sought. Although some compensation will be warranted, it should be short term in nature in order to avoid ongoing distortions in resource allocation.
5. Modernization and improvement of equipment and of labor-management relations should be encouraged.
6. Maintenance and refinement of quality standards and grading will serve long-run interests of industries, including meat packing.
7. The western agricultural processing sector would gain by paying the Crow's Nest Pass statutory freight rate to the producer. However, this change alone, without finding markets for expanded production, may not be fruitful.
8. Tripartite-style stabilization programs should be encouraged in order to reduce provincial subsidy competition and intercommodity competition and to allay foreign trade concerns.
9. Continued support for research in all areas is essential. Research based on the Prairies is preferred.

From the foregoing, it is apparent that the future of agriculturally related value-added activities in the Prairie region could be bright. Many such activities already exist, often in the face of severe constraints. Opportunities for value-added development exist in many areas, and some of these surely exist in industries outside the scope of this study. Many of these have been documented in the report by Grains 2000. Still, more available data would help all those involved in decisions about expanded ventures. In many areas marketing is the key along with physical production and processing. Unfortunately, this has been lacking.

Notes

[1] For price elasticities, see West *et al.*, Table 18, p. 72; for expenditure elasticities, see West *et al.*, Table 16, p. 70.
[2] For general overviews, see, for examples, Hathaway and also Carter, McCalla, and Schmitz.
[3] Personal communication with Canada Malting Co.
[4] This section draws heavily upon Bruce and Kerr.

References

Agricultural Institute of Canada. *Agrologist*, 1985.
Agriculture Canada. *Policy Issues and Alternatives Facing the Canadian Hog Industry*. A report for the National Workshop on Hog Marketing Alternatives. Ottawa, Ontario, February 1987.
Agriculture Canada, Agriculture Development Branch. *Potential Impact of Biotechnology on Agriculture in Alberta*. Edmonton, Alberta: Agriculture Canada, June 1987a.
_____. *Capacity and Competitiveness of the Western Canadian Red Meat Slaughtering Industry*. Edmonton, Alberta: Agriculture Canada, September 1987b.
Agriculture Canada, Livestock Development Division. *Registered Establishments & Tenants, Approved Inedible Provincial Agreement*. Ottawa, Ontario: Agriculture Canada, 1987.
Agriculture Canada, Market Information Service. *Livestock Market Review*. Ottawa, Ontario, various years.
Alberta Agriculture, Statistics Branch. "Estimates of Value Added for the Alberta Pork Industry." Edmonton, Alberta, Worksheet, First Quarter 1988.
Alberta Agriculture. *Canada-United States Trade Negotiations Analysis: Livestock and Red Meats*. Edmonton, Alberta: Agriculture Canada, 1986.
Bruce, C. J., and W. A. Kerr. "A Proposed Arbitration Mechanism to Ensure Free Trade in Livestock Products." *Canadian Journal of Agricultural Economics* 34(1986): 347-360.
Canada Grains Council. *Structure of the Feed Grain Market in Western Canada*. Winnipeg, Manitoba: Canada Grains Council, March 1985a.

_____. *Canada-E.E.C. Trade and Tariffs: Grains, Animals and Related Products 1980 to 1984*. Winnipeg, Manitoba: Canada Grains Council, October 1985b.

_____. *Canada-U.S. Trade and Tariffs: Grains, Animals and Related Products 1980 to 1984*. Winnipeg, Manitoba: Canada Grains Council, October 1985c.

_____. *Canadian Grains Industry Statistical Handbook, '87*. Winnipeg, Manitoba: Canada Grains Council, 1988, p. 61.

Carter, Colin, Alex F. McCalla, and Andrew Schmitz. *Canada and International Grain Markets: Trends, Policies and Prospects*. Ottawa, Ontario: Economic Council of Canada, 1989.

Chase-Wilde, L., L. D. Cornell, N. L. Sorenson, and J. Roy Black. "World Grain Trade." In *Evaluation of Factors Affecting Net Import Demand for Wheat and Coarse Grains by Selected Countries*. Agricultural Economics Report No. 487. East Lansing: Michigan State University, July 1987.

Clements, Douglas J., and Colin A. Carter. *Nontariff Barriers to Interprovincial Trade in Swine*. Extension Bulletin No. 84-1, Department of Agricultural Economics and Farm Management, University of Manitoba, Winnipeg, March 1984.

Dawson, Dau, and Associates, Ltd. "Impact of Government Programs on the Regional Location of Meat Production in Canada." Prepared for Dr. H. M. Horner and the Minister of Economic Development, Alberta, 1982.

Deloitte, Haskins and Sells. *Canada-U.S. Free Trade in Agriculture*. Edmonton, Alberta: Deloitte, Haskins and Sells, 1986.

Department of Regional Industrial Expansion. *The Canadian Flour Milling Industry*. Ottawa, Ontario: DRIE, 1984.

_____. *The Canadian Malting Industry*. Ottawa, Ontario: DRIE, 1984 and 1987.

_____. *Industry Sector Profile of the Canadian Red Meat and Meat Slaughtering Industry*. Ottawa, Ontario: DRIE, January 1987.

Evans, T. "Diversification by Value Added in Alberta." Paper presented to the Annual Meetings of the Alberta Agricultural Economics Association. Red Deer, Alberta, April 14 and 15, 1988.

Grains 2000. *The Road Not Taken: An Opportunity for the Canadian Grains and Meat Industry*. Ottawa, Ontario: Agriculture Canada, 1988.

Harling, K. F., and R. L. Thompson. "The Economic Effects of Intervention in Canadian Agriculture." *Canadian Journal of Agricultural Economics* 3(1983):153-176.

Hathaway, Dale E. *Agriculture and the GATT: Rewriting the Rules.* Washington, D. C.: Institute of International Economics, 1987.

Industry, Science, and Technology Canada. *Industry Profile—Agricultural Machinery, Industry, Science, and Technology.* Ottawa, Ontario: Industry, Science, and Technology Canada, 1988.

Intercambio, Ltd. "Net Financial Benefits from Government Programs for Red Meat Producers in Canada, 1981/82 to 1985/86." Guelph, Ontario, April 1988. Mimeographed.

Klein, K. K., C. A. Webber, and J. D. Graham. "Regional Impacts from Selected Policies on Freight Rates for Prairie Grains and Oilseeds." University of Lethbridge, Lethbridge, Alberta, 1986. Mimeographed.

Loyns, R. M. A., and Colin A. Carter. *Grains in Western Canadian Economic Development to 1990.* Discussion Paper No. 272. Ottawa, Ontario: Economic Council of Canada, 1984.

Menzie, E. L. *Interprovincial Barriers to Trade in Agricultural Products.* Economic Working Paper. Ottawa, Ontario: Agriculture Canada, 1981.

Milling and Baking News. 67(1988):1, 13, 16, and 17.

Mitchell, D. O., and J. W. Ross. "A Forecast of Grain and Soybean Exports to the Year 2000." Michigan State University Agriculture Model, Special Report, July 1981.

Organization for Economic Cooperation and Development. *Ministerial Mandate on Agricultural Trade.* Document No. C/MIN(87)4, 1987.

Oehmke, James F. "Persistent Underinvestment in Public Agricultural Research." *Agricultural Economics* 1(1986):53-65.

Peat, Marwick and Partners. *Preliminary Profile of the Canadian Pasta Industry.* Edmonton, Alberta: Peat, Marwick and Partners, 1986.

Saskatchewan Wheat Pool. *Research Reports.* Regina, Saskatchewan, 1986.

Statistics Canada. *Slaughtering and Meat Processors.* Cat. 32-221. Ottawa, Ontario: Minister of Supply and Services, 1980.

_____. *An Overview of Canadian Grain Milling.* Cat. 22-502. Canada Agriculture and Natural Resources Division. Ottawa, Ontario: Minister of Supply and Services, 1985.

_____. *Alcoholic Beverage Industry.* Cat. 32-232. Ottawa, Ontario: Minister of Supply and Services, various years.

_____. *Beverage and Tobacco Products Industries.* Cat. 32-251. Ottawa, Ontario: Minister of Supply and Services, various years.

_____. *Biscuit Industry.* Cat. 32-202. Ottawa, Ontario: Minister of Supply and Services, various years.

_____. *Bread and Other Bakery Products Industry.* Cat. 32-203. Ottawa, Ontario: Minister of Supply and Services, various years.

_____. *Flour and Breakfast Cereal Products Industry.* Cat. 32-228. Ottawa, Ontario: Minister of Supply and Services, various years.

_____. *Grain Trade of Canada.* Cat. 22-201. Ottawa, Ontario: Minister of Supply and Services, various years.

_____. *Manufacturing Industries of Canada: National and Provincial Areas.* Cat. 31-203. Ottawa, Ontario: Minister of Supply and Services, various years.

_____. *Miscellaneous Food Processors.* Cat. 32-224. Ottawa, Ontario: Minister of Supply and Services, various years.

Tung, F. L., and G. Strain. "Research, Technology and Productivity Change in Canadian Agriculture." *Canadian Farm Economics* 21(1987):37-42.

Ulrich, A. J. "Beer and Malting Barley: A Case of Jointly Funded Public and Private Research." M.Sc. thesis, University of Saskatchewan, Saskatoon, Saskatchewan, 1983.

U. S. Department of Agriculture, Foreign Agricultural Service. *Grains, World Grain Situation and Outlook.* FAS Circular, December, 1986.

Veeman, T. S., and Veeman, M. M. *The Future of Grain.* Toronto, Ontario: Canadian Institution for Economic Policy, 1984.

Western Producer. "Cattle Industry Restructuring." Saskatoon, Saskatchewan, March 3, 1988.

West, D. A., C. Lacasse, and S. Weedle. "Structure of the Grain and Oilseed Based Manufacturing Industries." Background paper of the Eastern Grain Production Seminar, Montreal, October 28-29, 1982.

5

Freer Trade in the North American Beer and Flour Markets

Colin Carter, Jeffrey Karrenbrock, and William Wilson

Introduction

The purpose of this chapter is to discuss the economics of the brewing and flour milling industries in the United States and Canada. Topics will include trends in international trade, trade barriers, market structure, and the impact of the U. S.-Canada trade liberalization agreement on these industries. Breweries and flour mills were chosen for three principal reasons. First, both are an important part of the North American food manufacturing industry. Second, the brewing and flour milling industries in Canada and the United States have many characteristics in common; they are industries which are highly concentrated, they receive both subsidies and protection from imports, and they are high "value added" industries. Third, it is expected that the U. S.-Canada Free Trade Agreement (FTA) will affect the brewing and flour milling industries in different ways because the brewing industry was excluded from the FTA and the flour milling industry was not. We argue in this chapter that both industries will be significantly altered because of the FTA. This chapter will analyze the two industries separately, focusing on the brewing industry first before turning to flour milling. The summary will compare and contrast the two industries.

Brewing Industry

Although the brewing industry was left out of the FTA, market forces continue to push the industry toward freer trade. A variety of factors on both sides of the border led to the exclusion of the brewing industry allowing both countries to continue practices that inhibit beer trade. Recent events, however, suggest that the exclusion of the industry may have only prolonged the opening of the market. A merger between Canada's second and third largest brewers and Canada's agreement to comply with a General Agreement on Tariffs and Trade (GATT) ruling calling for a reduction in discriminatory beer sales practices suggest that a rationalization of Canada's brewing industry may be underway.

The organization of the first half of the chapter is as follows. It begins with a discussion of world beer trade and the roles of the United States and Canada in the world beer market. Next, trade barriers restricting beer trade in Canada and the United States are outlined. A brief description of the markets that have developed behind the trade barriers in both countries is given. Then, we examine some of the most important reasons as to why the brewing industry was excluded from the FTA. Finally, we review some of the recent developments that have occurred in the brewing industry since the passage of the FTA.

World Beer Trade

The amount of beer traded on world markets has exploded during the past 20 years (Figure 5.1). The value of beer exports has jumped from $149 million in 1965 to $1.7 billion in 1986. (Dollar figures in this chapter are Canadian dollars unless otherwise specified.) Of course, a general expansion in world trade and inflation have both played a part in accounting for these trade figures. In constant dollars the beer trade in 1986 was over 10 times its 1965 value.

The largest exporters of beer, by value, are the Netherlands, West Germany, Canada, Mexico, and Denmark. In 1986, the Netherlands exported 26 percent of the total amount, by value, while the top five exporting countries accounted for 66 percent of world beer exports. The largest importers of beer, by value, are the United States, the United Kingdom, France, Italy, and West Germany. In 1986 the United States alone made up 46 percent of the import market, and the top five countries accounted for 70 percent of the import market.

United States as a World Player. In 1987, the United States imported over 9.35 million barrels of beer, which accounted for about 7.6 percent of total U. S. beer consumption.[1] In 1970, imports accounted for less than 1 percent of total U. S. beer consumption. On a quantity basis, U. S. beer imports have jumped to 10 times their 1970 level.

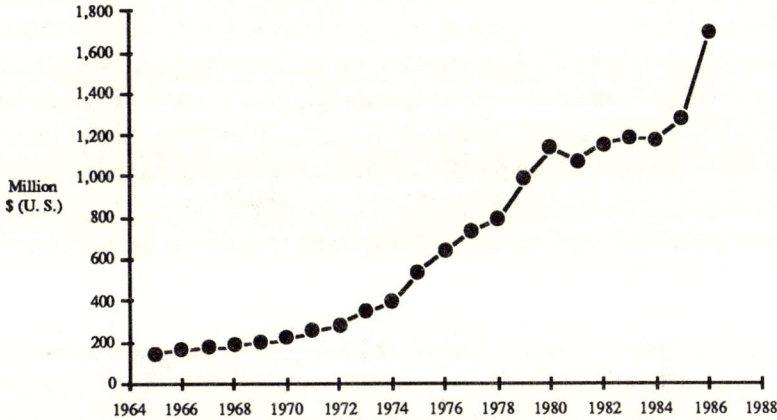

FIGURE 5.1. World Beer Exports, Selected Years, 1964-1988

Source: United Nations, Food and Agriculture Organization, Trade Yearbook.

The three largest exporters of beer to the United States in 1987, and their respective market shares, by volume, were the Netherlands, 27 percent; Mexico, 26 percent; and Canada, 22 percent. The top five exporters, which include the above countries plus West Germany and Australia, accounted for about 90 percent of all beer exports to the United States in 1987.

Although the United States is a large player on the import side, its role as an exporter is limited as it accounts for about 2 percent of world exports, by value. By 1986 the nominal value of U. S. beer exports had climbed to more than 30 times its 1970 value. The largest importers of U. S. beer and their respective market shares, by volume, in 1987 were Canada, 26 percent; Hong Kong, 19 percent; and Japan, 15 percent. The top five export markets for U. S. beer in 1987, which included the above countries plus Taiwan and Panama, accounted for 76 percent of total U. S. exports. In 1986 the United States exported about $39.5 million worth of beer, while importing approximately $786.3 million worth of beer.

Canada as a World Player. Canada's role in the world beer market is just the opposite of that of the United States. While the United States is a major importer and a minor exporter, Canada had less than 1 percent of the world beer import market but had more than 8 percent of the world beer export market (by value) in 1986. Beer imports in 1986 accounted for about 1.6 percent of total Canadian beer sales (by volume) compared to .4 percent in 1970. Beer imports have accounted for as much as 4.4 percent of Canadian beer consumption during years in which labor strikes have decreased the output of Canadian breweries. Between 1970 and 1986, the nominal value of Canadian beer imports have increased to more than 7 times their 1970 level (1986 was a nonstrike year). [In quantity terms, Canadian beer imports have increased from about 47,500 barrels in 1970 to about 287,000 barrels in 1986.]

The largest exporters to Canada and their respective market shares by volume in 1986 were the United States, 55 percent; the Netherlands, 13 percent; and the United Kingdom, 10.6 percent. The top five exporters of beer to Canada, which includes the above countries plus West Germany and Australia, accounted for 88 percent of all Canadian beer imports in 1986.

Canada was the third largest exporter of beer by value in the world in 1986. Essentially all of Canada's beer exports go to the United States. Since 1975, the United States has purchased about 99 percent of Canadian beer exports. The nominal value of Canadian beer exports in 1986 was 32 times its 1970 level. Overall, Canada has a substantial trade surplus in beer; in 1986 it exported about $136 million (U. S.) and imported only $14 million (U. S.) worth of beer.

U. S.-Canada Beer Trade. U. S.-Canadian beer trade was relatively constant until the early 1970s when shipments between the two countries began to grow rapidly. U. S. imports from Canada have shown steady growth but with slower growth in the most recent years. Canadian imports of U. S. beer have been increasing in general but have been very irregular due to strikes in the Canadian brewing industry. U. S. total beer imports and beer imports from Canada are shown in Figure 5.2, while total Canadian beer imports and beer imports from the United States are shown in Figure 5.3.

Canada's market share of total U. S. beer imports hit a high of 33.5 percent in 1980 but has since fallen to 21.5 percent in 1987. Canadian beer accounted for about 1.1 percent of total U. S. beer consumption in 1987. As shown in Figure 5.2, U. S. exports to Canada have been sporadic since 1977. Exports to Canada have accounted for as

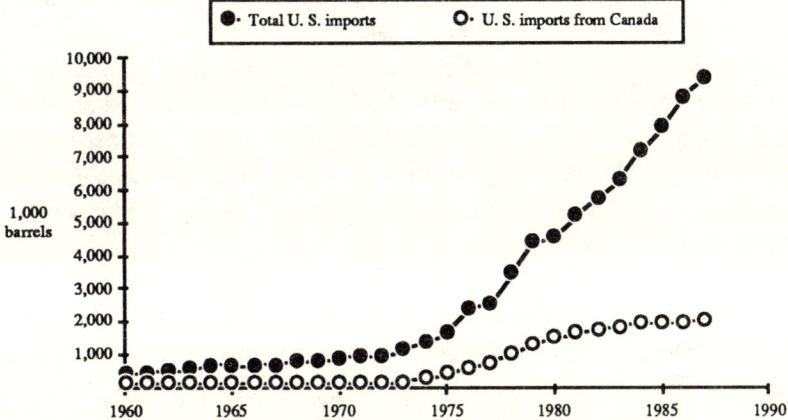

FIGURE 5.2. U. S. Beer Imports, Selected Years, 1960-1990

Source: National Association of Beverage Importers.

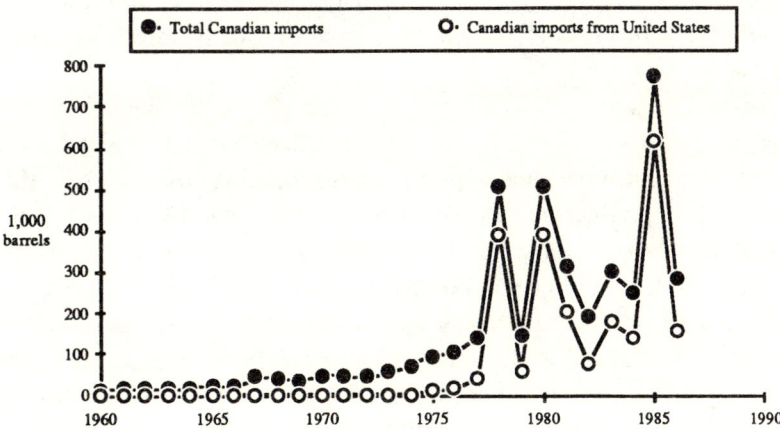

FIGURE 5.3. Canadian Beer Imports, Selected Years, 1960-1990

Source: Brewers Association of Canada.

much as 63 percent of total U. S. beer exports, by volume, but have averaged about 26 percent in recent years.

Canada imported very little U. S. beer during the 1960s but began to increase imports during the early 1970s. Then, as the Canadian brewing industry began to experience labor strikes, imports of U. S. beer increased substantially. U. S. beer imports as a percent of total Canadian consumption has been as much as 3.5 percent during strike years but was .9 percent in 1986.

Role of Licensing Agreements. The beer trade figures between Canada and the United States understate the true quantity of U. S. brand beer consumed in Canada. This is because three of the largest U. S. brewers have signed licensing agreements with the three largest Canadian brewers. A licensing agreement allows one brewer to make and sell another brewer's brand. The licensee then must pay the licenser a royalty fee for being allowed to sell its brand. The licenser is in essence exchanging the knowledge of how to brew a specific beer and its brand name for a royalty payment. In 1987, there were 30 licensing agreements between various producers all over the world (Finnegan). The U. S. companies licensing production in Canada and their respective Canadian licensers in 1987 were Anheuser-Busch and Labatt, Coors and Molson, and Miller and Carling O'Keefe. Brands produced in Canada under license to U. S. brewers may account for as much as 15 percent of sales in Canada (Conference Board of Canada, p. 9).

Several reasons promote the use of licensing agreements. First, the physical qualities of beer enhance the use of licensing agreements. Beer is about 90 percent water, and expensive transportation costs can be reduced through local production. In addition, beer has a shelf life of only about three to four months, of which two to three weeks can be taken up during overseas shipments. Second, companies often find it cost efficient to use existing plants and distribution systems when entering new markets. Third, companies purchasing the right to produce other firms' products sometimes do so in order to decrease excess capacity in their plants. Economies of scale exist in the brewing industry so, if brewers can increase their beer output, then their average cost of producing beer will fall.[2] Finally, brewers use licensing agreements to circumvent both tariff and nontariff trade barriers. For example, U. S. brewers who have licensing agreements with Canadian brewers can sell their brands in Canada without having to pay the import tariffs and do not have to face price markups at the retail level that discriminate against foreign beers. Of course, the brewers are not actually physically exporting beer to Canada, but they are receiving a royalty payment for their brands' sales. Brewers who do not have a licensing agreement must face the import tariffs and the nontariff barriers in Canada. In addition, nonlicensing brewers are at a competitive disadvantage since they do not have access to established distribution systems as the licensed brewers do.

Trade Barriers

There are several important barriers to trade in beer between the United States and Canada. Both countries have a variety of restrictions that inhibit beer imports and reduce economic welfare. The increase in retail prices caused by the barriers force consumers to pay more for beer, while the barriers also set the stage for the development of an inefficient domestic brewing industry.

U. S. Trade Barriers. The United States has both federal and state regulations that hinder the importation of Canadian beer.[3] At the federal level, a tariff of $1.86 per barrel is imposed on all ale, beer, porter, and stout entering from Canada. State-level restrictions come in the form of labeling and packaging requirements. While only 18 of the 50 states have such restrictions, the costs of making small production runs to meet the labeling requirements can significantly increase unit costs of exported beer for Canadian brewers. These state restrictions also affect domestic brewers.

Finally, Canadian brewers also face the problem of attempting to enter U. S. distribution systems. Many U. S. beer wholesalers are committed to distributing for one of the major U. S. brewers and will only carry that firm's products. This leaves little access to the most efficient means of distribution for the Canadian brewers.

Canadian Trade Barriers. Canadian restrictions on beer imported from the United States are somewhat more varied and complicated. Similar to the United States, restrictions exist at both the federal and provincial levels. However, the provincial barriers are not just targeted at foreign beer; they also restrict the consumption of other Canadian beer brewed outside the province.

Canadian federal restrictions include a tariff of $3.30 per hectoliter on all imported beer as well as setting a federal sales tax about 4 percent higher for imported beer. Provincial barriers are so numerous and varied that only the barriers considered to be the most significant are discussed here. It is important to realize from the outset that liquor sales in the provinces are controlled by liquor authorities established by the provincial and territorial governments. These authorities act in a manner that attempts to maximize the revenue raised from selling alcoholic beverages.

The first type of provincial barrier is known as a listing requirement. The provincial authorities require a brewer to obtain a listing for any brand he wishes to sell in the province. Discriminatory uses of the listing requirement include refusing to list out-of-province beer and requiring proof of demand before listing, minimum sales quota to remain listed, and frequency between listing decisions. Note that most of these are practices which private sellers would also employ but perhaps on a less formal basis. The fact that the provincial authorities have these restrictions does not imply that they necessarily restrict trade, but the ability to use them as restrictive practices certainly exists. Take the frequency between listings, for example. The provinces of Quebec and Alberta consider new beer listing applications on an as-needed basis for domestic products brewed in the province but only once a year for foreign imports. Since foreign beer applications are only considered once a year, failure to prove sufficient demand could delay market entrance for another year.

The second type of provincial nontariff barrier comes in the form of marketing practices in two of the largest Canadian beer markets, Ontario and Quebec (65.9 percent of total beer sales in 1987 occurred in these two provinces). In Ontario, U. S. brewers face the same problem Canadian brewers face in the United States—the most efficient means of distribution are reserved for beer brands associated with the brewers/distributors. In Ontario, about 99 percent of the beer is sold through the retail outlets of the Brewers' Warehousing Company, Ltd. (BWC), which is essentially owned (98 percent) by three Canadian brewers. The BWC distributes only beer produced by its owners and distributes for other Ontario brewers for a fee. It does not distribute out-of-province beer. Therefore, all foreign beer and out-of-province beer must be sold at the province's Liquor Control Board (LCB) outlets. In Quebec, only in-province brewed beer can be sold through private grocery stores and corner stores and account for the majority of beer sales.

Marketing practices of the provincial liquor authorities also have a negative influence on the sale of imported beers. In Ontario, the LCB does not cool any of the beer it sells (domestic or foreign) and often carries smaller, more expensive packages. These practices tend to make the products less appealing to the consumer. More importantly, since most of the beer is sold through the private outlets, the foreign beers are less accessible to consumers than domestic beers. The situation is similar in Quebec.

In addition to being marketed poorly by provincial outlets, foreign beers also face discriminatory pricing practices at the liquor authorities' outlets. Liquor authorities in all provinces have higher markups for imported beer than for domestic beer. In some cases, the government outlets also have higher markups for beer brewed outside of the province. These markup *differences* range from 2 percent to 100 percent across the Canadian provinces (see Table 5.1). Selling imports at a premium to domestic beers may be reasonable, as consumers often perceive imported beer to be of higher quality and there are greater costs associated with obtaining foreign beer. It is the amount of the price differential that is in question. Discussions with U. S. distributors suggest that most U. S. retailers markup domestic and foreign beer at *the same rate* of about 20 percent to 25 percent.

The final provincial barrier of significance comes from packaging laws. Four of the provinces have laws either prohibiting or limiting the amount of sales of canned beer or require a deposit on canned beer sales. For environmental reasons, recyclable bottles are preferred over cans. Cans are cheaper to ship over long distances than bottles, and there are no costs involved with recycling the discarded product. On the other hand, bottles are cost effective for markets near the breweries. Thus, breweries that are close to these provinces are favored over breweries that are located at a distance. However, as Canadians begin to accept the recyclable aluminum cans, this will likely become less of a barrier.

In fairness to the Canadian brewers, it should also be pointed out that some industry observers believe that some of the provincial barriers actually discriminate in favor of U. S. brewers, over Canadian brewers. For example, one observer believes that the U. S. brewers are not charged the full rate for distribution of their products by the provincial distributors.

Market Structure

The barriers to trade described above have influenced the market structure that has developed in the two countries. The U. S. industry, having relatively little protection, has undergone a substantial rationalization since the repeal of prohibition. The Canadian industry, hindered by the

TABLE 5.1
Provincial Markups on Brewery Products by Source

| Province | Markups on beer brewed in Canada ||| Markups on beer brewed outside Canada ||
	In province	Outside province	Outside province by an in-province brewer	United States	Other countries
Newfoundland	$0.33/can	a		$0.50/can	$0.50/can
Prince Edward Island	74 percent	71 percent			
Nova Scotia	67 percent	76 percent		80 percent	85 percent
New Brunswick	62 percent	65 percent		86 percent	86 percent
Quebec	Private stores set markups				124 percent
Ontario	21 percent		21 percent	80 percent	80 percent
Manitoba	50/75 percent[b]		50/75 percent[b]	50/75 percent[b]	50/75 percent[b]
Saskatchewan	58 percent		60 percent		60 percent
Alberta	50 percent			65 percent	
British Columbia	50 percent		83 percent	83 percent	83 percent

[a]Blanks indicate not applicable.

[b]Markup is 50 percent when sold by a vendor (the main retail outlets in Manitoba) and 75 percent when sold through the liquor board. Imports are also subject to 12 percent provincial sales tax rather than the 6 percent rate on domestic beer.

Source: Brewers Association of Canada, 1986.

provincial barriers, has not undergone as much rationalization. A brief discussion of each countries' market structure follows.

The U. S. brewing industry has undergone an almost continual concentration of production since 1945. The number of breweries operating in the United States fell from 471 in 1946 to 92 in 1983. Much of this concentration was brought on by advancements in production, storage, and transportation. As these improved, brewers could store their products longer and ship them faster and cheaper, eliminating the need to have breweries located in every major market. Another factor that has encouraged the reduction in the number of breweries is economies of scale. As brewers reduced the number of plants in operation, output increased at the remaining plants, thereby lowering the average cost of production. In the United States, the average output per plant went from about 412,000 barrels in 1960 to over 2 million barrels in 1983. Some plants, however, have substantially more capacity than the average plant. For example, Anheuser-Busch is building a new plant that will have a capacity of about 6 million barrels. Since 1983, the number of breweries operating in the United States increased to 147 in 1987. Much of the increase came from extremely small breweries, or microbreweries, that are attempting to serve a very specific market.

The Canadian industry has also seen some rationalization but not to the same degree as in the United States. The number of establishments producing beer in Canada in 1929 was 78 and, by 1987, the number had fallen to 57 establishments. Of these 57 plants, 37 were conventional breweries, the rest were microbreweries. The average output of Canadian brewers has grown from about 150,000 barrels in 1960 to 376,000 barrels in 1984.

As the number of breweries declined, so did the number of firms operating in the brewing industry; and sales became concentrated among a few firms. In the United States, the five largest brewers account for over 90 percent of all beer production. In 1987, Anheuser-Busch held 40.5 percent of the U. S. market, by volume. Miller, Stroh, Heileman, and Coors held 21 percent, 11.7 percent, 8.8 percent, and 8.4 percent of the U. S. market, respectively (Beverage Industry, p. 40). In 1987 the Canadian industry was dominated by three brewers who held over 90 percent of the market. Labatt Brewing Co., the largest in 1987, held about 41 percent of the market. As discussed below, Labatt is also one of the largest flour millers in Canada.

U. S.-Canada Free Trade Agreement

Thus far, our discussion has centered around the U. S.-Canada beer trade, barriers to beer trade, and the market structure which developed behind these barriers. Next, we need to take a look at some of the more important reasons why the brewing industry was all but excluded from the FTA.

The FTA calls for the elimination of federal tariffs placed on beer imports by both the United States and Canada in 10 annual steps, starting in 1989. However, the federal tariffs do not represent a significant barrier to trade. (The U. S. tariff amounts to about one-half cent per 12-ounce can of beer.) The more important need to reduce provincial barriers to beer trade under the FTA was sidestepped in Chapter 12 of the FTA: *Exceptions for Trade in Goods, Article 1204—Beer and Malt Containing Beverages*. The FTA will allow provincial liquor authorities to continue all existing nontariff trade barriers. It is interesting to note that provincial barrier reductions were negotiated for wine and spirits under the FTA in Chapter 8: *Wine and Distilled Spirits*. Issues pertaining to listing practices, price discrimination, distribution, and blending requirements were all addressed in this section of the Agreement.

The question remains then, why were the brewing industry's most important trade barriers exempted from the FTA? Factors on both sides of the border led to the exclusion. The Canadian case is analyzed first. Canadian brewers made a strong argument against the agreement on the grounds that they have higher costs than American brewers and, by allowing free trade, the U. S. firms could make sizable inroads into the Canadian market. This, in turn, could cause large layoffs in the brewing industry; and lower trade barriers would cause a loss of tax revenue to the provincial liquor authorities. Lastly, the Canadians claimed that the capital investment that would be needed to allow the industry to rationalize to a point where it would become competitive with the American brewers would be extremely high. Each point is analyzed briefly below.

Canadian Relative Cost Disadvantage.[4] The Brewers Association of Canada (BAC) estimates for the relative costs of producing canned and bottled beer in the United States and Canada are shown in Tables 5.2 and 5.3. The Canadian brewers claimed the cost disadvantage they faced stemmed from higher input costs and lack of economies of scale, both of which can be traced to government regulations. On the input side, barley

TABLE 5.2

Relative Costs of Producing Canned Beer: Canada and the United States, 1986

	United States		Canada	
	Brand A[a]	Brand b[b]	Ontario/ Quebec[c]	Other[d]
	dollars per hectoliter			
Brewing ingredients	3.55	5.92	7.50	8.00
Cans and other packaging	37.87	41.53	44.40	48.00
Total materials cost	41.42	47.45	51.90	56.00
Labor[e]	5.50	7.10	8.00	9.50
Variable product cost	46.92	54.55	59.90	65.50
Freight[f]	5.00	5.00	2.50	2.50
Delivered variable cost	51.92	59.55	62.40	68.00
Overhead and administration[g]	11.00	16.20	14.00	18.00
Distribution[h]	20.00	20.00	20.00	20.00
Total cost (excluding marketing)	82.92	95.75	96.40	106.00

[a]Typical costs for U. S. popular priced brand produced by firm with low costs.

[b]Typical average costs for top five brewers with primarily premium and superpremium brands.

[c]Typical costs for Ontario and Quebec plants.

[d]Typical costs for plants outside Central Canada.

[e]Direct and immediate supervisory labor.

[f]U. S. plant assumed to be twice the distance from distribution center.

[g]U. S. costs reflect much higher depreciation costs on newer plants and lower administration costs; Canadian costs show considerable variance across firms.

[h]Assuming U. S. brewers gain access to Canadian distribution systems; average for all provinces used in both cases.

Source: Brewers Association of Canada, 1986.

TABLE 5.3

Relative Costs of Producing Beer in Refillable Bottles: Canada and the United States, 1986

	United States		Canada	
	Brand A[a]	Brand b[b]	Ontario/ Quebec[c]	Other[d]
	dollars per hectoliter			
Brewing ingredients	3.55	5.92	7.50	8.00
Bottles and other packaging	e	13.08	13.75	15.00
Total materials cost		19.00	21.25	23.00
Labor[f]		13.55	12.25	14.00
Variable product cost		32.55	33.50	37.00
Freight[g]	9.50	9.50	3.70	6.50
Delivered variable cost		42.05	37.20	43.50
Overhead and administration[h]	11.00	16.20	14.00	18.00
Distribution[i]	22.00	22.00	22.00	22.00
Total cost excluding marketing)		80.25	73.20	83.50

[a]Typical costs for U. S. popular priced brand produced by firm with low costs.

[b]Typical average costs for top five brewers with primarily premium and superpremium brands.

[c]Typical costs for Ontario and Quebec plants.

[d]Typical costs for plants outside Central Canada.

[e]Blanks indicate no data available.

[f]Direct and immediate supervisory labor.

[g]U. S. plant assumed to be twice the distance from distribution center.

[h]U. S. costs reflect much higher depreciation costs on newer plants and lower administration costs; Canadian costs show considerable variance across firms.

[i]Assuming U. S. brewers gain access to Canadian distribution systems; average for all provinces used in both cases.

Source: Brewers Association of Canada, 1986.

malt accounts for about 70 percent of the total expenditure on raw material and ingredients (excluding packaging). Canadian brewers face a substantial cost disadvantage in procuring this input due to the monopoly selling right granted to the Canadian Wheat Board (CWB). The BAC estimates that American brewers had a cost advantage on malting barley of about $1.80 per bushel. As Tables 5.2 and 5.3 show, the difference between the average cost of brewing materials in the United States and Canada runs about $1.50 to $4.50 per hectoliter in favor of U. S. brewers, depending upon the type of beer brewed and brewery location.

The BAC also estimates that packaging costs are significantly different. They note that some of the large U. S. brewers roll their own cans and save roughly $3.50 per barrel. But Canadian firms enjoy lower average packaging costs because they use mostly recyclable glass bottles. Total packaging costs in the U. S. and Canada are about the same for refillable bottles, but U. S. brewers can package canned beer for about $3 to $10 per hectoliter less than the Canadians.

With respect to labor costs, the average hourly wage in 1984 for Canadian workers was about $14.47 per hour, while the average U. S. worker earned an equivalent of about $23.39 per hour. Some of the Canadian gain in labor costs is lost through lower average worker productivity. The BAC estimates that, in 1984, average output per Canadian brewery employee was only 30 percent of American workers. This is a result of less mechanization in Canada and the use of more labor-intense bottled beer. In 1984, the U. S. brewers had a wage bill of $5.33 per hectoliter, while Canadian brewers paid about $11.10 per hectoliter.

When all variable production costs are considered, U. S. brewers have a cost advantage ranging from $5 to $19 per hectoliter on canned beer, depending on brand of beer and location of production. The U. S. variable cost advantage on producing bottled beer was substantially less, ranging from $1 to $5 per hectoliter.

After including transportation and administrative costs, the total cost of selling canned beer in Canada could be as much as $23 more per hectoliter for Canadian brewers. However, most U. S. beers that compete in the Canadian market are premium beers, which means the higher of the two U. S. cost estimates is more appropriate to use. In addition, Ontario and Quebec account for most of Canada's beer sales, so this is the most likely target market for U. S. exporters under a free-trade situation.[5] Canadian beer production in these two provinces runs about $96.40 per hectoliter, while U. S. costs for producing premium beer are about $95.75 per hectoliter. Essentially, when all costs are considered, the U. S. brewers do not have a substantial cost advantage ($0.65 per hectoliter) over the Canadian brewers. Similarly, for bottled beer it would seem most reasonable to use the more expensive cost estimate for the United States and the lower estimate for Canada which, in this case, gives the Canadians a $7 per hectoliter cost advantage over U. S. brewers.

Although canned packaged beer sales have been increasing relative to total sales, bottled beer sales accounted for over 80 percent of Canada's beer sales in 1987. Given this consumer/provincial government preference for bottled beer and the Canadian producer's cost advantage, or at the very most a small disadvantage ($3.25 per hectoliter), there seems to be little evidence for the Canadian brewers' claim that the relative cost of producing beer in Canada would prevent them from competing with U. S. brewers. Although the Canadians do have a substantial production cost disadvantage, when transportation costs are included, the difference between Canadian and U. S. costs are not so great. The Canadian brewers also point out, however, that American brewers could view the Canadian market as a marginal one and use aggressive price-cutting measures to make inroads. While this is a possibility, it is not an inevitable outcome of free trade.

Stemming from their belief that the Canadian brewing industry was at a relative cost disadvantage to U. S. brewers, the Canadians estimated that free trade in beer would have widespread negative economic impact on the Canadian brewing industry. The BAC economic impact model estimated that free trade in the brewing industry would, in the long run, have the following impacts.[6]

- Lower domestic brand beer prices by 0 percent to 15 percent, depending on the province.
- Increase beer consumption by 0 percent to 5 percent, depending on the province.
- Allow imports to gain about a 32 percent share of the beer market in Canada. (In 1986, imports accounted for 1.6 percent of total Canadian beer sales.)
- A direct and indirect loss of over 19,000 jobs across Canada, including about 16,000 manufacturing jobs. (The Canadian brewing industry employed over 13,000 people in 1985.)
- A loss of almost $190 million in wages, salaries, and benefits and over $180 million in raw material purchases paid annually by the

brewing companies (Canadian brewers paid $255.3 million in wages and salaries in 1984).
- A direct and indirect loss in federal and provincial commodity, corporate and personal income tax revenue of over $270 million annually.
- A nearly $1.1 billion reduction in the Gross National Product (GNP) contribution made by the production and sale of beer in Canada. (Canada's GNP in 1986 was $489.9 billion.)
- An estimated $2 billion (1984) could be required in new capital to fund the rationalization of the industry.

Clearly, the Canadian brewing industry felt that the negative aspects associated with free trade in the brewing industry would far outweigh any positive impacts. Some in the Canadian brewing industry supported the FTA in general but felt that, first, provincial barriers to trade needed to be reduced to allow rationalization of the industry. Holding such beliefs, it is obvious why they did not support the FTA.

U. S. Brewing Industry's Mixed Stance on the FTA. The reaction of the U. S. brewing industry to the FTA was less obvious than the reaction of the Canadian brewers. Most of the large U. S. brewers either publicly supported or took a neutral position on the FTA, while some of the smaller firms were in favor of the Agreement. Licensing agreements played a major role in discouraging the large U. S. brewers from pushing for free trade.

Recall that three of the largest brewers in the United States (Anheuser-Busch, Miller, and Coors) had licensing agreements with the three largest Canadian brewers. While they still faced the interprovincial barriers, they did not have to face barriers that hindered foreign-brewed beer sales. Even the out-of-province restrictions were not so great because the Canadian brewers have breweries in each of the major selling provinces.

An executive with one U. S. brewer with a licensing agreement best summed up the dilemma. He noted that the U. S. firms were innovative and aggressive enough to break into the highly protected Canadian market. This is the type of activity you would expect the U. S. government to applaud and encourage. However, after these firms had sunk considerable manpower and money into developing a licensing agreement with the Canadian firms, the U. S. government wanted to give all the U. S. brewers equal access to the Canadian market. What incentive did the licensing firms have to support such a position? Allowing other

U. S. brewers into the Canadian market would only erode their product's performance. But, from a public image standpoint, it could be unwise to take a negative stance on free trade, especially if you were interested in entering other protected foreign markets yourself.

Of the brewers not having licensing agreements in Canada, some supported the FTA and others did not. The Stroh Brewing Company positioned itself to be able to take advantage of either an inclusion or exclusion of the brewing industry from the FTA. By buying partial interest in a small Canadian brewery, Stroh had an economic incentive to keep up the trade barriers. But, since it also has U. S. operations, it could have also gained had the brewing industry been included in the FTA. Some small brewers who were located close to Canada, such as Genesee in New York, supported a free-trade position with hopes of expanding sales through exports to Canada. But, with the trade negotiators knowing that the largest brewers in both countries either opposed or showed little public support for freer trade, they had little incentive to attempt to negotiate trade barrier reductions in this industry when efforts in other industries could provide more positive results.

Post FTA Agreement Activities.[7] Since the passage of the FTA, several events have occurred in the Canadian brewing industry that suggest it is moving closer to free trade despite its exclusion from the FTA. Shortly after the FTA was passed by the Canadian Parliament and it was clear that provincial barriers were going to remain in place, the Canadian brewers decided that, with some restructuring, they could compete with the U. S. brewers. Molson and Carling O'Keefe brewers announced that they were merging their brewing assets to form one company called Molson Breweries. Carling O'Keefe is owned by an Australian brewer, Elders IXL. The newly formed company will hold about 53 percent of the Canadian beer market.

The new firm plans to close 7 of its combined 16 breweries throughout Canada which will cause layoffs of between 1,400 and 2,000 employees. The new brewer is also planning to spend about $220 million to update its remaining plants. One of the objectives of the new firm is to more than double its exports to the United States.

At the same time, Labatt announced it would hire an additional 1,000 employees and spend about $270 million over the next three years to update its plants. Its plans are to expand its production capacity so that it can take advantage of the restructuring period of its new competitor and to grab market share from Molson Breweries. It also plans to double its exports to the United States.

While the Canadian industry is far from its final shake-out, it appears that the industry has realigned itself to where it believes it can compete with American brewers. If the new Molson Brewery and Labatt both intend to double their exports, they must believe that, at least in some aspects, they are capable of being price competitive with the American brewers. The cost of the new capital investment going into the industry is about $490 million—substantially less than the estimated $2 billion needed.

A second factor pushing the Canadian industry toward free trade is that in 1987 a GATT panel ruled that the Canadian provinces discriminate against European alcoholic beverages in their pricing and distribution practices. The suit brought by the European Community (EC) was targeted mainly at wine and spirits but also included beer. At first, the Canadian government said that it would comply with the GATT committee's recommendations to liberalize trade practices in the wine and spirits industry but that it did not plan to change its beer marketing practices. Later, however, a letter from the Canadian ambassador to the EC said that Canada recognizes its obligation under GATT to correct beer pricing practices and will move to do so once interprovincial barriers affecting beer are removed (Greenspon and Waddell). Under GATT rules, trade rights offered to one member must be extended to all GATT members. As a GATT member, the United States could push for compensation from Canada if U. S. brewers were not afforded the same rights as European brewers.

Finally, perhaps the most important sign of possible freer trade is the recognition by provincial governments that interprovincial barriers need to be reduced if the brewing industry is to remain viable in Canada. Since the passage of the FTA, meetings have been held by provincial authorities at which they agreed, at least verbally, that barrier reductions are necessary in the future. Presently, meetings are being scheduled at which provincial authorities will try to decide what needs to be done and set up a timetable to achieve these goals.

Flour Milling Industry

This section on flour milling is organized along similar lines to the above analysis of the brewery industry. However, because some of the issues are different and because the data are not as readily available for the same years as chosen for the breweries, there is not exact correspondence in the presentation.

First of all, the world flour trade is discussed as well as the role of the United States and Canada in this market. Then, we discuss trade in wheat, flour, and products between the United States and Canada, as well as tariffs and institutions impacting this trade, and salient features of the U. S.-Canada FTA as it pertains to this industry. Next, the structure and conduct of this industry is analyzed.

World Flour Trade

The world trade in wheat flour has been growing slowly compared to the growth in the world wheat trade. Over the past 15 years, flour exports have grown at an annual compound rate of 2.5 percent compared to 4.8 percent for wheat. Figure 5.4 shows the exports of flour by each of the principal exporting countries. Imports by region of the world are shown in Figure 5.5. The principal wheat flour market is North Africa, which is just less than one-half of the world market. This is followed by sub-Saharan Africa. Other markets are the Mideast, the U.S.S.R., and Latin America, each of which is declining in volume.

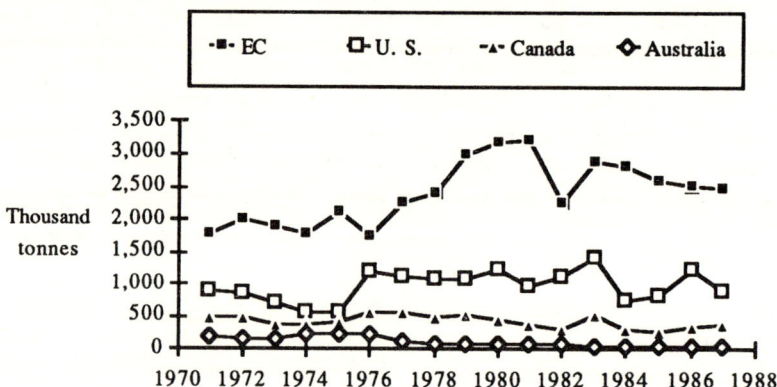

FIGURE 5.4. Flour Exports to World by Origin, Selected Years, 1970-1988

Source: International Wheat Council.

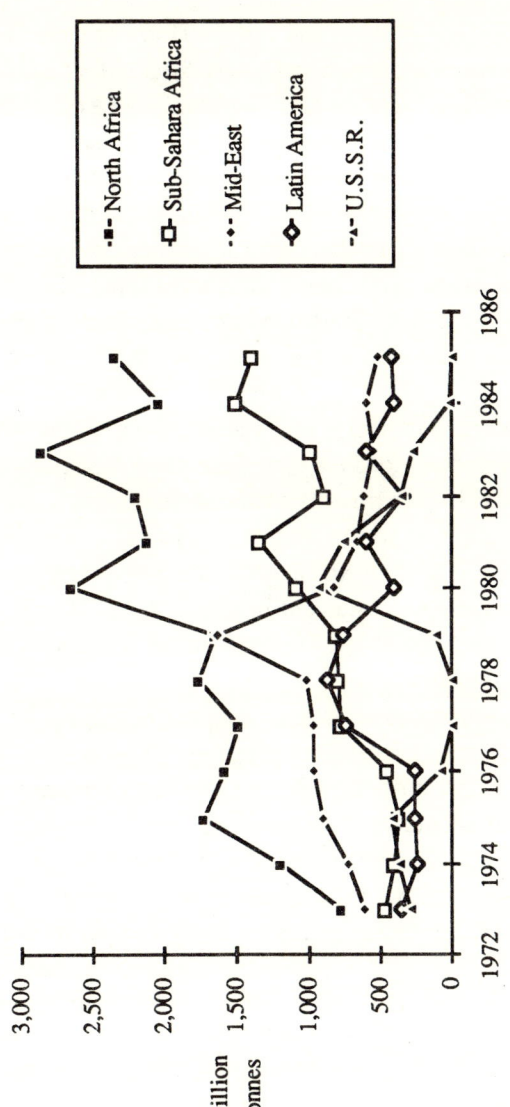

FIGURE 5.5. Regional Flour Imports from All Exporters, Selected Years, 1972

Source: U. S. Department of Agriculture, Foreign Agricultural Service.

The EC is the largest exporter of wheat flour, with domineering positions in each of the principal markets. Compared to the other exporting countries, flour exports are of greater importance to the EC. In the 1970s, as much as 60 percent to 70 percent of wheat exports from the EC were in the form of flour; this illustrates the important tradition of flour exports versus exports of wheat (Wilson and Hill). This has been facilitated by important commercial relationships and by the flour export subsidy program of the EC. In recent years, the relative importance of exports from the EC has declined (i.e., relative to wheat) and comprises 22 percent of exports in recent years. The U. S. market share for export flour has ranged between 20 percent and 30 percent in the world market, making the United States the second largest exporter in the world. Through the extensive use of export subsidies in recent years, the U. S. market share increased in 1985-86, but it has since declined. Canadian and Australian exports are nearly inconsequential and have each decreased during the 1980s. The primary market for Canadian flour is Cuba which has imported about 110,000 tonnes in recent years and accounts for about one-third of Canadian exports. Exports to Cuba have fallen from about 400,000 tonnes in the late 1970s. These are primarily food-aid shipments.

In general, market shares are quite sporadic. The EC has had 60 percent to 70 percent of the North African market, with the U. S. share ranging from 11 percent to 54 percent in recent years. The U. S. share of the sub-Saharan market, however, has been increasing since 1981 and that of the EC has been decreasing.

A large part of the retention or growth in market shares by the EC and United States can be attributed to programs which allow for export subsidies. The EC has exported grains and products for many years using export subsidies or restitutions. These are essentially refunds paid to exporters of a specific commodity or product to specific third countries, the purpose of which is to allow EC exports to be competitive in the world market given that EC domestic prices exceed those of the world. In the administration of the restitution program, differentials can be established to encourage or discourage the export of processed versus unprocessed commodities. As an example, in April 1989, the restitution on wheat to "other third countries" was about $25 per tonne, whereas that for wheat flour was $96 to $102 per tonne.

The traditional program in the United States is the Public Law (PL) 480 program which provides for special long-term credit at low rates to developing countries. Egypt is the largest recipient country for PL 480 shipments, accounting for about two-thirds of all PL 480 shipments.

In 1985, the United States initiated the Export Enhancement Program (EEP) which was implemented to combat the export subsidy programs of the EC and other exporters. The program provides a mechanism for exporters to receive "bonuses" for exports of specific commodities to specific countries. At its inception, the EEP was supposed to be used on both bulk and processed agricultural commodities. Bulk wheat has been by far the largest recipient commodity for EEP with 70.3 mmt offered through June, 1989. Wheat flour has also been an important recipient commodity with 2.2 mmt being offered through June, 1989. Like the EC, this program has differentials between bulk and processed commodities. As an example, in May, 1989, the bonus for flour exports to Yemen was $74 per tonne and that for wheat exports to Zaire was $4.70 per tonne.

The EEP came under political pressure in early 1989, but the question of using the EEP for bulk versus processed commodities has been argued since its inception (Green and Wilson; Tierney). The Millers National Federation, under the auspices of the "Export Processing Industry Coalition," has recently helped fund promotional material and lobbying activities in an effort to demonstrate the advantages of subsidizing processed versus bulk commodities. However, in a recent speech, the president of Cargill Flour Milling, L. DeWitt, indicated that a major threat (one of three glitches) to the recent prosperity of the U. S. flour milling industry is the likely reduction in exports due to reduced support for export subsidies on flour. "It is greatly tied up in the politics of wheat. It is tied up in the politics of the budget and the deficit. ... Opposition to export flour, in both the private and public sectors, is well organized." In response to the export flour markets, capacity utilization has moved up from about 80 percent to almost 95 percent and profits have improved (DeWitt, p. 17). Thus, reductions in subsidies on export flour implies a reduced capacity utilization, reduced margins, and closure of some plants—most likely interior, older mills.

Other countries, most notably Canada and Australia, have not been as aggressive in providing subsidy mechanisms for export of processed commodities. The only explicit subsidies used in Canada are credit and food-aid programs, and they are used only minimally. Sometime during the mid- to late-1970s there was a change in Canadian policy (either explicitly or implicitly) toward less aggressiveness in flour exports—in general this also applies to other processed products, such as malt, in which a similar phenomena was observed.

Market Shares. As shown in Figure 5.4, U. S. flour exports have increased over the last 15 years, in contrast to declining Canadian exports. In order to further understand factors behind these trends, we have decomposed changes in export sales into three effects. This analysis will help explain changes in a country's export position. This market-share approach was originally developed by H. Tyszynski; and it has been subsequently applied to the study of trade flows by Richardson (1971a), Rigaux, and Sprott.

The constant market-share model partitions export growth into three effects: (1) size of market, (2) market distribution, and (3) competitive effect. The size of market effect refers to an increase or a decrease in the volume of the total world demand for flour, and the distribution effect refers to the relative size of a particular importing country. The competitive effect is a residual factor which picks up changes in an exporter's competitive strength. Differences in competitiveness may be due to such factors as productivity changes, price and exchange rate changes, and changes in marketing and pricing policies. The first two effects (i.e., size of market and distribution effect) are calculated under the assumption that each exporter's share in each import market remains the same as in some base period. This means the model's results are sensitive to the choice of base period. In addition, Richardson (1971b) has shown that the relative magnitudes of the three "effects" are sensitive to the sequence in which they are calculated. We therefore recognize some of the drawbacks of a market-share analysis. While it may be arbitrary in describing the nature of competitiveness, it does help isolate it from the effects of changes in market share and distribution. In other words, it does provide a useful breakdown of explanations for market-share shifts.

Crop years 1973-74 through 1984-85 were selected as the time period for analysis of flour exports from Canada and the United States. Since the results depend on the base year chosen, four-year averages were selected for the market-share analysis. The four-year averages (1973-74/1976-77, 1977-78/1980-81, and 1981-82/1984-85) were designated Periods I, II, and III, respectively. Eleven importing regions were chosen, and the data were obtained from the U. S. Department of Agriculture and the International Wheat Council.

The results of a market-share analysis for the United States are reported in Table 5.4, where it is shown that the U. S. market share increased from 15.7 percent to 24.4 percent over the three periods studied. First, consider the size of market effect. This is calculated by subtracting actual U. S. exports from the hypothetical level of exports at constant, overall market share. From Period I to Period II, the size of market effect is positive (220,100 tonnes), but it is negative (-162,000 tonnes) from Period II to Period III. Second, consider the distribution effect. It is calculated by subtracting the level of U. S. exports that would have occurred under the Period I overall market share (15.7 percent) from the hypothetical level of U. S. exports that would have occurred at constant Period I market shares in each individual market. It is negative from Period I to Period II and positive from Period II to Period III. Finally, the competitive effect is equal to the actual level of exports in the current period, less the hypothetical sum of exports to individual markets in the current period but calculated using the previous period's individual market shares. In both periods the competitive effect is positive for the United States. In period III, the total world flour trade declined in volume and the negative size of the market effect for the United States was -162,000 tonnes (representing 40.1 percent of the change). However, the competitive effect worked in favor of U. S. exports, and it dominated the negative impact of the market effect. Over the entire period studied, the effective selling ability of the United States (i.e., its competitiveness) contributed most to its gain in flour exports.

A similar market-share analysis was conducted for Canada, and the results are reported in Table 5.5. Canada's flour exports increased from Period I to Period II, but then they dropped off sharply in Period III from 15 percent of the world market to only 7.4 percent.

From Period I to Period II, the size of market effect and competitive effect contributed in a positive way to Canada's increased exports, while the distribution effect was negative. Going from Period II to Period III, Canadian flour exports fell from 1.1 mmt to 494,800 tonnes, a decline of 56.1 percent. If Canada had maintained its 15 percent share of the world market, its exports would have been about 1 mmt in Period III. In contrast, if Canada had maintained its individual market shares for Period III that it had in the second period, its flour exports would have been only 738,500 tonnes. The difference between these two figures represents the distribution effect of -266,900 tonnes and reflects changes in the relative size of various markets. The distribution effect represents

TABLE 5.4

Components of U. S. Wheat Flour Export Gain, 1973-74 to 1984-85

	Period I 1973-74/1976-77 average	Period II 1977-78/1980-81 average	Period III 1981-82/1984-85 average
	thousand tonnes		
Total world exports	6,109.8	7,492.8	6,702.7
Total U. S. exports	956.3 (A_1)	1,556.2 (A_2)	1,636.0 (A_2)
U. S. market share	15.7%	20.8%	24.4%
Hypothetical U. S. exports in period II at same overall market share as period I		1,176.4 15.7% of 7,492.8 (B)	
Hypothetical U. S. exports in period II at period I market shares in each individual market		1,126.2 (C)	
Hypothetical U. S. exports in period III at same overall market share as period II			1,394.2 20.8% of 6,702.7 (B)
Hypothetical U. S. exports in period III at period II market shares in each individual market			1,427.1 (C)

Gain for United States	Period I to II		Period II to III	
	thousand tonnes	percent[a]	thousand tonnes	percent[a]
Total gain ($A_2 - A_1$)	599.9	100.0	79.8	100.0
Size of market effect ($B - A_1$)	220.1	31.4	-162.0	40.1
Distribution effect (C - B)	-50.2	7.2	32.9	8.2
Competitive effect ($A_2 - C$)	430.0	61.4	208.9	51.7

[a]The sum of the percent column disregards the signs of the three effects: size of market, distribution, and competition.

TABLE 5.5

Components of Canadian Wheat Flour Export Gain and Loss, 1973-74 to 1984-85

	Period I 1973-74/1976-77 average	Period II 1977-78/1980-81 average	Period III 1981-82/1984-85 average
		thousand tonnes	
Total world exports	6,109.8	7,492.8	6,702.7
Total Canadian exports	798.7 (A_1)	1,127.5 (A_2)	494.8 (A_2)
Canadian market share	13.1%	15.0%	7.4%
Hypothetical Canadian exports in period II at same overall market share as period I		981.6 13.1% of 7,492.8 (B)	
Hypothetical Canadian exports in period II at period I market shares in each individual market		914.8 (C)	
Hypothetical Canadian exports in period III at same overall market share as period II			1,005.4 15% of 6,702.7 (B)
Hypothetical Canadian exports in period III at period II market shares in each individual market			738.5 (C)

Gain (loss) for Canada	Period I to II		Period II to III	
	thousand tonnes	percent[a]	thousand tonnes	percent[a]
Total gain (loss) ($A_2 - A_1$)	328.8	100.0	-632.7	100.0
Size of market effect (B - A_1)	182.9	39.6	-122.1	19.3
Distribution effect (C - B)	-66.8	14.4	-266.9	42.2
Competitive effect (A_2 - C)	212.7	46.0	-243.7	38.5

[a] The sum of the percent column disregards the signs of the three effects: size of market, distribution, and competition.

42.2 percent of the total loss, the largest of the three. The size of market effect resulted in a loss of -122,100 tonnes and accounted for 19.3 percent of the loss. This is due to the overall decline in the global volume of the wheat flour trade. The competitive effect of -243,700 tonnes represented 38.5 percent of the total loss.

Canada's competitive position appears important in explaining its gains in one period and its losses in another. Canadian export flour prices are negotiated based on the price of wheat set by the CWB. Export flour programs from Canada depend to a large extent on negotiations between the Canadian millers and the CWB. It is believed by some that Canada has been losing its flour export business because the CWB is unwilling to supply a low-priced, low-quality wheat that allows the mills to offer wheat flour at prices competitive with exporters such as the EC and the United States.[8] Others attribute export losses to the handsome subsidies and incentives implemented by the EC and the United States that Canada simply cannot compete with.

Market Structure of U. S. and Canadian Milling Industries

In this section, a comparison of the market structures in the two countries is made. Major trends are identified, and a comparison is made of flour prices in the two countries. About 10 percent of the annual Canadian wheat crop is processed domestically into flour. This contrasts sharply with the United States where, in recent years, the domestic industry consumed up to 40 percent of the wheat production. In the past, the Canadian flour milling industry was not subject to foreign competition since flour imports into Canada were prohibited as described earlier. On the other hand, the U. S. industry was subjected to import competition, but most of this was in the form of product.

The structural evolution of the milling industry in the United States and Canada is shown in Table 5.6. In both countries there has been a decline in the number of mills over the years and an increase in the average size of mill. There has been an almost doubling of the number of large mills in the United States (i.e., with a capacity exceeding 10,000 cwt per day), from 24 to 42. The four-firm concentration ratios (i.e., concentration ratios in capacity) shown in Table 5.6 illustrate that, in each case, the industries are becoming more concentrated. There are three dominant firms in Canada who alone control over 70 percent of the capacity in the industry. The four-firm concentration ratio in Canada is 76, compared to 57 in the United States. The disparity in the concentration ratio between percent of mills and

TABLE 5.6
Structural Evolution of U. S. and Canadian Milling Industries Selected Years

Wheat milling industry	Number of mills	Daily capacity[a]	Average mill size	Top four firms	
				Percentage of mills	Percentage of capacity[a]
		tonnes	tonnes per day	percent	
United States					
1972	305	45,727	150	15.1	33.4
1978	281	51,240	182	14.2	35.8
1987	239	55,255	231	28.0	50.3
1989	231	56,143	243	32.0	57.1
Canada					
1972	42	8,617	205	38.0	76.0
1978	38	8,150	241	44.7	77.6
1987	37	9,097	246	51.4	77.4
1989	36	8,917	247	50.0	75.9

[a] All capacities include durum.

Source: Derived from data published by Sosland Companies.

percent of capacity in each country confirms that the four largest mills have disproportionately larger capacity.

The milling capacities of the top firms are shown in Figures 5.6 and 5.7. The largest in the United States is ConAgra, followed by Archer Daniels Midland, Cargill, and Pillsbury. It is interesting that, in 1972, Pillsbury was the leader and Cargill was not even in the business. The four largest firms in the Canadian industry are Maple Leaf Mills, Ltd.; Ogilvie Mills, Ltd.; Robin Hood; and Dover. Two of these major Canadian milling companies are owned by non-Canadian firms. Maple Leaf Mills is owned by Hillsdown Holdings of Britain; and Robin Hood is owned by the U.S. firm, International Multifoods. (It is interesting that International Multifoods sold its U. S. milling interests but retained those in Canada in which they held a more domineering position). Ogilvie, on the other hand, is owned by Labatt, the major Canadian brewer.

An important difference between these two countries is that capacity utilization in the United States is upward of 93 percent in recent years, an increase from 81 percent to 86 percent capacity utilization during the 1970s (Harwood, Leath, and Heid). This has been due to a number of concurrent phenomena including export sales programs and a sharp increase in per capita consumption. On the other hand, the Canadian milling industry, as an industry, has been operating at about 70 percent to 75 percent of rated capacity.

Flour milling companies in the United States have been classified in four categories by Goldberg. These are:

1. Vertically integrated food processors—e.g., Pillsbury, General Mills, and International Multifoods.
2. Multiunit flour millers diversified into other grain operations—e.g., ConAgra, Cargill, and Archer Daniel Midland.
3. Medium-sized firms that are primarily regional flour producers—e.g., Bay State Milling and Mennel.
4. Small millers with one or two mills in local market niches—e.g., Amber Milling and Miller Milling.

This categorization was made in the early 1980s. Examination of historical data on firm ownership and capacity indicates that important structural changes have been occurring in the United States. In particular, the first and third categories have been declining in importance, where those in the second and fourth categories have been increasing in importance. For

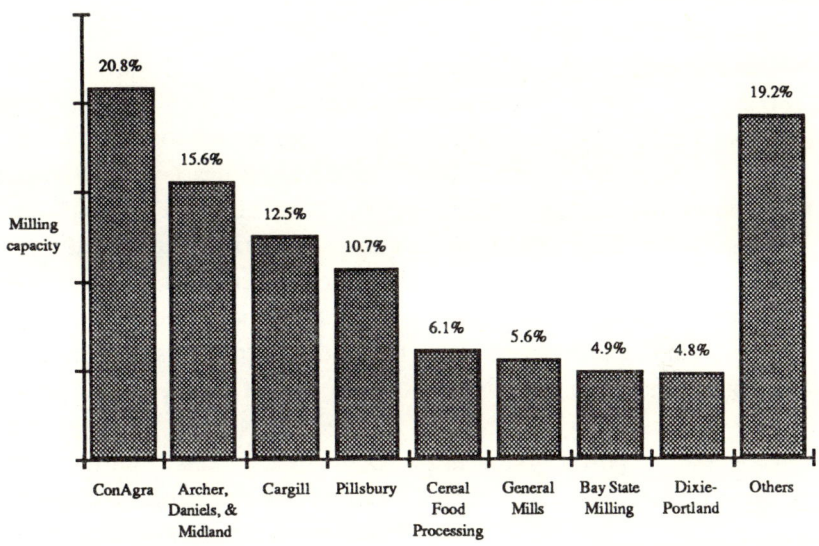

FIGURE 5.6. Milling Capacity of the Top Eight U. S. Firms in 1987

Source: Sosland Companies.

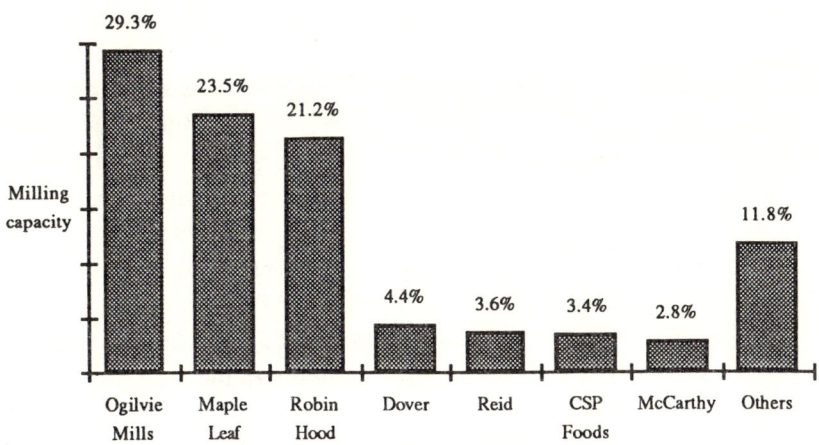

FIGURE 5.7. Milling Capacity of the Top Seven Canadian Firms in 1987

Source: Sosland Companies.

example, a number of traditional vertically integrated mills in the United States have exited (e.g., International Multifoods) or have decreased in importance (e.g., Pillsbury and General Mills) during the last two decades. On the other hand, there has been an increase in the number of new entrants which appear to be oriented to service local market niches. As an example, in the last 2 years, construction has been announced on eight new mills. Three of these, and more than half of the new capacity, are for durum (World Grain). The important overriding trend has been the increased dominance, from virtually nil, of larger diversified grain firms. As a result, one of the important longer term trends which has occurred in the flour milling industry has been the tendency for firms in the industry to be more commodity oriented rather than part of an integrated processed product industry.

Another important trend in the U. S. industry has been the apparent gradual shift in location that generally favors milling at the point of consumption. In the past, transit- and other rail-rate relationships generally had a neutral impact on the location of mills. As a result, much of the traditional milling capacity was at milling centers such as Minneapolis, Kansas City, etc. However, due to a number of changes, it appears that milling at destination is being increasingly favored. These changes include an elimination or reduction of the transit relationships and, more importantly, the introduction of unit-train shipping rates on wheat with substantial discounts relative to single-car rail shipments. Shipments of flour, requiring specialized equipment and destined to bakers not capable of handling unit trains, have become increasingly less appealing, thereby favoring destination milling. Though this is preliminary, evidence that virtually all of the new milling capacity has been at consumption points supports this longer term trend (e.g., in the past two years, Cargill has announced the construction of two new mills in California which would take advantage of unit-train wheat rates on the inbound movement).

Comparison of Flour Prices in Canada and the United States. As an indication of the lack of competitive arbitrage in the U. S.-Canadian flour market, it is instructive to compare retail flour prices in the two countries. According to the Prices Division of Statistics Canada, the 2.5 kilogram bag is the most popular size at retail food stores. This is the size most frequently purchased. Retail flour prices from 1976 to 1985 were obtained from the U. S. Bureau of Labor Statistics and Statistics Canada. These prices are based on 2.5 kilogram bags of flour, and they are plotted in

Figure 5.8 in U. S. cents per kilogram. According to this data, there was not much difference between the U. S. and Canadian prices in 1976; however, the gap quickly widened and, by 1985, the Canadian price increased to over $0.50 per kilogram above the U. S. price. These data indicate that the Canadian consumer was paying roughly twice as much as the U. S. consumer. Statistics Canada data also shows that normally over 50 percent of the flour milled in Canada is purchased at the retail level and/or by small bakeries. Therefore, this "overpricing" at the retail level in Canada is not an insignificant issue. Furthermore, there is evidence to suggest the Canadian millers overcharge at the wholesale level as well. However, this is more difficult to substantiate because the data on wholesale prices are not available. In private communication, a senior executive in one of Canada's largest bakeries reported to us that his company was being charged 60 percent over and above the U. S. price of flour in the mid 1980s.

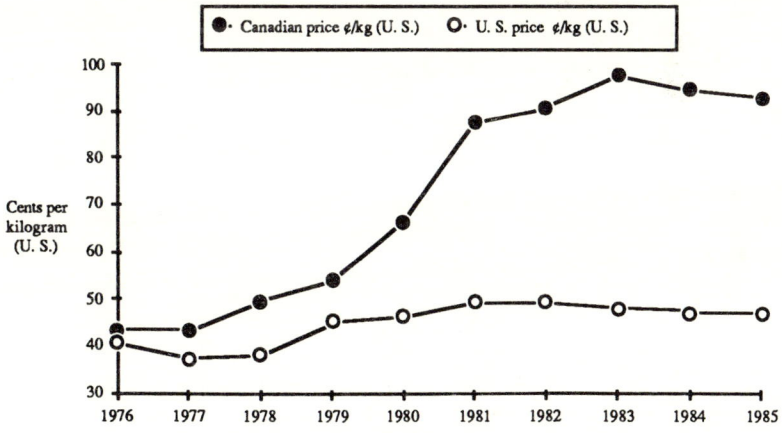

FIGURE 5.8. Canadian/U. S. Price of Flour, Selected Years, 1976-1985

Note: Prices based on 2.5 kg bags at retail level.

Source: U. S. data—U. S. Bureau of Labor Statistics. Canadian data—Statistics Canada.

An Explanation for the Protection Afforded Canadian Millers. Consider the simplified model of the market for bread in Figure 5.9. The demand for the final product, bread, is represented by D_b. Bread is produced from flour and other inputs represented by labor.[9] The supply functions for flour and labor are represented by S_f and S_l, respectively. Assuming that one unit of bread is produced by using one unit of flour and one unit of labor, the derived demand functions for flour and labor are then D_f and D_l, respectively. The demand for flour is obtained by vertically subtracting S_l from D_b. This graphic representation does not allow for input substitution possibilities but the model could be generalized to accommodate this. Under a competitive market for bread, the equilibrium price is P_1 and the

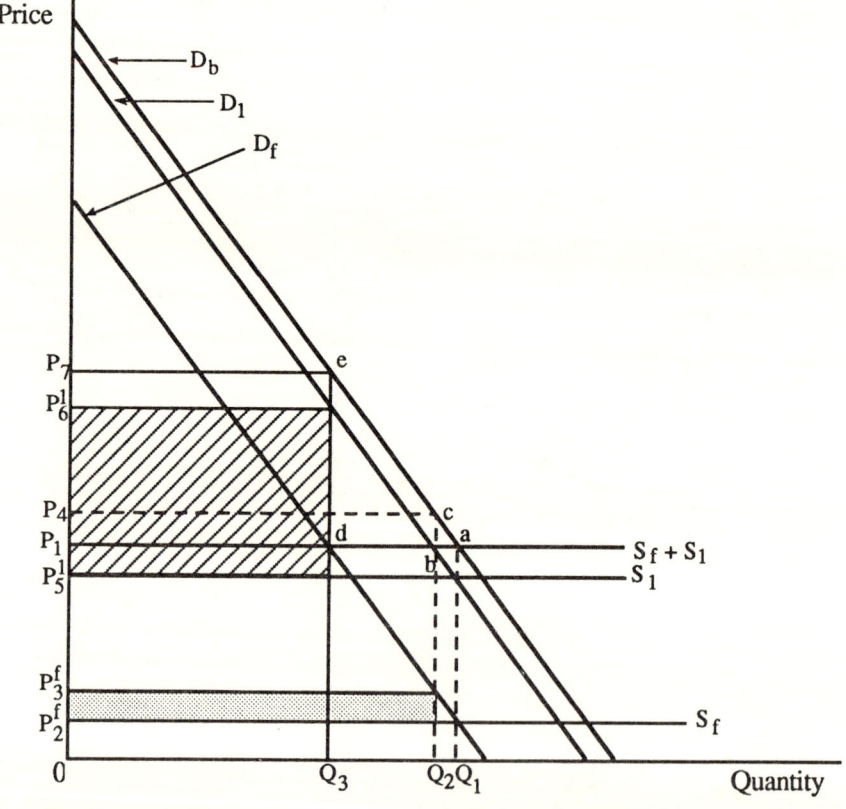

FIGURE 5.9. Market for Bread and Flour and Gains from Protection

output is Q_1. In the intermediate product market, the price of flour is P_{f2} and the amount demanded is Q_1. Now, suppose the two input groups that supply flour and labor independently lobby for import protection that will double their respective supply prices. Which group will be most successful in obtaining protection?

Clearly, if tariff protection increases the supply price of the input, the price of the final product rises. From Figure 5.9, we see that increasing the price of flour, through tariff protection, from P_{f2} to P_{f3} only results in a small price increase of the final product. The price of bread increases from P_1 to P_4 with the higher price of flour. Even though the price of flour has doubled in the model drawn in Figure 5.9, the price of bread increased by only about 13 percent. In addition, the new equilibrium quantities demanded of bread, flour, and labor are only marginally reduced from Q_1 to Q_2. This means the deadweight social loss associated with tariff protection to flour is relatively small. It is represented by the triangular area, *abc*.

Alternatively, consider the likely outcome of lobbying efforts of the other input for protection. If the price of labor rises by 100 percent from P_{l5} to P_{l6}, the price of bread rises dramatically from P_1 to P_7—an increase of approximately 83 percent. As a consequence, the equilibrium quantities demanded falls from Q_1 to Q_3 and the deadweight social loss is *ade*, which is much larger than *abc*.

Rather than investigating the outcome of a 100 percent increase in the input price, we could alternatively consider the impact on the final product of generating equal income transfers to flour and labor through import protection. If we set the cross-hatched and shaded area in Figure 5.9 equal to each other, we obtain the same result as above. Namely, transferring $x to the flour input results in a smaller increase in the final product price and a lower deadweight loss than does a similar protectionist scheme designed to transfer $x to the labor input. These results arise because, at any given output level for bread, the demand elasticity is lower for flour than labor; and the share of flour costs in total product costs is smaller. This model, therefore, predicts that the group representing the flour input would have a larger chance of obtaining protection than would the group representing the labor input.[10]

This explanation fits in with the literature on the political economy of and is especially related to the "Pressure Group Model" (Olson; Stigler; Brock and McGee). The Pressure Group Model hypothesizes that various interest groups in society compete with one another for economic rent from protection. This class of models predicts that the level of protection an industry receives is inversely related to the number of firms and/or the concentration ratio in the industry. Protection is also related to demand conditions.

We believe the Canadian flour millers have been successful in receiving protection because the industry is highly concentrated and because of the special demand characteristics for flour, an intermediate good. The three key parameters are the final product demand elasticity, the elasticity of input substitution, and the share of flour in total bread costs. The demand elasticity for bread in the United States has been estimated to be about -1.5 (George and King), and it seems plausible the elasticity in Canada would be similar. The elasticity of substitution between flour and other inputs is essentially zero. Finally, the value of flour represents only about 15 percent of the retail price of bread. Given that these three parameters are all relatively small in value, the bargaining power of the millers ought to be high.

To estimate the extent to which the Canadian industry gains from the import protection, profit margins were estimated over a six-year period. These margins were then compared to those in the United States where the industry is more competitive and is not protected from imports. The financial data for the Canadian millers is reported in Table 5.7 and, for the United States, in Table 5.8. There are some striking differences between the two countries but the basic finding is that net profits are considerably higher in Canada. Over the 1977-1982 period, the Canadian industry earned an estimated economic profit of $15.20 (U. S.) per tonne of wheat ground (milled). In comparison, the U. S. millers earned $6.36 per tonne of wheat milled. The Canadian profits would be even higher if it were not for the tremendous amount of "X-inefficiency" in the industry. For instance, in 1982 the sales and administration costs were $50 per ton of flour produced in Canada compared to only $12 per ton produced in the United States.

As an estimate of the excess profits earned by the Canadian millers due to the import protection, consider the net profit differentials in Tables 5.7 and 5.8. If we simply multiply this annual differential by the tons of wheat ground and then average across the six years, we come up with $22 million. This assumes the U. S. industry is competitive.

The question that naturally arises is, why doesn't the U. S. flour milling industry also receive protection from its government? Flour is a small component of the final cost of bread in the United States as well as in Canada. There are two significant differences between the Canadian and

TABLE 5.7
Financial Data for Canadian Flour Milling Operations, 1977-1982

Year	Net sales	Cost of material	Gross profit	General sales and administration costs	Net profit[a]	Net profit as a percentage of sales[a]	Wheat ground	Net profit per ton of wheat[a]
	thousand dollars (U. S.)					percent	million tons	thousand dollars (U. S.)
1977	427,005	335,133	91,872	55,804	36,068	8.45	2.60	13.87
1978	415,424	328,616	86,808	57,211	29,597	7.12	2.60	11.38
1979	506,055	411,997	94,058	66,142	27,916	5.52	2.70	10.34
1980	590,712	481,401	109,311	71,288	38,022	6.44	2.60	14.62
1981	700,541	573,111	127,430	77,430	50,000	7.14	2.50	20.00
1982	636,112	500,783	135,329	84,953	50,376	7.92	2.40	20.99

[a]Before income taxes.

Source: Compiled from Statistics Canada data (published and unpublished). Published data appear in Statistics Canada, *Flour and Breakfast Cereal Products Industry*, Cat. No. 32-228.

TABLE 5.8

Financial Data for U. S. Flour Milling Operations[a], 1977-1982

Year	Net sales	Cost of goods sold	Gross profit	General sales and administration costs	Net profit[b]	Net profit as a percentage of sales[b]	Wheat ground	Net profit per ton of wheat[b]
	thousand dollars (U. S.)					percent	million tons	thousand dollars (U. S.)
1977	1,722,512	1,582,351	140,161	55,345	84,816	4.9	10.3	8.23
1978	1,524,811	1,402,260	122,551	53,822	68,729	4.5	10.4	6.61
1979	1,806,886	1,688,355	118,531	59,276	59,255	3.3	10.4	5.70
1980	2,242,163	2,095,579	146,584	73,045	73,539	3.3	10.4	7.07
1981	2,538,511	2,374,296	164,215	85,206	79,009	3.1	10.7	7.38
1982	2,469,153	2,331,617	137,536	94,012	43,524	1.8	11.5	3.78

[a]These data represent approximately 65 percent of the total U. S. flour volume.
[b]Before income taxes.

Source: Ray Goldberg (ed.), *Research in Domestic and International Agribusiness Management*, JAI Press, 1983.

U. S. industries, however. The first is mentioned above and relates to the fact that, in Canada, the millers receive protection through the CWB and this enhances the power of the CWB. In other words, the Wheat Board supports the millers and the millers support the Wheat Board. The second difference lies in the degree of concentration. Where three firms control over 70 percent of the industry's capacity in Canada, in the United States this percentage of capacity is shared by eight firms. However, there is reason to believe that U. S. millers have received government support, but in the form of export subsidies rather than as import barriers. The U. S. support may have been as great as that in Canada but it is not as transparent.

U. S./Canada Trade in Wheat, Flour, and Products and the FTA. There are important institutions (or nontariff barriers) impacting trade flows of wheat and products between the United States and Canada, in both directions. Most important for wheat imports into the United States is Section 22 of the Agricultural Adjustment Act of 1933 which allows the U. S. Secretary of Agriculture to impose quotas on imports if it is determined that such imports are threatening U. S. price-support programs. Though this has always been a contentious threat, it has not been activated in the case of grains since the early 1970s. In Canada, imports of wheat and flour (and unpackaged products) are subject to licensing by the CWB. Licenses can be granted if a shortage exists or if an extreme special circumstance exists. In addition, there are numerous nontariff barriers related to packaging, labeling, transport, etc.

Trade in wheat and wheat products are also impacted by tariffs and quotas as shown in Table 5.9. The important point is that wheat and flour trade into Canada is restricted due to import licenses as discussed above. Wheat and flour trade into the United States is subject to tariffs (and subject to Section 22 and sales strategies of the CWB). Thus, in the past there has been relatively little movement of these commodities. On the other hand, selected products are subject to tariffs to varying degrees in each country. In general, the tariffs are not highly restrictive (compared to wheat and flour) and are usually greater going into Canada.

There was a sharp rise in U. S. imports of Canadian wheat beginning in 1984. Due to the import licensing mechanisms, there are virtually no exports of wheat into Canada except for seed. However, there is a brisk trade for products, in both directions. In general, the United States imports more wheat products from Canada than vice-versa. The one exception is

TABLE 5.9

Current Tariffs and Quotas Impacting Trade

	Sales into:	
	Canada	United States
Wheat	Import license	21 cents per bushel
Flour	Import license	52 cents per hundredweight
Selected wheat products:		
Pasta	Free-10%	Free
Cereals	10%	2.5%
Bakery mixes	7.5-10%	10%
Bread: leavened	Free	Free
other	7.5-10%	Free
Biscuits	Free	Free

Source: U. S.-Canada Free Trade Agreement.

that of breakfast cereals where Canada imports more from the United States than vice-versa (Wilson).

The FTA has potentially important impacts on trade between these two countries in bulk and processed commodities. Key elements of the FTA affecting wheat and product trade include:

General points:

1. Each party shall take into account the export interest of the other party in the use of any export subsidy on any agricultural good exported to third countries.

2. The United States retains the right to exercise Section 22 if imports interrupts operations of farm programs. If imports are a result of a substantial change in either party's support program, these rights of reintroducing quantitative restrictions are on an individual country basis.

Specific points:

1. Elimination of all tariffs over a period of 10 years.
2. Elimination of Canadian rail subsidy on exports to the United States shipped through Canadian west coast ports.
3. Canadian import licenses for wheat and wheat products are removed when grain-support levels in the two countries, as measured by producer subsidy equivalents (PSEs), are equal. However, the CWB retains the right to require end-use certificates for wheat imports.

Crucial to bilateral trade is the equivalency of the PSEs. From 1982 to 1986, the most important component of the PSE for U. S. wheat was the deficiency payment—on average, 39 percent of the total PSE. The most important component of the Canadian PSE on wheat is the transportation subsidy—also 39 percent of the PSE during the 1982-1986 period. Historic levels of PSEs in the two countries are shown in Table 5.10. As indicated, the PSE for the United States has exceeded that for Canada in each year except 1984. At the point in time that the two-year moving average of these PSE values are equal, the licensing capabilities governing exports of wheat from the United States into Canada are lost. However, end-use certification can still be used in some form which has yet to be defined.

One of the most important overriding impacts of the FTA relates to Canadian wheat pricing. Prior to 1988, the CWB had a two-priced system whereby the price to domestic mills exceeded that of the world market. Beginning in September, 1988, in anticipation of the FTA, this policy was changed so that sales would be made to domestic mills at the "North American" price, however defined. Prices were offered for up to two to three months which more closely approximated the U. S. market price. It is anticipated that this policy will change again, in general allowing for more frequent price changes. However, the important point is that the effective purchase price for Canadian mills was reduced to more closely reflect the price paid by U. S. mills.

TABLE 5.10

Comparison of U. S. and Canadian Wheat PSEs, 1982-1986

Year	PSEs per unit value	
	Canada	United States
	percent	
1982	16.8	14.1
1983	38.1	20.2
1984	28.5	34.2
1985	37.9	33.9
1986	62.9	49.9
Average	36.5	30.5

Source: U. S. Department of Agriculture, Economic Research Service.

The FTA is perhaps one of the more important institutional changes impacting the North American flour industry in many years. As a result there are numerous changes occurring concurrently as discussed above. It is premature to fully identify the impacts of the FTA on these industries; however, there is little doubt that there will eventually be a greater amount of trade and competition between these sectors, eventually evolving toward a North American market. Several individuals have indicated likely changes. Wilson identified a number of important impacts. Canadian mills will have access to U. S. wheat which has a greater diversity in quality and, in some cases, is cheaper; this point was later supported and emphasized by Sosland (1989). To be competitive, and for strategic purposes in procurement negotiations, Canadian mills will likely originate at least a portion of their wheat from the United States. At least, using historic prices, there have been times when arbitrage could occur between some qualities of U. S. and Canadian wheats. Also because of the reduced domestic wheat price, Canadian mills will improve their competitive position and, at least in the short run, will be protected from U. S. competition. As discussed earlier, U. S. mills already have access to Canadian wheat but for a number of reasons the flows will likely increase. One is that, though the CWB has never been precluded from selling into the U. S. market (i.e., the United States being viewed as a residual market), the FTA provides a more formal means of legitimizing sales into the United States with less threat of imposition of Section 22. Second, and perhaps of greater importance, is that through U. S. export subsidies on wheat the international market becomes depressed relative to the U. S. market. In other words, the U. S. market has become a premium market relative to the rest of the world, and natural arbitrage pressures would force more Canadian wheat into the United States (Wilson).[11] DeWitt expects that, due to the large amount of excess capacity in Canadian mills relative to the United States, there is a threat of greater flow of flour into the United States than vice-versa. Ultimately, whether there is an increase in wheat trade or flour trade between these countries ultimately depends on the logistics of moving bulk versus processed products, locations of plants, location of consumption, and border area competitive environment. Some Canadian milling firms have already taken strategic positions, likely in response to the opportunities or threats embedded in the FTA. For example, Maple Leaf Mills has announced plans to expand and modernize its flour mill and build a new bakery near Toronto; in addition, they have purchased a small specialty flour milling company located in New Mexico.

Conclusion

The brewing industry was exempted from the U. S.-Canada FTA because Canadian provincial laws have forced Canadian brewers to build plants that are relatively inefficient. Because of this inefficiency, Canadian brewers would have seen their profit margins shrink as beer prices would have fallen if free trade had been allowed. With profits falling, Canadian brewers would have laid off workers. On the U. S. side, most of the large breweries have found ways to effectively operate in the protected industry and, therefore, have had little incentive to remove the barriers.

The passage of the FTA has bought the Canadian brewing industry some time to begin restructuring. The industry has realized that the trade barriers are not a long-run solution to their inefficient operations. Therefore, Canadian brewers have decided to retool while their profit margins are artificially inflated by the trade barriers. Once rationalization has occurred, their average cost of production will be lower and they will be better able to compete with European and U. S. brewers. Perhaps the provincial governments will be less opposed to tearing down their barriers since the rationalization process has already started without them changing their policies and with the additional pressure of the GATT countries looming in the background monitoring their progress. While the Canadian brewing industry is still highly protected, international market forces are moving it in the direction of freer trade.

The process of liberalization is more forceful yet less voluntary in the case of flour milling in Canada. For years, the Canadian flour milling industry has been highly protected from foreign competition. As a result, Canadian consumers and bakers were paying substantially more than the world price for flour. Under the new FTA the industry in Canada will come under competition from their U. S. counterparts. In the past, the U. S. flour milling industry has gained from export subsidies. The international flour market will not provide many commercially attractive opportunities for demand growth in the foreseeable future and export subsidies may be reduced. Therefore, the North American industry will have to address the challenge of producing for a stagnant domestic market. This is already happening in the United States with the construction of larger, more efficient plants; and, because of the FTA, it will also take place in Canada. The brewery and flour milling industries are two clear examples of where freer trade will generate economic gains.

Notes

[1] A barrel of beer contains 31 gallons of beer.

[2] Evidence of economies of scale in the brewing industry is provided by Elzinga; Scherer; Fuss and Gupta; and Keithman.

[3] Information on U. S. and Canadian trade barriers was obtained from the Brewers Association of Canada's *Analysis of the Competitive Position of the Canadian Brewing Industry*.

[4]For a more detailed discussion, see the Brewers Association of Canada publications (1985, 1986, 1987) listed in the Reference section.

[5]In 1987, the provinces of Alberta and British Columbia accounted for 68 percent of total Canadian beer imports. In these provinces, the differential between U. S. and Canadian production costs are larger, as Tables 5.2 and 5.3 show.

[6]For a more detailed discussion, see the Brewers Association of Canada publications (1985, 1987).

[7]Much of the information in this section was obtained from articles appearing in the *Toronto Globe and Mail*.

[8]According to "Assessment of the World Flour Trade." *World Grain*, 4(April, 1988).

[9]Labor may be taken to represent the group of all other inputs excluding flour.

[10]This idea of the "importance of being unimportant" is not new. Marshall applied it to the theory of labor unions and their ability to extract rents from their employers. He suggested that, ceteris paribus, labor unions would have more bargaining power the smaller their share of costs in total costs.

[11]U. S. wheat grower groups such as the Durum Growers Association are lobbying already for changes in the FTA to reduce the amount of Canadian wheat exported to the United States.

References

Beverage Industry. "Beer Trends." *Annual Manual 1988/1989*. Cleveland, Ohio: Edgell Communications, Inc., 38-47.

Brewers Association of Canada. *Analysis of the Competitive Position of the Canadian Brewing Industry*. Ottawa, Ontario: Brewers Association of Canada, December 1986.

_____. *The Impact of a Canada-U. S. Free-Trade Agreement on the Brewing Industry and Its Contribution to the Canadian Economy*. Ottawa, Ontario: Brewers Association of Canada, June 1987.

_____. *Perspectives on Canada-United States Free Trade*. Ottawa, Ontario: Brewers Association of Canada, May 1985.

Brock, W. A., and S. P. McGee. "The Economics of Special Interest Politics." *American Economic Review* 68(May 1978):246-250.

Conference Board of Canada. *The Canadian Brewing Industry: Historical Evolution and Competitive Structure*. Ottawa, Ontario: Conference Board of Canada, February 1989.

DeWitt, L. "Milling must prepare for end of 'good ride', Cargill's DeWitt tells A.O.M." A speech to the Association of Operative Millers, May 21, 1989; and in *Milling and Baking News* (June 6, 1989):17-22.

Elzinga, Kenneth G. "The Restructuring of the U. S. Brewing Industry." *I.O. Review* 1(1973).

Finnegan, Terri. "International Licensing Pacts on the Rise." *Modern Brewery Age* (July 13, 1987).

Fuss, Melvyn A., and Vinod K. Gupta. "Cost Function Approach to the Estimation of Minimum Efficient Scale, Returns to Scale, and Suboptimal Capacity." *European Economic Review* 15(1981):123-135.

George, P. S., and G. A. King. *Consumer Demand for Food Commodities in the U. S. with Projections for 1980*. Giannini Foundation Monograph No. 26, University of California, Berkeley, March 1971.

Goldberg, Ray. *Research in Domestic and International Agribusiness Management*. Greenwich, Connecticut: JAI Press, 1983.

Green, Paul, and Ewen Wilson. "A Choices Debate: Export Subsidies on Value-Added Products." *Choices* (Second Quarter 1988):4-8.

Greenspon, Edward, and Christopher Waddell. "Canada Gives Pledge to EC on Beer Pricing." *Toronto Globe and Mail*. Toronto, Ontario (January 26, 1989).

Harwood, J. L., M. N. Leath, and W. G. Heid, Jr. "Structural Change in the U. S. Wheat Milling Industry." *Wheat Situation and Outlook*, Report WS-283, Washington, D. C.: U. S. Department of Agriculture, Economic Research Service, November 1988.

International Wheat Council. *Market Report*, London, England, October 27, 1988.

Keithman, Charles. *The Brewing Industry*. Staff Report of the Bureau of Economics. Washington, D. C.: U. S. Federal Trade Commission, 1978.

Marshall, A. *Principles of Economics*. 8th ed., London: Macmillan and Co., 1948.

National Association of Beverage Importers, Inc. *NABI Annual Statistical Report*. Washington, D. C., various issues.

Olson, Mancur. *The Logic of Collective Action: Public Goods and the Theory of Groups*. Cambridge: Harvard University Press, 1965.

Richardson, J. D. "Constant-Market-Shares Analysis of Export Growth." *Journal of International Economics* 2(1971a):227-239.

_____. "Some Sensitivity Tests for a 'Constant-market-Shares' Analysis of Export Growth." *Review of Economics and Statistics* 53(1971b):300-304.

Rigaux, L. R. "Market Share Analysis Applied to Canadian Wheat Exports." *Canadian Journal of Agricultural Economics* 19(1971):22-34.

Scherer, F. M. "The Determinants of Industrial Plant Sizes in Six Nations." *Review of Economics and Statistics* 55(1973):135-145.

Sosland Companies, Inc. *Milling Directory/Buyers' Guide*. Merriam Kansas: Sosland Publishing Company, various issues.

Sosland, M. "U. S.-Canada Breadstuffs Trade May Expand in Wake of Pact." *Milling and Baking News* 68(June 27, 1989).

Sprott, D. C. "Market Share Analysis of Australia Wheat Exports Between 1950-51 and 1969-70." *The Wheat Situation*. Canberra, Australia: Bureau of Agricultural Economics, 1972.

Statistics Canada. *Flour and Breakfast Cereal Products Industry*. Cat. No. 32-228. Ottawa, Ontario: Minister of Supply Services, various years.

Stigler, G. J. "The Economic Theory of Regulation." *The Citizen and the State*. Chicago: The University of Chicago Press, 1975.

Tierney, W. "The Case for an Increase in EEP Funding for U. S. Flour Exports." Manhattan: Kansas State University, September 1, 1988.

Tsyzynski, H. "World Trade in Manufactured Commodities, 1899-1950." *The Manchester School* 19(1951):272-304.

U. S.-Canada Free Trade Agreement. Communication from the President of the United States, July 26, 1988.

U. S. Department of Agriculture. Economic Research Service. "Estimates of Producer and Consumer Subsidy Equivalents: Government Intervention in Agriculture, 1982-86." Washington, D. C. Agriculture and Trade Analysis Division Staff Report No. AGES880127.

U. S. Department of Agriculture. Foreign Agricultural Service. "Grains: World Grain Situation and Outlook." FG 15-86, Washington, D. C.: December, 1986.

United Nations. Food and Agricultural Organization. *FAO Trade Yearbook*, various issues.

Wilson, William. "Potential Impact of the FTA on the Wheat Foods Sector." A speech to the 87th annual meeting of the Miller's National

Federation, Washington, D. C., May 2, 1989; and in *Milling and Baking News* (May 16, 1989).

Wilson, William W., and L. D. Hill. "The Grain Marketing System and Wheat Quality in France." North Dakota Research Report No. 110. North Dakota State University Agricultural Experiment Station, Fargo. June, 1989.

World Grain. "New U. S. Mills Built to Serve Domestic Farmers, Pasta Markets." Shawnee Mission, Kansas: Sosland Publishing Company (May 1988):8-10.

6

Irrigation and Prairie Agricultural Development

Surendra N. Kulshreshtha

Introduction

Parts of the Canadian Prairies are located in a semiarid climatic zone. The Palliser Triangle—the area that extends from the southern foothills of the Rockies to the southern part of Manitoba—was described by the Canadian explorer Captain John Palliser in 1857 as too dry to support agriculture. Most parts of the southern Prairies are also susceptible to frequent droughts.[1] Converting dryland areas into irrigation, during years following dry conditions, has been of prime interest among farmers as well as among various levels of the government. As a result, irrigation farming dates back as far as 1877. By 1986, some 530 thousand hectares of Prairie land were under irrigation. Recently, the government of Canada adopted a strategy for diversification of the Prairie economy. Since agriculture is a large and important sector of this economy, the agricultural industry may also support this strategy through its own diversification. The question that remains is whether irrigation development can help diversify Prairie agriculture.

The analysis presented centers around three issues related to irrigation and diversification of Prairie agriculture.

1. Does irrigation development help diversify Prairie agriculture by reducing its dependence on traditional grains and oilseeds?

2. Has there been a significant accomplishment in terms of the magnitude of diversification resulting from irrigation development?

3. Where does the future lie for irrigation development on the Prairies and for diversification of the Prairie economy?

These questions are analyzed for the Prairie region as a whole and for individual provinces, where possible.

In this chapter the discussion of irrigation development and diversification of Canadian Prairie agriculture is somewhat limited. No attempt is made in this chapter to carry out a rigorous cost-benefit analysis of irrigation development on the Prairies, neither from the farmers' perspective nor from society's perspective. Also, the notion of opportunity cost of water being used for irrigation is not discussed. Regional development impacts from irrigation and diversification are not considered in any detail, although some reference is made to these issues. Demographic changes and other social impacts, as well as environmental quality issues related to irrigation development and/or diversification, are not discussed. A complete investigation of stability at the farm level and at the regional level was not included.

The objectives of this chapter are to (1) describe the historical development of irrigation in the three Prairie provinces; (2) address the links between irrigation and diversification of agriculture, defining "diversification" as it is referred to in this chapter, and discussing the key issues of how irrigation assists in on-farm diversification; (3) explore the diversification of regional economy; (4) describe the actual progress made toward diversification of Prairie agriculture, contrasting it against a hypothetical situation where no irrigation development occurred; and (5) examine the future of irrigation development on the Prairies, taking into account the factors that may affect this development and then translating these factors into prospects for diversification on the Prairies. The summary and conclusions are presented at the end.

Signing of the U. S.-Canada Free Trade Agreement can also have significant implications for irrigation development and resulting diversification on the Prairies. Although the issues deserve a careful and more detailed analysis, the free trade agreement may lead to a slower rate of change in adoption of irrigation by farmers and may also retard new investment in a primarily agricultural water project development. This is based on the premise that, under the free trade agreement, subsidies to agricultural producers are reduced or eliminated. This would lead to a different impact upon various commodities. For example, Canada and the United States trade only in small quantities in wheat and feed grains. Increase in imports of corn in Eastern Canada may reduce use of Western Canadian feed grains, some of which may be grown on irrigated farms. Canada currently imports a large quantity of fruits and vegetables; however, very small quantities of these products are produced on irrigated farms. Currently, Canada enjoys a large surplus in red meat product trade. With the agreement, providing assured access to the U. S. market, beef exports to the United States may increase. However, the extent to which this would lead to improved economics of producing feed grains and forages on irrigated farms needs further investigation.

Background to Irrigation Development

The climate of the Prairies is classed as semiarid due to low precipitation. Annual precipitation is a minimum of 28 centimeters in the southern region along the boundary of Alberta and Saskatchewan, and increases to the east, north, and west to an average of 56 centimeters in eastern Manitoba, 41 centimeters on the northern fringe, and 64 centimeters along the foothills of the Rockies (Topham). Irrigation occurs mainly in South Alberta and southwestern Saskatchewan where approximately 400,000 hectares were irrigated in 1985.

The first recorded irrigation in Alberta was a small private scheme on Fish Creek, just south of Calgary in 1877 (Craig). Major infrastructure was developed during the late 1800s. The first irrigation canal, measuring 115 miles, was dug during 1898 and 1900. The project was located in the Raymond-Magrath-Lethbridge area, and water was diverted from St. Mary's River.

Irrigation developed gradually but continually since the 1890s in three distinct phases (Topham, p. 16). The first phase was the company-built or commercial phase which lasted until about 1920. Financing of irrigation was done by the railways in order to increase the value of the land sales. The second phase in irrigation development in Alberta was the irrigation district phase which followed passage of the Irrigation Districts Act in 1915. Under this phase, farmer-owned and operated projects were established under the supervision of the provincial government. The final phase could be called the government-developed phase, which began after World War II. During this phase, both levels of the governments (federal and provincial) accepted direct responsibility for the establishment of new developments and the expansion and maintenance of older irrigation schemes. At present, there are 13 irrigation districts, of which St. Mary Irrigation District and the Eastern Irrigation District are the largest.

In Saskatchewan the irrigation of crops dates back to the 1890s. In 1894 a small number of private irrigation schemes existed in the Cypress Hills district. Some of these were the result of the droughts of 1885 and

1889, particularly in the southwest corner of the province. In Saskatchewan, as compared to Alberta, irrigation proceeded at a much slower rate. It was not until the severe drought of the 1930s that both the federal and provincial governments undertook irrigation development. The largest concentration of irrigation is located around Lake Diefenbaker, a lake created by damming the South Saskatchewan River. In addition, several small-scale water projects were developed by the Prairie Farm Rehabilitation Administration (PFRA) in southwest Saskatchewan. In fact, since 1936, PFRA has developed 26 storage reservoirs and six irrigation projects (Topham, p. 41).

Irrigation has not developed in Manitoba to the extent it has in Alberta and Saskatchewan. This is primarily because moisture deficiency in Manitoba is not as severe. In fact, there are no large-scale irrigation developments in Manitoba. This is not because of a lack of water for irrigation; in some areas irrigation is done by withdrawing water directly from the river or, where available, from underground sources.

Major irrigation projects in the Prairie provinces are shown in Figure 6.1. The largest concentration of these projects is in Alberta, and the smallest is in Manitoba. In fact, as shown in Figure 6.2, until 1950 there was very little irrigation outside Alberta. During 1988, there were 691 thousand hectares of irrigated land in the Prairies. Of this total, 81.8 percent was in Alberta, 15.2 percent in Saskatchewan, and the rest in Manitoba. Even in the province of Alberta, irrigated areas constituted only 3.2 percent of the improved farmland. This proportion is even lower (at 1.4 percent) for the Prairies as a whole (Table 6.1). Thus, relative to the land base of the Prairie region, irrigation constitutes a very small portion of the agricultural economy.

The importance of irrigation can also be assessed in economic terms. The contribution of irrigation, in terms of the value of product sold, is estimated at 6.62 percent of total agriculture's gross value of production. These range from 0.6 percent in Manitoba to 1.7 percent in Saskatchewan and to 13.6 percent for Alberta.[2] These numbers suggest that regional economy irrigation constitutes a significant contribution to the economic well-being of the provincial and certain subprovincial economies.[3]

Irrigation development in the three Prairie provinces during 1950 and 1985 is shown in Table 6.2. In Alberta, by 1950, there were 180.9 thousand hectares irrigated, most done by surface methods. By 1980, there was a reversal in the relative area irrigated by sprinkler; now about three-quarters of all irrigated areas was irrigated using sprinklers. In

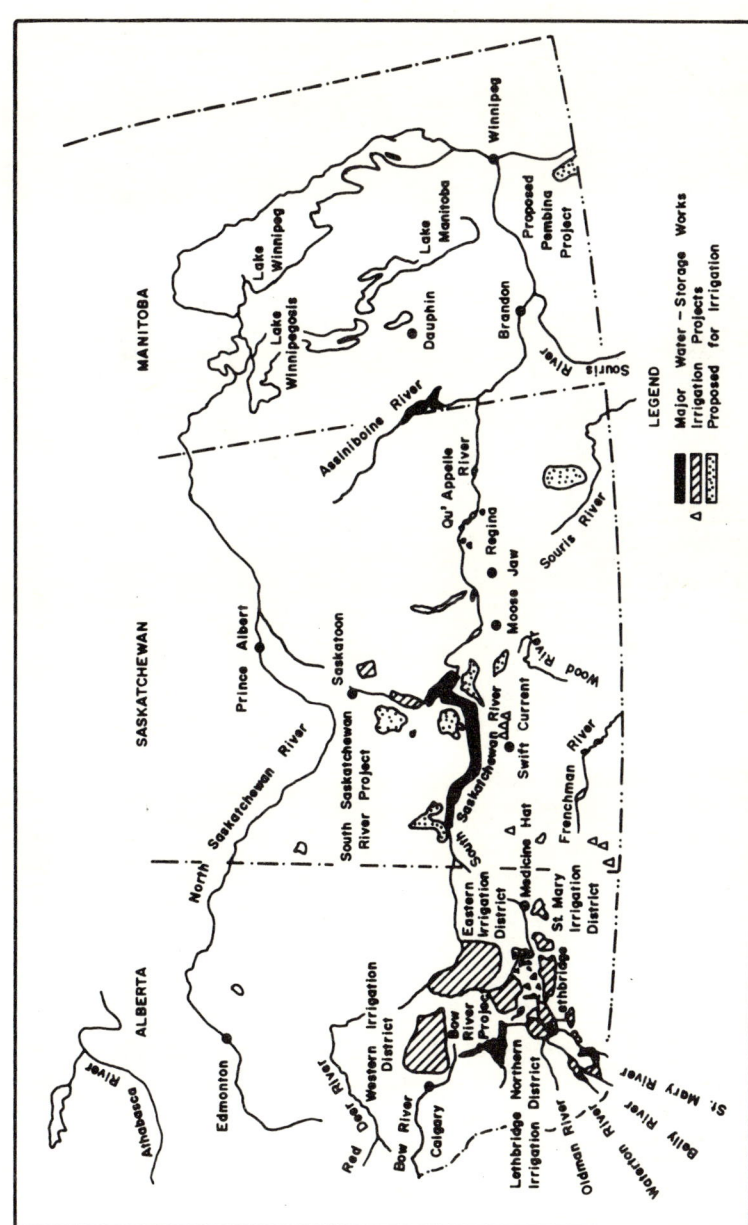

Figure 6.1. Irrigation Projects in the Prairies, Canada
Source: Adopted from S. Dubetz and D.B. Wilson, with recent developments added.

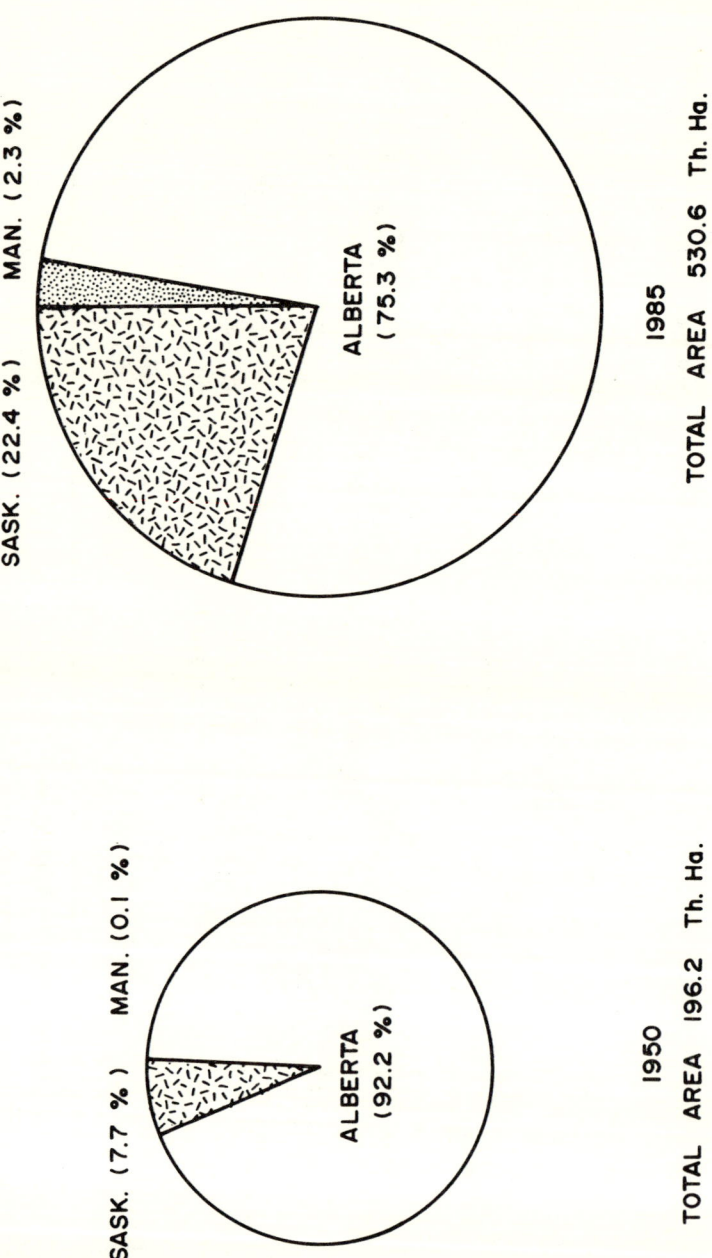

Figure 6.2. Irrigation Development in the Prairies, 1950 and 1985

TABLE 6.1

Irrigated Area Relative to Total Farmland
The Prairies, 1985

	Farmland			Irrigated as a percentage of improved
	Total	Improved	Irrigated	
	thousand hectares			percent
Alberta	19,109	12,525	400	3.2
Saskatchewan	25,947	19,684	119[a]	0.6
Manitoba	7,616	5,504	12	0.2
Prairies, total	52,672	37,713	531	1.4

[a] 1986 level.

Source: Estimated.

Saskatchewan, although the relative size of sprinkler irrigation increased by 1985, there was still 53 percent of the total area irrigated by surface methods (backflood or spring flood). For Manitoba, estimates for 1980 are not available. However, according to the 1979 Prairie Provinces Water Board Study, there was a total of 4,600 hectares under irrigation by some 68 farmers. Of these farmers, 57 percent used a hand-move method of irrigation, whereas some 19 percent used a center pivot method (Prairie Provinces Water Board, p. 59).

Most of the earlier irrigation developments in Alberta were organized under district irrigation; only 4 percent of the total irrigated area was developed through private schemes. By 1978, the proportion of private

TABLE 6.2

Irrigated Area in the Prairies
Selected Years, 1950-1988

Prairies	Year	Total area	Proportion of total area under:			
			Sprinkler	Surface and other	District	Private
		hectares		percent		
Alberta	1950	180,910	2	98	96	4
	1960	219,600	9	91	97	3
	1970	217,450	24	76	a	
	1978	373,140	55	45	84	16
	1980	393,969	74	26		
	1988	565,000				
Saskatchewan	1950	15,180	3	97	48	52
	1960	19,840	8	92	64	36
	1970	31,360	8	92		
	1978	76,890	37	63	33	67
	1988	105,000	47	53		
Manitoba	1950	99				
	1960	500				
	1970	2,970	22	78		
	1978	4,450	83	17		
	1980	6,935				
	1988	21,000				

[a]Blanks indicate no data available.

Source:

    For 1950-1978: Topham.

    For 1988: Agriculture Canada and Shady.

schemes grew to 16 percent of the total. In Saskatchewan, a majority (67 percent) of the total areas is under private irrigation.

In light of the small irrigated area in Manitoba, in subsequent discussions we focus only on Alberta and Saskatchewan.

Irrigation and Agricultural Diversification

Prior to presenting the linkages between diversification and irrigation development on the Prairies, a definition of diversification is needed. According to the *MIT Dictionary of Modern Economics*, diversification can be defined either in the context of a region or a firm (Pearce). With regard to a firm, diversification refers to a policy of spreading the range of products marketed and has the analogous motive of reducing reliance upon the markets for a small number of products. In the context of an industry, diversification has a similar meaning. Regionally, diversification can be defined in terms of the variety of industry within a given region. A region is more diversified if its industrial structure includes a number of industries which serve as major employers in the region.

In this study, the first concept of diversification—one with respect to a firm—is used.[4] Diversification is primarily viewed in terms of the ability of a farmer to grow and successfully market a variety of products. Two types of linkage between irrigation and farm diversification are examined: diversification of the crop mix and diversification of the sources of farm income. Let us examine each of these one at a time.

Choice of Crops on Irrigated Lands

Every successful irrigation project could be considered an oasis in the desert, a focal point around which prosperity grows. In summary, irrigation offers a wider variety of crops that can be chosen by the farmer than it is possible to have under dryland conditions. For example, in Southern Alberta about 35 different crops are grown on irrigated land only some of which are grown under traditional dryland conditions (*Canadian Environmental Control Newsletter*, p. 7.1).

Soft wheat is a preferred crop under irrigation. This type of wheat appears to have better livestock feeding qualities than hard wheats.[5] Sunflower crops also appear to have considerable merit for use in Prairie irrigated areas since they are resistant to early spring as well as to fall frosts. Production of field peas, lentils, and faba beans is also possible under irrigated conditions. A number of specialty crops are also possible: Sugar beets in South Alberta, potato production, and vegetables for both fresh market and canning are some of the examples that have been successfully grown under irrigation. Generally, irrigation offers a greater diversity in crop selection than does dryland farming. Under irrigation, farmers have a wider choice of crops: Soft wheat, durum wheat, canola, and flax are planted by more farmers under irrigation than under dryland conditions. Furthermore, crops such as potatoes, fresh vegetables, faba beans, and corn silage can only be grown under irrigation in dryland areas.

Although irrigation may widen a farmer's choice of crop, it may not necessarily translate into a benefit. In order for this diversity in crops grown to become beneficial, it must lead to enhanced farm income, more economic stability over time, and more opportunities for growth of the firm. Thus, irrigation may or may not provide opportunity to farmers, through a wider crop selection to adjust to fluctuations in world markets and to grow higher value crops.

Linkages Between Irrigation and Livestock Production

Linkages between irrigation and livestock production are less since livestock uses very little water directly. Although they are surrounded by a certain degree of controversy, these linkages are significant, particularly on the issue of line of causation between cattle numbers and irrigation. Benefits of irrigation include the increase in net incomes of industries processing the increased agricultural production, provided these increases would not have occurred in the absence of the irrigation (Howe, p. 39). However, it has been suggested that the expansion of the livestock and the processing industries, anticipated as a result of proposed irrigation projects, are forward-linked activities; their inclusion in the calculation of direct benefits is incorrect (Stabler, p. 11). It is the author's contention that livestock is an important enterprise under irrigation, leading to further diversification at both the farm and the regional levels. For example, in the southern part of Alberta,[6] cattle numbers have been on the increase during the period of irrigation development. Cattle on farms in this region increased from 342 thousand in 1941 to 817 thousand in 1986. The numbers of cattle per thousand hectares of improved land for Alberta and South Alberta, as shown in Figure 6.3, show that the irrigated regions have a higher cattle population per unit of land. Linkages between livestock production and irrigation can be exemplified as follows:

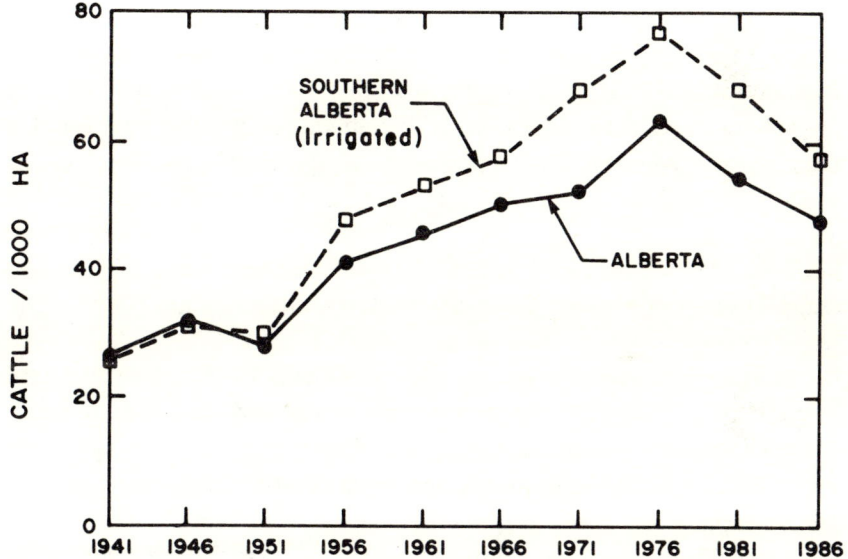

Figure 6.3. Cattle on Irrigated Regions of Alberta and in Alberta, Selected Years, 1941 – 1986

- Irrigation farms have more livestock than dryland farms under similar situations.
- It is more economical to feed livestock irrigated forage than that produced under dryland conditions.
- Irrigation provides by-products from specialty crops (for example, sugar beet tops and crushed canola) used often as livestock feed.
- A larger proportion of income from irrigation farms is derived from livestock production.

Opportunity for expanded livestock production on irrigation farms is shown (Table 6.3). Although the proportion of irrigated farms with livestock is only marginally higher than that for dryland farms, irrigated farms have more livestock per farm. The ratio is easily higher than 2 to 3 in favor of irrigated farms. For example, on average an irrigated farm had 173 head of cattle, about 2.7 times as many as on a nonirrigated farm. Breaking down the cattle by type, irrigated farms have twice as many milk cows and heifers and 2.4 times as many beef cows and heifers as their dryland counterparts. For steers and cattle on feed, the ratios are 6.1 and 7.6, respectively.

Similar situations exist for Saskatchewan (Table 6.4). About 47 percent of all irrigated farms had cattle and calves as against 26.7 percent on dryland. At the same time, the average number of cattle per acre was twice as large on irrigated farms as on dryland farms.

An explanation of why irrigated farms have more livestock lies in terms of economic benefits. To determine this, one can observe the comparative advantage of forage grown under irrigation versus dryland. To produce a tonne of hay costs $18.92 less under irrigation than under dryland conditions (Table 6.5).[7] (Dollar figures in this chapter are Canadian dollars.) This translates into a saving of $32 per head for cow-calf, $19 per head for background cattle, and $11 per head for feeding to finish operations. Given the favorable economics of livestock enterprise under irrigation, farmers in irrigated regions will opt for diversification through livestock enterprises. As much as 60 percent of irrigation farmers' gross income in Alberta is derived from livestock production (Anderson and Associates, Ltd. 1978). A study by Carter and Schmitz addresses the issue of major economic factors contributing to the establishment of the feedlot industry in Alberta rather than in Saskatchewan or Manitoba. According to their study, "the production of cereal silage under irrigation provides Alberta feeders with a cost advantage over Saskatchewan" (p. 134). These findings

TABLE 6.3

Livestock on Farms: Irrigated and Nonirrigated Alberta, 1981

Type of livestock	Number of livestock per farm		Percentage of all farms responding	
	Irrigated	Non-irrigated	Irrigated	Non-irrigated
Cattle	173	64	68	61
Milk cows and heifers	8	4	15	15
Beef cows and heifers	73	30	53	51
Steers	49	8	32	29
Calves	40	21	59	22
Cattle on feed	69	9	25	22

Source: Statistics Canada.

TABLE 6.4

Livestock on Farms: Irrigated and Nonirrigated South Saskatchewan River Irrigation District, 1986[a]

	Livestock on farms	
	Irrigated	Non-irrigated
Farms reporting cattle and calves (percent)	47.0	26.7
Cattle and calves per 100 acres (head)	2.20	1.08

[a]SSRID No. 1.

Source: Based on data by Manning.

TABLE 6.5

Comparative Economics of Livestock Enterprises on Irrigation Farms, Saskatchewan, 1986

	Cost of forage		
	Dryland	Irrigation	Difference
Cost per acre (dollars)			
Total variable	34.57	76.47	a
Total fixed	72.26	120.02	
Total per acre	106.83	196.49	89.66
Yield per acre (tonnes)	1.33	3.20	
Average cost per tonne (dollars)	80.32	61.40	-18.92
	Cost of feed per head		
	Cow-calf	Back-ground	Feed to finish
	dollars		
Dryland			
Grazing and other	106	16	68
Feed: Forage	134	79	47
Grain		40	195
TOTAL	240	135	310
Irrigated			
Grazing	106	16	68
Feed: Forage	103	60	36
Grain		40	195
TOTAL	209	116	299
Marginal difference	31	19	11

[a]Blanks indicate not applicable.

Source: Personal communication with Dr. K. D. Russell, Underwood McLellan and Associates, Ltd., Lethbridge.

support the contention that irrigation leads to more farm level diversification through the addition of livestock on farms along with irrigated crops, although more research is needed in order to make definitive statements.

Irrigation and Regional Diversification

A diversified crop mix, as well as the emergence of livestock, have impacts that extend far beyond the farm gate. This is because irrigation adds to the demand for farm inputs, and it may spur further processing because of increased output thereby increasing the level of secondary industries in the region. This results in a somewhat more diversified nonagricultural sector and, as a result, a more diversified regional economy.[8]

Some of the input-supplying industries that are affected by irrigation development are fertilizer manufacturing; irrigation equipment manufacturing, distribution, and retail; and other machinery and equipment, electrical goods, construction industries among others. Similarly, the output from irrigated farms, particularly livestock and specialty crops, is used by agricultural processing firms. Each of these industries requires goods and services produced by still many other industries. Eventually, this leads to a multiplier effect. For the province of Alberta, Kulshreshtha *et al.* estimated such a multiplier to be in the neighborhood of 3.37. Thus, for every dollar's worth of agricultural gross domestic product (related to irrigation production), another $2.37 is contributed by other industries in the province of Alberta.[9] A comparison (based on Johnson and Kulshreshtha) of various agriculture subsectors in Saskatchewan also suggested that, with the exception of poultry, the subsectors have very similar backward linkages and, thus, comparable multiplier effects. However, the forward linkages of irrigation are significantly higher through more livestock raising and processing.

Progress of Diversification on Irrigated Farms

What is the actual record of diversification on Prairie farms? The crop mix of Saskatchewan and Alberta irrigated farms is shown in Figure 6.4. In Alberta, grains still predominate in various cropping rotations followed on irrigated farms. About half the total irrigated cropland is under wheat, oats, barley, mixed grain, and rye. These crops occupied about 192 thousand hectares of land in the province by 1982 (Table 6.6). The other major category of crops include tame pasture, hay, and other fodder.

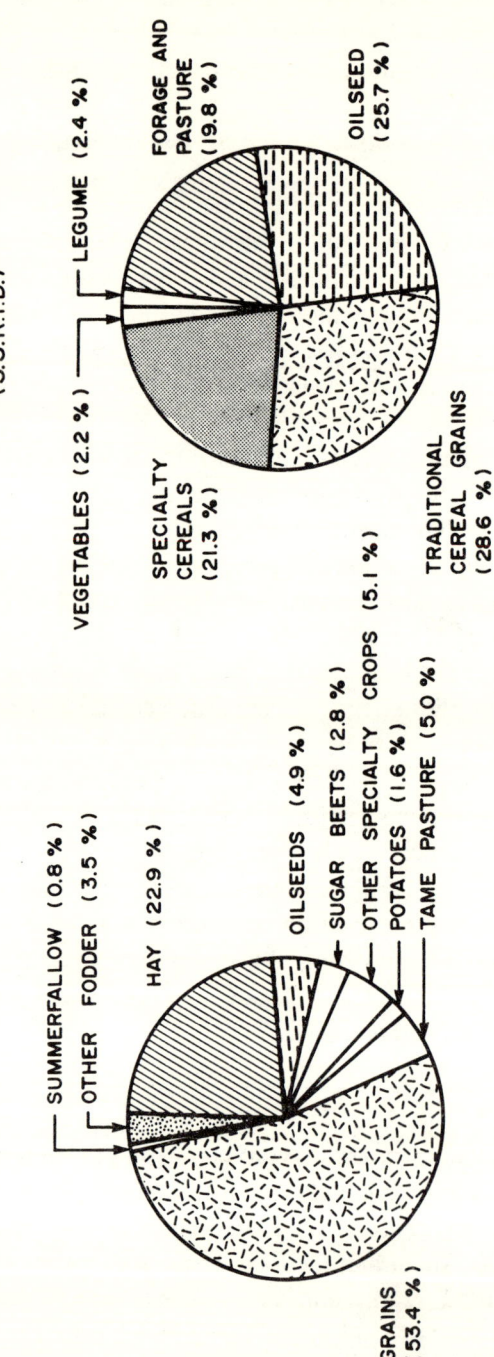

Figure 6.4. Crop Mix on Irrigated Farms in Alberta and Saskatchewan, 1986

TABLE 6.6
Crop Area in the Major Irrigation Districts Alberta, by Crop, 1979 and 1982[a]

	1979		1982	
	Total	Percentage of total	Total	Percentage of total
	hectares	percent	hectares	percent
Grains[b]	166,664	46.3	192,346	54.8
Oilseeds[c]	33,252	9.2	15,519	4.4
Hay[d]	71,071	19.7	67,098	19.1
Tame pasture[d]	29,869	8.3	20,555	5.8
Other fodder[e]	13,361	3.7	9,513	2.7
Sugar beets	13,141	3.6	12,737	3.6
Potatoes	4,612	1.3	5,394	1.5
Other specialty crops	22,617	6.3	24,521	7.0
Irrigated crops	354,587	98.4	347,683	99.0
Summer fallow	5,588	1.6	3,424	1.0
Irrigated cropland	360,175	100.0	351,107	100.0

[a]Does not include data for the Irrigation Districts of Aetna, Leavitt, Mountain View, and Ross Creek; in 1982, data were not available from the Western Irrigation District which had over 46,000 irrigated acres in 1979.

[b]Wheat, oats, barley, mixed grain, and rye.

[c]Flaxseed, canola/rapeseed, sunflowers, and mustard.

[d]Alfalfa hay and sweet clover hay.

[e]Greenfeed, silage corn, silage, and grass hay.

[f]Peas, canning corn, timothy, brome grass, alfalfa seed, sweet clover seed, small seeds, and miscellaneous crops.

Source: Alberta Agriculture.

In 1986 these crops constituted about 31.4 percent of irrigated cropland. In 1982 there were 42,652 hectares of specialty crops. Major specialty crops include potatoes and sugar beets. Other specialty crops include peas, canning corn, timothy and brome grass, alfalfa seed, sweet clover seed, and other miscellaneous crops. In order to obtain a valid measure of increased diversification on irrigated farms, one should compare the above crop mix with that under dryland conditions. Using 1981 Census of Agriculture data, almost three-fourths of the total dryland area under crops (including summer fallow) is devoted to production of grains. Thus, the reduction of this proportion to 53 percent on irrigated farms indicates that a significant progress has been made toward diversification.

In Saskatchewan, data for all irrigated cropland are not available. However, data for the South Saskatchewan River Irrigation District (SSRID) No. 1 was taken as a proxy for all intensive irrigation districts. In this region, cereals (both traditional and special) constitute roughly half of the total irrigated cropland. Oilseeds account for another 25 percent of the area. Specialty crops are not very widely cultivated in this region; they constitute only 348 hectares of irrigated cropland (Table 6.7).

Progress towards crop diversification in SSRID No. 1 is shown in Table 6.8. Irrigated area in the district increased from 2,566 hectares in 1969 to 15,952 hectares in 1985. The area under potatoes and other specialty crops increased from 34 hectares to 268 hectares—an increase of almost seven times.

Irrigated land use in southwest Saskatchewan is slightly different from that in the SSRID No. 1; there, the type of irrigation followed is spring flood and most of the land irrigated is devoted to forage (Table 6.8). Almost three-quarters of the total irrigated land is given over to native hay, tame hay, tame pasture, and native pasture.

The other dimension of farm level diversification is livestock. As irrigated lands produce more forage, and as the carrying capacity of pastures is enhanced through irrigation, more and more farmers find it profitable to raise livestock. Data presented in Table 6.3 and Figure 6.3 for Alberta and in Table 6.4 for Saskatchewan support this thesis. The importance of South Alberta as a major North American cattle center is likely to increase in the future, particularly in light of aquifer depletion in other major cattle centers, such as Texas and Colorado, resulting in reduced forage availability (based on Alberta Agriculture 1983).

In summary, irrigation has already resulted in significant change in the land-use pattern in South Alberta. Diversification has taken two forms:

TABLE 6.7

Irrigated Land Use
South Saskatchewan River Irrigation District[a]
Selected Years, 1969-1985

Crop and forage	Irrigated land use				Percentage of total, 1985
	1969	1979	1982	1985	
	hectares				percent
Traditional cereals[b]	1,356	4,763	8,069	4,569	28.6
Special cereals[c]	200	420	1,551	3,396	21.3
Forage	556	2,290	2,512	3,104	19.5
Pasture and other forage	14	287	154	53	0.3
Oilseeds[d]	291	2,465	1,924	4,104	25.7
Legumes[e]	0	1,667	310	378	2.4
Potato	34	194	277	268	1.7
Other vegetable crops	0	53	52	80	0.5
Miscellaneous	113	0	0	0	f
Total irrigated crops	2,564	12,139	14,849	15,952	100.0

[a]SSRID No. 1.

[b]Includes hard wheat, durum, barley, oats, utility wheat, and rye.

[c]Includes triticale, buckwheat, soft wheat, corn, winter wheat, and canary seed.

[d]Includes flax, canola, mustard, and sunflower.

[e]Includes lentils, peas, faba beans, field beans, and Chinese broad beans.

[f]Blank indicates no data available.

Source: Saskatchewan Water Corporation.

TABLE 6.8

Irrigated Land Use, Southwest Saskatchewan, 1980

Forage and crop	Percent of total	
Cereals	16.8	
Oilseeds	0.7	
Other crops	4.2	
Total crop		21.7
Native hay	1.1	
Tame hay	54.8	
Pastures	15.1	
Total forage		71.0
Summer fallow	7.3	7.3
TOTAL	100.0	100.0

Source: Prairie Farm Rehabilitation Administration.

a variety of new crops and livestock as a major source of farm income. Significant strides have been made on both of these farms in Alberta. In Saskatchewan progress has been much less relative to Alberta.

Future of Irrigation in Prairie Provinces

The future of irrigation in the Prairie provinces is difficult to forecast. A complex set of economic-social-political factors affects the future supply of and demand for irrigation. Some of these considerations are now reviewed.

Technical Feasibility

Technical constraints play a role in two ways: they dictate the magnitude of irrigable land and they determine the supply of water for irrigation. At present, a magnitude of 520 thousand hectares of irrigation, when judged against either of these criterion, suggests that there is considerable scope for future expansion. In Alberta, irrigated acreage could be doubled to 810 thousand hectares using internal basin supplies (Veeman 1985, p. 26). Estimates based on Alberta Agriculture (1983) suggest an increase of 40 percent (648 thousand hectares). In addition, another 2.4 million hectares in the South Saskatchewan River basin in Southern and East Central Alberta is believed to be potentially irrigable; however, this would require that water be imported from other basins. In Saskatchewan the forecasted level of irrigable area has been estimated at 197 thousand hectares by the year 2020.[10] In the case of Saskatchewan, irrigation potentials suggest a possible increase in the present irrigated area by 66 percent. Thus, one cannot avoid the conclusion that technically irrigation can be expanded both in Saskatchewan and Alberta.

Economic Feasibility

Given the large potential for irrigation on the Prairies, an obvious question is "why are farmers not adopting irrigation on their farms?" This raises the possibility that the major issue in determining the future of irrigation may not be technical feasibility but economic feasibility. The following are the top 10 reasons for adopting irrigation, based on Manning (pp. 89-109), ranked in descending order of importance:

- Increased yields
- Profitability of irrigation
- Reduced effect of drought on yields
- Ability to diversify the crop mix
- Availability of capital for the investment
- Availability of water

- Cost of irrigation equipment
- Reduced interest rates
- Grant of $100 per acre
- Subsidization of water costs.

The profitability of irrigation ranked high as a factor conducive to adopting irrigation. Over 60 percent of those surveyed ranked profitability as "very important" with regards to their original decision to adopt the technology.

A similar listing of reasons for not adopting irrigation by dryland farmers is provided below, again arranged in order of importance:

- Cost of irrigation equipment
- Low crop prices
- Salinity concerns
- Availability of capital for the investment
- Availability of markets
- Risk associated with changing to irrigation
- Increased time and labor requirements
- Suitability of land (topography)
- Cost of water
- Investment in irrigation not compatible with goals.

Economic reasons selected by both irrigators as well as nonirrigators support earlier contentions that the future of irrigation would be dictated by the economic feasibility of irrigation. Three aspects of irrigation are now examined.

Crop Mix. Irrigated farms still depend upon traditional low-valued cereals and oilseeds for a major part of its income. High-valued specialty crops constitute a rather small proportion (5 to 10 percent) of the total cropland under irrigation. Unless markets for these products expand, the Prairie provinces would not be devoting a higher proportion of area than at present under these crops (Veeman and Veeman). The markets for traditional grains and oilseeds have been suggested to be modest to optimistic in the coming decade.[11]

Financial Requirements. Converting a dryland farm into an irrigated farm is a capital-intensive proposition. Costs, however, vary depending upon the irrigation system chosen. For example, if a farmer chooses a surface irrigation system, his major cost is land leveling. These costs may be in the neighborhood of $900 to $1,300 per hectare (in 1987) as shown in Table 6.9. In Saskatchewan in 1987, the cost for a border dyke system has been estimated to be between $850 and $1,300 per hectare (Table 6.10).

TABLE 6.9

Economics of Alternative Irrigation Systems on Farms Alberta, 1979 and 1987

Type of system	Total cost	
	1979	1987
	dollars per hectare	
Border dikes	1,297	897
Furrows with gated pipe and pump	1,574	1,438
Wheel move: 4 laterals	665	914
Center pivot		
Towable: high pressure	557	941
low pressure	a	825
Nontowable: high pressure	908	1,329
low pressure		1,250
Traveling volume guns	728	1,235

[a]Blanks indicate no data available.

Sources:

    For 1979: Susko and Andruchow.

    For 1987: Alberta Agriculture.

TABLE 6.10

Cost of Converting Dryland into Irrigated Saskatchewan, 1987

Type of system	Initial cost	Cost per hectare
		dollars
Center pivot	70,000	1,300
Wheel move: 2 laterals	53,000	935
4 laterals	70,000	1,080
Linear system	150,000	1,160
Traveling volume gun	a	1,000 and up
Handmove		500-600
Solid set		3,700-4,900
Drip		4,000
Border dike		850-1,350

[a]Blanks indicate no data available.

Source: University of Saskatchewan.

One major problem with surface irrigation is that it is not applicable to all land topography; if the topography is highly undulating, it is preferable to apply a sprinkler system. A variety of such systems is available, and costs vary according to the method of irrigation. The initial investment is a major deterrent to the adoption of irrigation. Irrigation is more appealing to farmers in better financial situations. Furthermore, the cost of capital plays an important role in deciding whether or not a farmer adopts the irrigation.

Relative Returns. Returns from irrigation relative to dryland cultivation fluctuate from year to year as well as from region to region. Using data for Alberta for the period from 1982 to 1984, net returns to irrigation over dryland farming are positive. In the South Saskatchewan River basin (Alberta portion), the increase in net returns is estimated to vary between $213.27 to $554.25 per hectare, depending upon the irrigation climatic zone (Table 6.11). One must be cautioned against generalizing from these results as these are area and time specific.[12]

For Saskatchewan, using 1986 data, the economics of irrigation over dryland is shown in Table 6.12. Under 1986 economic conditions, the benefits to farmers from converting dryland area into irrigation are not positive. For example, to grow wheat on fallow (under dryland conditions) results in a net return of -$106.50 per hectare. Similarly, to grow soft wheat under irrigation results in a net return of -$156.14 per hectare. Thus, there is a decrease in the net return of $49.67 per hectare by irrigation over dryland. However, if that area were converted into growing alfalfa, there would be a net positive return of $208.95 per hectare. One should again be cautioned against generalizing from these figures since they are based on a single year and are for a period when prices were relatively low.

To show the effect of increasing product prices, examine the data in Table 6.13. If prices were to increase by 25 percent above the 1986 level, returns from irrigation over dryland farming would be $84.65 higher per hectare over cash costs and $39.32 higher per hectare over total cost. This is an improvement of $101.36 per hectare over the 1986 level.

Besides prices, yield levels are equally critical in determining a favorable return from irrigation over dryland farming. Break-even yields for irrigated wheat would have to exceed 185 bushels per hectare under medium levels of yield on dryland (Table 6.12). For Manitoba, it is estimated that wheat yields must be over 62 bushels per hectare, grain corn over 86 bushels per hectare, and corn silage over 17.3 tonnes per hectare in order to attain a desirable rate of return (Kraft *et al.*).

TABLE 6.11

Annual Net Returns to Irrigated Crop Production Alberta South Saskatchewan River Basin Average, 1982 and 1984

Zone[a]	Dryland			Irrigation			Marginal increase
	Revenue	Costs	Net returns	Revenue	Costs	Net returns	
	dollars per hectare						
A1	91.61	82.88	8.73	1,071.46	508.48	562.98	554.25
A2	98.27	83.80	15.47	841.95	438.70	403.25	387.78
B	150.00	131.41	18.59	779.90	427.43	352.47	333.88
C	146.43	123.41	23.02	591.50	355.21	236.29	213.27
D	185.76	198.03	-12.27	706.12	410.81	295.31	307.58

[a]Refers to irrigation climate zones based on heat units, days of growing season, and moisture deficit. These extend from the eastern part of the basin (Zone A1) to the foothills of the Rockies (Zone D).

Source: Desjardins and Egglestone, p. 334.

TABLE 6.12

Relative Returns for Various Crops, Dryland and Irrigated Saskatchewan, 1986

Crop	Gross return	Variable cash	Fixed cash	Return over cash cost	Return over total cost[a]
	dollars per hectare				
Dryland					
Wheat: on fallow	172.91	88.61	154.86	- 60.56	-106.50
on stubble	154.12	113.45	123.82	- 83.15	-117.72
Irrigated					
Hard wheat	420.03	279.38	322.10	-181.45	-301.15
Soft wheat	447.09	229.71	268.80	- 51.42	-156.17
Durum	455.12	286.57	293.22	-124.67	-222.59
Canola	531.77	293.17	317.31	- 78.71	-191.58
Flax	386.09	270.01	383.71	-267.62	-426.33
Alfalfa	689.37	162.03	337.42	189.92	102.45

[a]Excludes operator labor and management charges.

Source: Brown, Kulshreshtha, and Linsley.

TABLE 6.13

Changes in the Relative Returns for Various Crops Under Varying Levels of Product Prices Dryland and Irrigated, Saskatchewan, 1986

Crops	Returns over:					
	Cash cost under increased product prices			Total cost under increased product prices		
	10 percent	25 percent	50 percent	10 percent	25 percent	50 percent
	dollars per hectare					
Dryland						
Wheat: on fallow	-43.27	-17.33	25.89	-89.21	-63.27	-20.05
on stubble	-67.74	-44.62	-6.09	-102.31	-79.19	-40.66
Irrigated						
Hard wheat	-139.45	-76.44	28.57	-259.15	-196.14	-91.14
Soft wheat	6.71	60.35	172.13	-111.46	44.40	67.38
Durum	-79.16	-10.89	102.89	-177.08	-108.81	4.97
Canola	25.53	54.23	187.18	-138.40	-58.64	74.31
Flax	-229.01	-171.10	-74.58	-387.72	-329.81	-233.29
Alfalfa	258.76	362.26	534.61	171.39	274.79	447.14
Marginal benefit of irrigation[a]	38.80	84.65	167.17	-9.91	39.32	88.22
From 1986 level	36.68	82.53	165.05	52.13	101.36	150.20

[a]Estimated using average share of dryland and irrigated crops in the South Saskatchewan River Irrigation District (SSRID No. 1).

The economic benefits of irrigation on farms may also result from its drought-mitigative effects. During periods of agricultural drought, irrigation provides a stabilizing effect on farms and thereby on the regional economy.[13] In addition, farmers' preferences with respect to changed lifestyle[14] will continue to be a major factor.

Future Policy Directions

In addition to the technical and economic feasibility of irrigation, another set of factors will influence the future of irrigation on the Prairies: This set includes the various policies of national and regional governments.

Evaluation Criteria. Evaluation of water projects can be carried out using the criteria of economic efficiency, income distribution, regional development, or other criteria relevant to the decision maker. The future of water development projects and of irrigation in general may depend upon the criteria used. Some past projects may not be found to be socially desirable solely on the basis of economic efficiency. For example, the development of the South Saskatchewan River Project was questioned by the Royal Commission with respect to its contributions to national economic efficiency.[15] Similarly, the study of the expansion of irrigation through the Oldman Dam in South Alberta suggests the benefit-cost ratio to be near, or even slightly less than, one. Furthermore, one could also question such developments on the basis that farmers would still produce crops subsidized by the public treasury. In light of limited local and international markets, it is doubtful that irrigation development in the near future could be justified on the grounds of economic efficiency.

An irrigation project does contribute toward development of regional economic activity. Water used through direct use, such as irrigation, recreation, power, and municipal water supply, does spur economic activity in other sectors that have economic linkages with them. Some 12 percent of economic activity in South Alberta is directly or indirectly related to irrigation (Kulshreshtha *et al.*). Similarly, for Saskatchewan (Kulshreshtha), it is estimated that, on a comparable land base, the multiplier effect of irrigation production is almost three times as high as dryland production. Thus, on account of a relatively higher multiplier effect of irrigation, such projects do contribute significantly to regional development. From a regional development perspective, one could justify irrigation development. However, even in this context, one should be cognizant of two types of income transfers: (1) from consumers and taxpayers on the Prairies and other regions to producers in the water project region and (2) from income distributions that are more unequal after irrigation is adopted in the region. Only 8 percent of nonirrigators had net worth exceeding $120,000 as opposed to 20 percent of irrigated farmers.

Large Water Development Project. Water development projects require a large initial capital investment. For example, the South Saskatchewan River Project cost $136 million (in nominal dollars), which translates into $827 million in 1986 dollars. The Oldman Dam in South Alberta is expected to cost $140 million in 1980 dollars. In addition to the cost of developing reservoirs, off-farm investments are fairly high. For example, in Saskatchewan, to bring water to the farmer's field costs the Saskatchewan Water Corporation $5,065 to $6,710 per hectare.[16] To what extent such investment outlays would be possible is doubtful in light of a sluggish economy, regionally and nationally, plus the public outcry for more responsible fiscal management of taxpayers' money.

Water Resource Utilization Policies. Policies that would affect the future of irrigation include those affecting water exports and interbasin transfers of water. Large-scale irrigation development in drier regions of the Prairie provinces cannot be sustained without some transfer of water from northern regions. However, the policy of the federal and regional governments is unclear in this regard; thus, future implications for irrigation cannot be drawn. Water export policies may also limit irrigation development on the Prairies. However, the interbasin transfer and water export policies are interrelated since many large interbasin transfer projects may involve water exports.[17]

Water Pricing. How will society view water in the future? At present there is no established price for water use.[18] As demand for water increases and society considers its opportunity cost in deciding its use for irrigation, irrigators will likely face stiff competition for available water supplies.

Irrigation and Environment Quality

The future of irrigation on the Prairies may also be determined by environmental concerns, particularly soil salinity. Due to water seepage, the area around canals and other distribution channels is prone to salinization.[19] This may lead to more drainage on irrigated lands, the use of pressured pipelines (instead of open canals), and better lining of canals. All of these add further to the cost of both public and private irrigation. Assessment should also be made as to whether irrigation and diversification lead to better land use and soil conservation.

Summary and Conclusions

The semiarid climate in much of the farming areas in the southern part of the Prairies makes irrigation development very appealing. As a result, irrigation has been developed over 530 thousand hectares of this area. Three-quarters of this irrigation development is located in South Alberta, 22 percent in South and West-Central Saskatchewan, and only 2 percent of the total in Manitoba. Although, in terms of the land base, irrigation accounts for only 1.4 percent of improved farmland, its contribution to the value of products sold by agriculture is estimated at 6.62 percent of the total.

Irrigation leads to the diversification of agriculture and the diversification of the wider regional economy through its multiplier effect. Farmers have a wider choice in terms of the variety of crops that can be grown. Production of forage leads to induced livestock activity. Both of these, through backward and forward linkages, induce development of other sectors and lead to regional development. When one looks at the record of diversification in Alberta, it appears that much progress has been made in diversifying the agricultural crop mix. Specialty crops account for 12 percent of the irrigated cropland. In Saskatchewan the emergence of specialty crops has not been as dramatic. Farmers continue to grow traditional crops.

About 20 to 28 percent of irrigated cropland is devoted to forage, which leads to higher cattle intensity in the irrigated areas. In Alberta irrigated farms have three times more cattle and about seven times more cattle on feed than in the dryland farms. Recent free trade agreement between the United States and Canada may lead to assured markets for Canadian beef in the United States and, thus, more cattle raising in irrigated areas. However, since irrigation development is heavily subsidized, the same agreement may reduce subsidies and lead to lack of future expansion in irrigated agriculture on the prairies.

Given this link between diversification and irrigation, the future of diversification may, to a significant extent, be determined by the future of irrigation. Forecasting irrigation development is a risky process because of the interplay of complex socio-economic-political factors. However, based on technical feasibility, there are no constraints to irrigation expansion.

Irrigated areas could be doubled with existing basin supplies in Alberta and increased by 66 percent of the present area in Saskatchewan. Much of this expansion would, therefore, depend upon farmers' adoption of irrigation. Aside from the living style preferences, the economics of irrigation is a very significant factor in a farmer's decision to adopt irrigation. At present, with low crop prices, a favorable economic climate does not exist in the Prairie region for the rapid expansion of irrigated farmland.

Notes

[1]In fact, this region was hard hit during the extremely dry years of 1936 and 1937.

[2]Based on S. N. Kulshreshtha and M. T. Yap. Estimates refer to the year 1979 but were based on data obtained from the 1981 Census of Agriculture. Contribution of irrigation to value of output was estimated as:

Manitoba	$ 7,096 thousand
Saskatchewan	$ 47,421 thousand
Alberta	$402,549 thousand.

Calculation of relative contribution is based on the definition of value of output as used in the input-output accounts.

[3]For subregions such as South Alberta, irrigated agriculture is one of the primary goods-producing industries. Although precise estimates are unavailable, it is stipulated that irrigation may account for 50 to 70 percent of direct output of the agriculture industry in that subregion.

[4]The second concept—regional diversification—requires a more thorough inquiry and, therefore, considered to be beyond the scope of present study. A brief discussion of this topic is, however, contained in a later section.

[5]Soft wheats have a very high energy content which makes them more desirable for cattle feed. Furthermore, because of their better amino acid balances, they have an advantage as feed in hog production.

[6]South Alberta was defined as an aggregate of census divisions 1 to 3 for the 1956 to 1986 censuses and of census divisions 1 to 4 for the earlier censuses.

[7]Evidence on the favorable economics of hay production under irrigation, over dryland, is far from being definitive. Information provided by Mr. C. D. Radke of Alberta Agriculture suggests that in 1987, in the

Lethbridge area, cost of providing a tonne of hay under dryland conditions was $11.14 lower than under irrigation.

[8] As noted in the beginning of this report, estimation of secondary impacts of irrigation is considered beyond the scope of this study. This is not to suggest that such impacts are neither important nor relevant to a discussion of irrigation.

[9] At this moment, such estimates are unavailable for the province of Manitoba.

[10] This forecast is based on several sources of information and data. It includes 118,620 hectares of existing irrigated area and adds to it 17,105 hectares of expansion in irrigated area around Lake Diefenbaker, plus 56,858 hectares of additional new development in the South Saskatchewan River Project. In addition, 4,855 hectares are included, which are planned to be developed in connection with the Souris River Basin Development. Data on the South Saskatchewan River Project were obtained from UMA Engineering, Ltd. (Kulshretha, Russell, and Klein).

[11] According to the Economic Council of Canada, the most probable outlook for the world grains market over the next eight years is for the production and consumption of wheat and coarse grains to converge during the first half of the 1990s. The report suggests that the productive capacity of the developed countries will continue to outstrip not only their own domestic demand but also the effective demand in both the developing countries and the centrally planned economies in the foreseeable future (pp. 15 and 16).

[12] For example, under 1987 conditions, Alberta Agriculture (Production Economics Branch) estimates that dryland farming may be more economical than irrigation farming in the Lethbridge area.

[13] Such benefits may not be present with full force if the agriculture drought coincides with a meteorological drought.

[14] Change in the life-style of a farmer after adopting irrigation may result from several sources: (1) managerial decision-making process may be more sophisticated on irrigation farms, leading to more stress-related changes in life-style; (2) diversification into livestock production may affect the farmers' choice of residence, particularly during winter months; and (3) farming would tend to be more labor intensive, resulting in less availability for other activities.

[15] The Royal Commission on the South Saskatchewan River Project (Chaired by Dr. T. H. Hogg, 1952) found that "... the economic returns to

the Canadian people on the investment in the proposed South Saskatchewan River Project . . . are not commensurate with the cost thereof . . ." (p. 6).

[16] Based on personal communications with Mr. Dave Richards, Saskatchewan Water Corporation.

[17] For example, the GRAND CANAL scheme is designed for both of these purposes.

[18] This is not to suggest that there is no charge for water. The users pay for the cost of services in connection with delivery and treatment of water. However, there is no direct charge for water itself.

[19] This is not to suggest that irrigation is the only cause of salinization. This issue, however, needs more objective evidence linking irrigation to salinization.

References

Agriculture Canada. *Irrigation in Canada*. A report to AIC Task Force on Water Use in Agriculture. Ottawa, Ontario, undated.

Alberta Agriculture, Statistics Branch. *Agriculture Statistics Yearbook, 1982*. Edmonton, Alberta, 1983.

Alberta Agriculture, Irrigation Branch. Personal Communication, 1989.

Anderson, M., and Associates, Ltd. *Oldman River Basin Study. Phase II: Economic Analysis of Water Supply Alternative*. Edmonton: Oldman River Basin Study Management Committee, 1978.

Brown, W. J., S. N. Kulshreshtha, and J. Linsley. *Economics of Irrigation in Saskatchewan: Results of the 1987 Farm Workshops*. Research Report No. 87-02, Department of Agricultural Economics, University of Saskatchewan, Saskatoon, 1987.

Canadian Environmental Control Newsletter. "Oil Money in Alberta to Be Used for Increased Irrigation Efforts." 79(1976):7.1.

Carter, C. A., and A. Schmitz. "Cattle Feeding in Canada: The Economics of Its Location." *Agribusiness* 2(1986):119-135.

Craig, K. R. "Irrigation in Alberta: Past, Present and Future." Paper presented to the APEGGA Annual Convention, June 1987.

Desjardins, R., and J. Egglestone. "The Economic and Financial Feasibility of Irrigation." In *Irrigation on the Prairies*. Saskatoon: Saskatchewan Water Corporation, Fourth Annual Western Provincial Conference, November 1985.

Dubetz, S., and D. B. Wilson. *Growing Irrigated Crops on the Canadian Prairies*. Ottawa, Ontario: Canadian Department of Agriculture, Publication No. 1400, 1969.

Economic Council of Canada. *Handling the Risks—A Report on the Prairie Grain Economy*. Ottawa, Ontario: Economic Council of Canada, 1988.

Howe, C. W. *Benefit-Cost Analysis for Water System Planning*. Water Resources Monograph 2. Washington, D. C.: American Geophysical Union, 1971.

Johnson, T. G., and S. N. Kulshreshtha. "Exogenizing Agriculture in an Input-Output Model to Estimate Relative Impacts of Different Farm Types." *Western Journal of Agricultural Economics* 7(1982):187-198.

Kraft, D. F., R. M. Josephson, M. J. M. Kapitany, and S. Geddes. *Economic Analysis of Irrigation Investment in Manitoba*. Department of Agricultural Economics and Farm Management, University of Manitoba, Winnipeg, 1981.

Kulshreshtha, S. N. "Implications of Alternate Levels of Water Charge on Agricultural Diversification in Saskatchewan." In *Water for World Development*. Volume III. Proceedings of the VIth IWRA World Congress on Water Resources, Ottawa, May 29-June 3, 1988.

Kulshreshtha, S. N., and M. T. Yap. *Economic Structure of the Prairie Provinces: The Prairie Regional Input-Output and Employment Model*. Regina, Saskatchewan: Prairie Farm Rehabilitation Administration, 1985.

Kulshreshtha, S. N., K. D. Russell, and K. K. Klein. *Evaluation of the Impacts and Effects of the South Saskatchewan River Project*. Ottawa, Ontario: Audit and Evaluation Branch, Agriculture Canada, June 1988.

Kulshreshtha, S. N., K. D. Russell, G. Ayers, and B. C. Palmer. "Economic Impacts of Irrigation Development in Alberta Upon the Provincial and Canadian Economy." *Canadian Water Resources Journal* 10(1985).

Manning, S. L. F. "A Socio-Economic Analysis of the Adoption of Irrigation in Saskatchewan." Unpublished M.Sc. thesis, University of Saskatchewan, 1988.

Pearce, D. W., ed. *The MIT Dictionary of Modern Economics*. Cambridge, Massachusetts: The MIT Press, 1986.

Prairie Farm Rehabilitation Administration. "Irrigation Water Uses Survey." Regina, Saskatchewan: Prairie Farm Rehabilitation Administration, Draft Report, 1980.

Prairie Provinces Water Board. *Agricultural Water Uses.* Appendix 3, Water Demand Study. Regina, Saskatchewan: Prairie Provinces Water Board, 1982.

Russell, K. D. Personal Communication.

Saskatchewan Water Corporation. *History of South Saskatchewan River Irrigation Project.* Moose Jaw, Saskatchewan, July 1987.

Shady, A. M. *Irrigation Drainage and Flood Control in Canada.* Ottawa, Ontario, June 1989.

Stabler, J. C. "Post Irrigation Assessment Study." A report prepared for the Saskatchewan Water Corporation, Moose Jaw, Saskatchewan, 1987.

Statistics Canada. *Census of Agriculture, 1981.* Ottawa, Ontario. Minister of Supply and Services, 1982.

Susko, R. J., and L. J. Andruchow. *Economic Evaluation of Irrigation Systems.* Edmonton, Alberta: Production Economics Branch, Alberta Agriculture, 1979.

Topham, H. L. *History of Irrigation in Western Canada.* Regina, Saskatchewan: Prairie Farm Rehabilitation Administration, 1982.

UMA Engineering, Ltd. "South Saskatchewan River Basin Planning Program, Agronomy Component, Irrigation Water Requirement Criteria Study." Lethbridge, Alberta, 1982.

University of Saskatchewan. *Guide to Farm Practice.* Saskatoon, Saskatchewan: University of Saskatchwan: 1987.

Veeman, T. S. "Water and Economic Growth in Western Canada." Discussion Paper No. 279. Ottawa, Ontario: Economic Council of Canada, 1985.

Veeman, T. S., and M. M. Veeman. *The Future of Grain.* Toronto, Ontario: James Lorimar & Co., 1984.

7

The Adoption of Modern Irrigation Technologies in the United States

Gary Casterline, Ariel Dinar, and David Zilberman

Introduction

Irrigated farming plays an important role in the United States. About one-sixth of the U. S. agricultural acreage is irrigated, and this land is producing about one-third of the product value grown in the United States. Obviously, the relative importance of irrigation varies among crops. For example, most specialty crops, as well as rice and cotton, are produced on irrigated land.

While irrigated farming has been practiced for thousands of years, irrigation technology has changed substantially over the last 40 years. We have seen substantial improvement in water pumping and conveyance technologies—an improvement that adapted much of the technology first introduced in oil exploration and conveyance. Moreover, modern irrigation technologies (in particular, sprinkler and drip irrigation) have been introduced. These technologies drastically affected water use and crop yields and changed crop-production patterns. Adoption of these technological changes has had substantial impact on trade and market conditions of many crops; cotton and corn are notable examples. Moreover, the availability and adoption of these technologies substantially extended the ability of U. S. agriculture to respond to environmental degradation problems as well as possible changes in weather patterns.

This chapter first presents a farm-level assessment of the characteristics and impacts of modern irrigation technologies such as drip and sprinkler irrigation. It argues that they serve to increase irrigation effectiveness and that their relative contribution is larger in areas where soil and water conditions are not very favorable for irrigation, e.g., water-holding capacity or water quality is low. Introduction of these new technologies is likely to increase yields, reduce water use, and reduce runoff and percolation. These technologies are more likely to be adopted in regions with high output revenue and water prices, low water quality, sandy soils, and rather uneven land. These general tendencies are illustrated with a simulation on data from California and supported by empirical evidence.

The second part of the chapter attempts to explain and analyze general patterns of irrigation technology uses in the United States over the last 40 years and to assess the changes that the introduction of modern irrigation technologies have made on the structure of U. S. farming. Results show that there is a significant difference between the development of irrigated agriculture in the Southwest and the Midwest.

Irrigated agriculture was introduced to the Southwest in the 1890s and has been growing constantly in acreage (for most states) since that time. Gravitational irrigation systems are dominant forms of irrigation in most states. Sprinkler irrigation was introduced in the 1950s and is used for much of the land of field crops, fruits, and vegetables. Drip irrigation was introduced in the 1970s. It is used with high-value vegetables and fruits and low-quality or high water cost regions, and it is starting to be used for cotton.

The West is facing water shortages that will lead to increases in water price and faster adoption of drip and sprinkler irrigation because empirical evidence shows a high likelihood to adopt modern irrigation technologies as water price increases. Moreover, transitions to modern irrigation technologies and the allocation of the requisite management skills may be accompanied by a transition to higher value crops (e.g., transitions from alfalfa or potatoes to specialty crops) as long as prices for specialty crops (fruits, etc.) are sufficiently high. Reduction in trade barriers and increased marketing and educational efforts will tend to accommodate the transitions.

Substantial irrigation of the Plains states started in the 1950s. It involved much use of groundwater and the application of gravitational irrigation technologies. In the 1970s, fast adoption of center pivot irrigation in the Midwest was seen with a resulting increase in output supply. The prices of corn and soybeans have been affected by the changes in supply and are likely to decline since the irrigated land base in the Midwest is increasing.

There is some evidence that governmental support programs have tended to encourage overadoption of modern technologies and exploitation of water in the Midwest. Reduction in support levels and possible reductions in output prices, associated with free-market conditions, may slow the adoption of irrigation technologies in the Midwest; and that will increase resource conservation and reduce the tendency to oversupply in agriculture.

The Economics of Irrigation Technology Choice

In arid (and many semiarid) areas, irrigation has been the primary source of water application in crop production. In other areas, irrigation has been used to supplement rainfall and to ensure an adequate water supply during dry years. Among other benefits, auxiliary irrigation tends to improve yields and allows double cropping.

In irrigation farming, there is total reliance on ground- and surface water. The use of groundwater has increased as pumping technology has improved. It is quite common for farmers to drill wells on their own property and to manage their groundwater independently. Most of the surface water is obtained by the diversion of water from rivers and lakes, and it frequently supplies a large number of farmers so that its utilization requires substantial investments in damming and conveyance systems. There are situations of conjunctive use of ground- and surface water. In these cases, the ground is used for water storage, and groundwater is used to stabilize water use in farming when surface water supplies are lacking (Burt).

The main interest of this chapter is on on-farm irrigation technology choices. In traditional irrigation technologies (furrow, border, and flood irrigation), water is delivered to the individual plants through gravitation. Irrigation of fields utilizing these traditional technologies is quite infrequent (once every two to three weeks, or even less). During irrigation, large amounts of water are deposited on the fields over a short period of time. However, with modern irrigation technologies (drip and sprinkler), energy and equipment are used rather than gravity to deliver water to the plant. This results in more frequent irrigation, improves irrigation uniformity, and reduces water losses to deep percolation and runoff. In essence, the output produced with a given amount of water is increased with these modern technologies. This improvement, however, involves higher equipment and energy costs.

Caswell and Zilberman introduced a methodological approach to analyze the choice of irrigation technologies. They distinguished between applied and effective water; the latter denotes the amount of water actually utilized by the plants. Irrigation effectiveness is defined as the ratio of effective water to applied water. There is likely to be lower irrigation effectiveness with traditional technologies than with modern technologies. For example, the irrigation effectiveness of furrow irrigation in many parts of California is 0.6 (while the irrigation effectiveness on the same land is about 0.8); whereas the effectiveness of drip irrigation is about .95. Irrigation effectiveness also varies with land characteristics. It is less on lands with lower water-holding capacity (sandy soils) or on steep slope (lands that for our purpose can be denoted low quality). The relative gain in irrigation effectiveness associated with the transition from traditional to modern technologies is greater in areas with lower land qualities than those with higher land qualities. The irritation effectiveness of traditional technologies is likely to be quite high on flat lands and heavy soil, and not much can be gained by transition to modern technologies. On the other hand, irrigation effectiveness can be substantially improved by a transition from furrow to drip or sprinkler irrigation on sandy soil.

Because of the higher irrigation effectiveness of modern technologies, the variable cost of one unit of effective water with a modern technology is less than with a traditional technology. Since the level of output is contingent upon effective water use, profit maximizing will lead to higher yields with the modern technologies than with the traditional. Despite the higher yields of the modern technologies, their higher irrigation effectiveness may result in lower application levels of water. Thus, under reasonable conditions, a transition from traditional to modern technologies yields increasing and water-saving effects.

The magnitude of these effects depends on the specific environment. When water price is high and profit-maximizing yield of the traditional technology is far below the maximum attainable yield, transition to a modern technology may have a relatively substantial yield effect. When water price is very low and the profit-maximizing yield under the traditional technology is close to the maximum attainable yield, transition to modern technology has a substantial water-saving effect.

It is suggested from this analysis that a transition from traditional to modern technology is more likely to occur as output price or water price increases. The implication is that modern technologies are more likely to be adopted in regions with higher water costs and/or high-value crops. Thus, research and development activities and extension efforts that will reduce the costs of modern technologies are likely to increase their adoption.

Technology choices are likely to be influenced by land characteristics; for example, regions with high-quality land (flatland or heavy soil) are less likely to adopt modern technology than regions with low-quality land (steep slopes or sandy soils) where the transition to modern technologies has a strong impact on irrigation effectiveness. Moreover, the introduction of modern technologies may extend the utilized land base, allowing profitable production in regions with sandy soils and uneven land surfaces.

Feinerman, Letey, and Vaux suggest that the gains from transition to modern irrigation technologies do not depend only on average-quality land but also on its distribution. Their results indicate that the likelihood of adoption of modern irrigation technologies increases as the variability of the water-holding capacity of the soil and slope of the land increases.

Environmental characteristics other than land quality are likely to affect irrigation technology choices. Letey, Dinar, Woodring, and Oster suggest that modern irrigation technologies are more likely to be adopted in locations with low water quality (high salinity). Moreover, introduction of new technologies may increase the ranges of water quality used in agricultural production; also, weather affects technology choices. Drip irrigation is more likely to be adopted in high-temperature areas with high evaporation rates. Sprinkler irrigation can be useful for frost protection, and this advantage may encourage its adoption in locations with frost problems.

Modern irrigation technologies tend to save inputs other than water. Sprinkler and drip systems are used to dispense chemicals (chemigation), such as fertilizers and pesticides. Their use tends to reduce application rates, and the higher costs of chemicals tend to increase the likelihood of adoption. Water costs are likely to be directly related to energy prices since energy is a major element of pumping costs. One may expect that adoption of modern technologies, especially those requiring relatively low pressure (drip or sprinkler) is likely to increase as energy prices increase. The impact of energy cuts on adoption is likely to be higher at locations which rely on deep groundwater.

Applied water that is not utilized by the crop may be a source of concern. In some regions, deep percolation may lead to waterlogging problems; in others, runoff water may be a source of environmental contamination. There tends to be less generation of deep percolation or runoff by technologies with higher irrigation effectiveness. Thus, taxing or regulation of these activities is likely to induce adoption of more modern technologies (Dinar, Knapp, and Letey).

Farmers may be influenced into adoption of modern technology, as well as changing of cropping patterns, by existing circumstances or a change in circumstances, such as an increase in water price (cost). The technological innovations and the introduction of high-yield varieties of crops may be the very incentives to bring about the adoption of modern irrigation and needed changes. Hayami and Ruttan document how the high-yield, high-quality variety of crops (wheat and corn) introduced during the Green Revolution have been accompanied by the introduction of irrigation and chemical fertilizers in many parts of the world. The same economic logic suggests that introduction of modern irrigation technologies or provision of new water supplies by, say, government agencies, may be accompanied by the introduction of new crops or crop varieties. In this sense, increased irrigation possibilities may lend themselves well to diversification of crop mix and an increase in production possibilities.

Numerical Example

The theoretical results of the previous sections are illustrated through a numerical simulation based on conditions prevailing in the northern segments of the west side of the San Joaquin Valley in California (Panoche fan area). Two major crops in this region are cotton (involving more than 50 percent of the planted acreage in the 1970s and 1980s) and tomatoes. Furrow irrigation is the dominant irrigation technology for both cotton and tomatoes in the region, but there is substantial use of sprinkler irrigation (especially with tomatoes). There have been experimentations in the use of both drip and low energy precise applications (LEPA) in the region. Mostly surface water is being used in the region; it is provided at low prices (from $1.00 to $2.00 per hectare-centimeter, or from $12.50 to $25.00 per acre-foot). However, the cost may be substantially increased in the future. The region is facing substantial drainage problems which ultimately may result in drainage taxation as well as requirements of drainage disposal. The simulation shows the impacts of changes in water prices, drainage charges, and drainage pumping costs on relative profitability, yield, and water use of the four alternative technologies in the region.

In the simulation, the Letey and Dinar crop-water production function model is applied to investigate profit-maximizing behavior. This application requires specification of the highest potential yields (Table 7.1), a measure

TABLE 7.1
Characteristics of Irrigation Technologies and Costs, Prices, and Yields for Cotton and Tomatoes Used for this Analysis, San Joaquin Valley, California, 1987

	Furrow	Sprinkler	LEPA[a]	Drip
Technology characteristics				
Energy cost (dollars per hectare per centimeter)	0	.95	.28	.40
Uniformity (CUC)	70	80	90	96
Cotton				
Potential yield (tons per hectare)	<----	1.6	---->	
Price (dollars per ton)	<----	1,665	---->	
Harvest cost (dollars per ton)	<----	230	---->	
Annual irrigation capital cost (dollars per hectare)	49	216	242	476
Production cost (dollars per hectare)	1,029	950	998	813
Tomatoes				
Potential yield (tons per hectare)	<----	85	---->	
Price (dollars per ton)	<----	50	---->	
Harvest cost (dollars per ton)	<----	18	---->	
Annual irrigation capital cost (dollars per hectare)	49	240	240	581
Production cost (dollars per hectare)	1,102	1,034	1,041	899

[a]Low energy precise applications.

Source: University of California, Committee of Consultants on Drainage Water Reduction. *Associated Costs of Drainage Water Reduction*, No. 2. Sacramento: University of California Salinity/Drainage Task Force and Water Resources Center, 1988.

of water-quality conditions and weather conditions similar to those in the southern San Joaquin Valley.

Table 7.1 presents the values of other key parameters used in the analysis. Its first row shows that (1) application with furrow does not require energy, (2) linear movement of sprinkler irrigation requires a relatively high amount of energy for application, and (3) energy requirements with drip irrigation, especially LEPA, are substantially less than with the sprinkler system. The second row reflects the substantial gains in irrigation uniformity as one changes from furrow irrigation to sprinkler, LEPA, or drip irrigation.

With drip irrigation, less labor and fewer chemicals are required, and that is reflected in its comparatively lower production cost. However, drip irrigation requires much higher annual irrigation capital costs, while furrow requires very little irrigation equipment; thus, production and irrigation capital costs are highest for drip and lowest for furrow irrigation.

For cotton and tomatoes under alternative technologies, comparison of profits, yields, water use, and drainage per acre are made in Tables 7.2 and 7.3 for six scenarios. The first three scenarios assume a very low water price (equivalent to $12.50 per acre-foot): In scenario A, drainage is not regulated; scenarios B and C require pumping and the disposal of drainage, along with a drainage penalty, which is quite severe in scenario C. The remaining three scenarios assume a high water price (equivalent to $100 per acre-foot): Scenario D assumes no regulation on drainage, while scenarios E and F assume moderate and severe drainage regulations.

Consider the first outcome for each crop separately. The results of Tables 7.2 and 7.3 demonstrate the yield-increasing and water- and drainage-saving effects associated with transition to a more capital-intensive irrigation technology. The yield-increasing effect is greater than the water-saving effect as water and drainage costs increase. For example, the transition from furrow to drip irrigation in cotton production has a small yield effect ($0.04/1.56 \approx 3$ percent) in scenario A and quite a substantial yield effect ($0.26/1.33 \approx 20$ percent) in scenario F. On the other hand, the water-saving effect is substantially larger ($73/159 \approx 45$ percent) under the low-cost case (scenario A) than under the high-cost case (scenario F) where water saving (associated with transition from furrow to drip irrigation) is about 4 percent (3/86).

The relative magnitude of yield increases and water savings associated with the transition to the more advanced technology is larger for tomatoes than for cotton because water is used more heavily for tomatoes. For

TABLE 7.2

Profits and Yields for Cotton and Tomatoes Irrigated by Various Irrigation Technologies Under Different Environmental Conditions[a] and Prices,[b] San Joaquin Valley, CA, 1987

Scenario	Cost			Irrigation technology[c]	Profit		Yield	
	Water	Draining	Pumping		Cotton	Tomatoes	Cotton	Tomatoes
	dollars per hectare per centimeter				dollars per hectare		tons per hectare	
A	1	0	0.0	1	1,726	2,014	1.56	72.4
				2	1,662	2,380	1.57	83.9
				3	1,723	2,638	1.58	84.0
				4	1,712	2,550	1.60	85.0
B	1	8	0.8	1	1,344	1,000	1.45	65.7
				2	1,526	1,950	1.52	79.2
				3	1,668	2,425	1.57	81.6
				4	1,712	2,546	1.60	84.8
C	1	12	0.8	1	1,225	651	1.41	60.2
				2	1,487	1,793	1.51	77.8
				3	1,651	2,360	1.57	81.6
				4	1,712	2,546	1.60	84.8
D	8	0	0.0	1	858	337	1.44	65.9
				2	992	1,161	1.51	79.2
				3	1,102	1,598	1.56	81.6
				4	1,117	2,080	1.59	84.8

(Continued on next page.)

TABLE 7.2—continued.

Scenario	Cost			Irrigation technology[c]	Profit		Yield	
	Water	Drain-ing	Pump-ing		Cotton	Tomatoes	Cotton	Tomatoes
	dollars per hectare per centimeter				dollars per hectare		tons per hectare	
E	8	8	0.8	1	661	-274	1.35	54.2
				2	934	888	1.46	74.4
				3	1,079	1,466	1.54	81.6
				4	1,117	2,080	1.59	84.8
F	8	12	0.8	1	584	-487	1.33	51.2
				2	914	782	1.45	74.4
				3	1,072	1,403	1.54	81.2
				4	1,117	2,080	1.59	84.8

[a] Seasonal pan evaporation is 165 and 140 centimeters for cotton and tomatoes, respectively; soil water-holding capacity is 60 percent, and salt concentration of the irrigation water is 4 mmhos/cm.

[b] Long-term crop prices for cotton and tomatoes are 75 cents per pound of lint and $60 per ton, respectively.

[c] 1 = Furrow irrigation; 2 = Sprinkler irrigation; 3 = Low energy precise applications (LEPA); and 4 = Drip irrigation.

TABLE 7.3

Applied Water and Drainage Volumes for Cotton and Tomatoes Irrigated by Several Irrigation Technologies Under Different Environmental Conditions[a] and Prices[b]

Scenario	Cost			Irrigation technology[c]	Applied water		Drainage water	
	Water	Draining	Pumping		Cotton	Tomatoes	Cotton	Tomatoes
	dollars per hectare per centimeter				hectares per centimeter			
A	1	0	0.0	1	159	250	77	136
				2	108	205	25	32
				3	96	154	12	32
				4	86	124	1	1
B	1	8	0.8	1	108	206	33	96
				2	89	160	11	41
				3	86	137	4	16
				4	86	123	1	0
C	1	12	0.8	1	99	180	27	74
				2	86	154	9	36
				3	86	154	4	32
				4	86	123	1	0
D	8	0	0.0	1	104	206	30	96
				2	86	160	9	41
				3	85	137	4	16
				4	83	123	0	0

(Continued on next page.)

TABLE 7.3—continued.

Scenario	Cost			Irrigation technology[c]	Applied water		Drainage water	
	Water	Drain-ing	Pump-ing		Cotton	Tomatoes	Cotton	Tomatoes
	dollars per hectare per centimeter				hectares per centimeter			
E	8	8	0.8	1	89	158	20	57
				2	79	142	5	26
				3	80	137	2	16
				4	83	123	0	0
F	8	12	0.8	1	86	149	19	50
				2	78	142	5	26
				3	79	136	2	16
				4	83	123	0	0

[a]Seasonal pan evaporation is 165 and 140 centimeters for cotton and tomatoes, respectively; soil water-holding capacity is 60 percent, and salt concentration of the irrigation water is 4 mmhos/cm.

[b]Long-term crop prices for cotton and tomatoes are 75 cents per pound of lint and $60 per ton, respectively.

[c]1 = Furrow irrigation; 2 = Sprinkler irrigation; 3 = Low energy precise applications (LEPA); and 4 = Drip irrigation.

example, in the lowest cost case (scenario A), the switch from furrow to drip will reduce water use by more than 50 percent.

Considering the profitability of technologies under each of the crops, note that, under conditions resembling the present—low water cost and no drainage controls (scenario A)—the low-yield effects in the case of cotton makes furrow irrigation the most desirable technology. In the case of tomatoes, the greater yield effects associated with the transition from furrow to sprinkler irrigation make it more profitable (by $366/2,014 \approx 18$ percent) to change. The gain in profit seems to be greater with drip irrigation (especially LEPA) than with sprinkler, but the relative gain from LEPA ($258/2,380 \approx 11$ percent) does not seem to be substantial enough to motivate farmers to go through the transition effort and a risk adoption that a new technology requires. Actually, the 18 percent increase in profit associated with the transition from furrow to sprinkler irrigation is likely to be insufficient incentive for such a transition for some of the farmers. Thus, the results of scenario A serve to explain the use of furrow irritation with cotton production and furrow and drip irrigation with tomato production in low-cost water regions in the Central Valley of California.

When water costs are high and drainage is still not a problem, the profitability of technologies is correlated with their capital intensity. In the case of cotton, however, the relative gains associated with changing from sprinkler to drip irrigation ($120/1,117 \approx 11$ percent) and even from furrow to drip irrigation ($259/858 \approx 28$ percent) are not very high and thus may not attract many farmers to attempt this "unknown" technology. Even high water prices are not likely to induce quick adoption of drip irrigation in cotton. The tendency of farmers to use sprinkler irrigation with cotton when water prices are high may be explained by its higher profitability relative to using furrow irrigation (by $134/858 \approx 16$ percent) and the fact that it is an established technology in the Valley.

When water price becomes high, the profitability of growing a water-using crop (e.g., tomatoes) with furrow irrigation declines drastically (by $1,687/335 \approx 500$ percent) as one moves from scenario A to scenario D). In this case, farmers are likely to use sprinkler, or even try drip irrigation—the relative gain in profitability associated with a change from sprinkler to drip irrigation is about 85 percent ($919/1,161$).

Scenario D in Table 7.2 is useful to illustrate another interesting result. When furrow is the only available technology, regions with highly priced water are likely to grow cotton and not tomatoes. Introduction of modern technologies to high water-cost regions (sprinkler, LEPA, or drip irrigation) may result in both technology adoption and change to another crop. (In scenario D, tomatoes are more profitable than cotton with all the modern technologies.) In this case, the introduction of a modern technology results in diversification of crop production.

Considering the impact of drainage regulation on technology adoption by cotton producers, note that introduction of these regulations by themselves provides less of an incentive to adopt drip irrigation or LEPA than an increase in water price. Evaluation of the impact of the introduction of a policy combining increase in water price and strict regulation of drainage (a change from A to F) for cotton producers suggests that such a policy is likely to induce many cotton growers to change from furrow to sprinkler irrigation (increase of $340/584 \approx 70$ percent in profitability). The introduction of such a policy makes growing tomatoes with furrow irrigation unprofitable, while the profitability of drip irrigation ($1,292/782 \approx 140$ percent increase in profitability relative to sprinkler irrigation) makes it especially attractive.

The Spread of Irrigation in the United States in the Post World War II Era

The area of irrigated agriculture in the United States has grown immensely over the last 40 years, and many modern irrigation technologies have been introduced and adopted. An understanding of the historic evolution of irrigated agriculture in the United States is essential for assessing its future development and impact. In this section we present and interpret data depicting developments in the use of irrigation in the United States from the 1950s to the present.

The primary source of data presented and analyzed in this section comes from the *Irrigation Journal* which published the results of its annual state-by-state survey of irrigation technology use from 1956 to 1985. Annual data on total-irrigated acres and its distribution by major technology categories and by major crop categories for each state are utilized in this section. Most of the analysis distinguishes only between gravitational technologies and modern technologies (sprinkler, center pivot, drip, etc.) that are presented under the heading, "sprinkler."

Using these data, we attempted to estimate the behavior of the total-irrigated acreage and the sprinkler-irrigated acreage as functions of time for all 50 states. We found that the simple linear function of the type,

$$A_{it} = a_i + b_i \cdot t,$$

has a very good statistical fit (high R^2 above .90) and a significant coefficient for almost all states for both total-irrigated acreage and sprinkler-irrigated acreage.[1] The variable, i, is a state indicator; and t is a time indicator, assuming the value of zero for 1950. Irrigated acreage (A_{it} can represent total- or sprinkler-irrigated acreage) at state i in 1950 + t is denoted by A_{it}, b_i is annual change in irrigated acreage, and a_i is irrigated acreage in year 0 (1950). Assuming that b_i also represents annual increase in irrigation prior to 1950, $g_i = 1950 - a_i/b_i$ denotes an estimate of a year with no irrigated acreage. If irrigated acreage increases over time ($b_i > 0$), g_i is an estimate of the year when irrigation began. If irrigated acreage is estimated to decline over time ($b_i < 0$), g_i is an estimate of the time when irrigation will end.

Table 7.4 presents the estimated values of annual changes and zero level years of both total-irrigated acreages and sprinkler-irrigated acreages for all 50 states. For most states, the estimates of a_i and b_i were highly significant, which implies that the estimated zero levels (g_i) are quite reliable. Indeed, the results of Table 7.4 suggest some reasonable patterns of irrigation development.

In many western and southwestern states, irrigated agriculture is substantial and well established. Irrigation with traditional methods began relatively early (late 19th Century and early 20th Century). The use of sprinkler irrigation began in the early 1950s; since that date, a high percentage of newly irrigated land has been irrigated by the modern technologies. This pattern is observed in California (CA), Colorado (CO), Hawaii (HI), Idaho (ID), Montana (MT), Nevada (NV), New Mexico (NM), Oregon (OR), Texas (TX), Utah (UT), Washington (WA), and Wyoming (WY). In California, for example, it is estimated that the use of irrigation began in 1872 and, since that time, has grown in increments of 89,000 acres. Thus, total-irrigated California acreage in 1985 is estimated to be around 9.85 million acres. Sprinkler irrigation is estimated to have begun immediately following World War II (1947) and to have increased at the rate of 61,000 acres annually, approaching 2.5 million

TABLE 7.4

Estimated Values of Annual Changes and Zero Level Years, United States

	All irrigation		Sprinkler			All irrigation		Sprinkler	
	Annual acre	Zero level	Annual acre	Zero level		Annual acre	Zero level	Annual acre	Zero level
State	change[a]	year[b]	change	year	State	change	year	change	year
AL	4.01	1954	4.52	1957	MT	56.48	1922	9.14	1946
AK	0.04	1941	-0.23	1986	NB	229.80	1949	141.27	1960
AZ	-0.41	4749	2.68	1959	NV	24.69	1927	2.15	1953
AR	42.44	1938	1.75	1937	NH	0.18	1945	0.15	1941
CA	88.65	1872	61.00	1947	NJ	3.67	1937	1.40	1883
CO	17.23	1799	28.96	1958	NM	22.47	1924	6.57	1956
CT	-0.27	2015	-0.39	2003	NY	0.23	1711	-0.14	2407
DE	1.49	1953	1.49	1953	NC	5.44	1950	5.40	1949
FL	68.70	1947	28.72	1951	ND	5.15	1949	6.02	1961
GA	33.51	1958	39.97	1960	OH	0.96	1936	0.93	1935
HI	2.78	1919	2.40	1960	OK	24.78	1946	15.42	1954
ID	63.79	1915	74.54	1957	OR	21.30	1886	36.38	1953
IL	4.87	1958	5.51	1959	PA	-0.11	2169	-0.23	2074
IN	3.21	1955	3.22	1956	RI	0.10	1945	0.11	1951
IA	7.74	1954	7.85	1958	SC	2.76	1949	2.64	1949
KS	111.12	1951	50.17	1960	SD	13.27	1949	15.89	1960
KY	0.72	1939	0.68	1937	TN	0.31	1907	0.34	1917
LA	1.97	1647	0.55	1954	TX	99.85	1895	71.64	1949
ME	0.11	1894	0.04	1767	UT	8.77	1815	12.06	1957
MD	1.24	1950	1.24	1950	VT	0.03	1909	0.03	1889
MA	0.70	1927	0.85	1937	VA	1.67	1940	1.68	1940
MI	11.50	1955	12.83	1956	WA	38.41	1933	41.14	1953
MN	19.42	1961	18.75	1961	WV	0.03	1898	0.01	1816
MS	13.78	1950	-0.92	1988	WI	8.84	1954	9.55	1956
MO	15.92	1957	6.72	1959	WY	17.00	1873	7.57	1957

[a]This and the other "change" columns are actually the slope coefficients from a simple regression of acreage (measured in thousands) on the date (1950-1985).

[b]The "zero level year" columns are minus the ratio of the intercept to the slope from each regression. This is the estimated "x intercept" or, in this case, the year during which irrigation first appeared in the state.

Source: *Irrigation Journal*, various issues.

acres in 1985. Moreover, in the recent past, more than two-thirds of newly irrigated acreage in California (61/88.7) is estimated to use modern sprinkler technology. These estimates appear to be very close to reality and demonstrate the usefulness of the linear trend model used in this chapter.

The estimated values for other western and southwestern states seem reasonable as well, with some noticeable exceptions (for irrigated agriculture in Colorado). For some states (Oregon, Colorado, and Idaho), annual increments in sprinkler irrigation are larger than increments in total-irrigated acreage—indicating substantial conversion from traditional irrigation to modern irrigation.

In some agricultural states, development of irrigated farming began after World War II with the introduction of sprinkler irrigation in the late 1950s. These states include Nebraska (NB), Kansas (KS), Oklahoma (OK), Florida (FL), the Dakotas (SD and ND), and Missouri (MO). The growth in irrigated acreage in Nebraska and Kansas has been staggering. Nebraska has the second largest irrigated acreage in the United States (following California).

The growth in production of center-pivot irrigated corn, soybean, and sorghum (particularly in Nebraska and Kansas) has drastically affected the markets for these products by greatly increasing their supply. The development of irrigated citrus and vegetable crop production in Florida has made this state the "garden state" of the East Coast. Due to the "late start" of traditional irrigation in these states, their sprinkler-irrigated land share is higher than that of the more arid western states.

In some midwestern and southern states, substantial irrigation began in the 1950s and the early 1960s with the introduction of sprinkler irrigation; traditional irrigation technology has been little used. These states include dairy and lake states such as Wisconsin (WI), Minnesota (MN), Michigan (MI), Indiana (IN), and Iowa (IA); and southern states such as Georgia (GA), Alabama (AL), and the Carolinas (SC and NC).

Sprinkler irrigation is the major irrigation technology in many central and eastern states. In many of these states, agriculture is not very important, and sprinkler irrigation is used in the production of vegetables and pasture growth. These states include Maryland (MD), Massachusetts (MA), Maine (ME), Kentucky (KY), Tennessee (TN), Ohio (OH), Vermont (VT), Virginia (VA), West Virginia (WV), Rhode Island (RI), New Hampshire (NH), and New Jersey (NJ).

In some states, irrigated acreage has declined or has not changed much over time. Water scarcity has resulted in the decline of irrigation acreage in

Arizona (AZ) while the sprinkler-irrigated acreage in this state is estimated to have been increasing since 1959.

Urban pressures and the decline of crop production have caused a reduction in irrigation acreages in both Pennsylvania (PA) and Connecticut (CT) and restrict the growth of irrigated acreage in New York (NY). In the case of Louisiana (LA), irrigated acreage has not changed. Lands on the Mississippi Delta continue to be irrigated with traditional technology. Irrigated acreage in Mississippi (MS) has been increasing; but furrow irrigation is mostly used—mainly because of the cheap price of the water in this state. It should be noted that in both Mississippi and Arkansas (AK), the acreage of sprinkler irrigation has appeared to decline. It may be that experimentation with this technology has proven it unprofitable relative to traditional technologies because of the low water price and heavy soils in these states.

The evolvement of irrigated acreage thus far in the United States is presented quite well by the linear model, but we are doubtful that the same trends will continue far into the future. It appears that we are reaching the end of an era during which irrigation sources have been expanded, allowing irrigated acreage to increase.

States like Arizona and Texas, and even California, are beginning to realize that they have reached or are approaching the limit of their water extraction capacities and that extending their water supply may be either infeasible or very costly. Many aquifers have been badly depleted. Water quality, which should be viewed as an exhaustible resource, has deteriorated; and its maintenance imposes constraints on pumping and irrigation practices. It seems that water utilization in the United States, especially in the western and southwestern states, is in the same developmental stage experienced by land utilization in the 1890s. According to Cochrane, until that time, land base expansion had been the major cause for increased agricultural production in the United States. By the 1890s, the West had been "won" and the Far West had been settled. Since then, the main source of increased production has been increased land productivity.

Technological changes and innovations, embodied in new crop varieties, chemical inputs, and new practices have contributed to this increased productivity. The government has contributed to the generation and diffusion of the new innovations by establishing and supporting public research and extension systems in agriculture.

By analogy, the history of land-use development in the United States suggests that, as agricultural water-use levels in the United States is reaching its physical limit, emphasis is shifting toward increasing water-use productivity through the development and adoption of new irrigation technologies and the improvement of water conveyance systems. Therefore, one can expect slower growth in total-irrigated acreage with substantial increase in the acreage irrigated by advanced systems such as sprinkler, LEPA, and drip. Moreover, the water productivity with all technologies is likely to improve as computer software for irrigation management improves in quality and accessibility and as better monitoring equipment needed for computerized irrigation is developed.

The results of Table 7.4 served to identify major regional differences in the patterns of irrigation over time and to provide the broad historic overview. Tables 7.5 and 7.6 and Figure 7.1 will be used to identify and explain recent developments in irrigated agriculture and their implications.

Tables 7.5 and 7.6 present the distributions of irrigated acreage by regions according to technology and crop for 5 years of the period, 1960-1985.[2] Some of the implications and interpretations of these results follow.

The acreage using traditional irrigation technologies grew by about 30 percent between 1960 and 1970, grew only slightly in the 1970s, and has been in slow decline since 1980. Much of the added acreage in the 1960s was used in feed grain and oilseed production (in Nebraska, Texas, Kansas, and Oklahoma). Irrigation was essential for the drastic increase in the production of oilseed in the United States in the last 30 years. Some of the added acreage of the 1960s was used in cotton production in Texas and the West. The increase in cotton acreage was affected substantially by the introduction of the cotton harvester in the 1960s. The acreage in gravity irrigation reached its peak in 1975 during a period of extremely high agricultural prices. The slight decline in gravity-irrigated acreage since that time can be the result of lower food prices and transition to sprinkler irrigation.

Sprinkler-irrigated acreage (this acreage includes drip irrigation) increased by close to 60 percent between 1960 and 1980 and has grown very slowly since. Sprinkler acreage increased by about 5 million acres in the 1960s and 10 million acres in the 1970s.

The sprinkler-irrigated acreages added in the 1960s were used in the production of feed grains and oilseed in Texas, Nebraska and the western states; fruits and vegetables in California, Florida, Washington, and Oregon; potatoes in the West; and small grains in the West and Texas.

TABLE 7.5
Technology Acreage

Region	Gravity					Sprinkler					Drip		
	1960	1970	1975	1980	1985	1960	1970	1975	1980	1985		1980	1985
						thousands of acres							
CA	7,100	7,200	7,200	7,999	7,500	900	1,450	1,559	2,025	2,589		305	350
HI	119	113	97	125	146	9	17	27	50	123		29	108
FL	520	1,040	1,097	1,452	1,058	228	545	821	816	837		36	298
GA	2	0	0	0	0	102	145	230	1,013	1,080		5	45
NB	2,415	3,563	3,973	4,509	4,404	250	866	1,642	3,128	3,793		.1	0
TX	5,152	6,600	6,700	5,620	4,617	610	1,700	1,918	2,197	2,141		20	0
WA; OR	690	2,100	1,922	1,444	1,365	354	1,200	1,609	2,367	2,577		6	13
SW[a]	2,772	1,985	3,304	3,461	3,536	37	180	207	294	294		4	42
West[b]	8,472	12,076	11,535	10,517	9,913	392	1,600	3,076	3,733	4,030		1	2
KS; OK	920	1,971	2,306	2,802	2,539	100	560	953	1,754	1,677		.7	3
MI	.2	.2	1	0	0	71	139	112	302	422		9	36
IA; IL	30	44	41	23	5	45	73	104	359	438		1	2
NE[c]	6	.7	.4	4	97	126	206	336	327	226		2	7
TOTAL	28,197	36,693	38,177	37,956	35,181	3,223	8,680	12,594	18,365	20,225		419	906

[a]Arizona, Nevada, and New Mexico.

[b]Idaho, Montana, North Dakota, South Dakota, Wyoming, Utah, and Colorado.

[c]Maine, New Hampshire, Vermont, Massachusetts, Rhode Island, Connecticut, New York, Pennsylvania, New Jersey, Maryland, and West Virginia

Source: *Irrigation Journal*, various issues

TABLE 7.6

Irrigated Crop Acreage

Region	Year	Potatoes, rice, and sugar	Orchards	Vegetables and legumes	Feed grains and oilseeds	All small grains	Alfalfa and cotton
				thousands of acres			
CA	1975	880.0	1,297.3	914.0	420.0	916.0	962.0
	1980	700.0	1,928.0	1,500.0	440.0	1,180.0	1,600.0
	1985	755.0	2,205.0	1,036.0	354.0	1,186.0	1,379.0
HI	1975	117.0	2.1	2.5	0.0	0.0	0.0
	1980	120.0	7.3	4.8	0.0	0.0	0.0
	1985	114.0	7.3	4.8	0.0	0.0	0.0
FL	1975	285.0	650.0	352.1	0.0	0.0	0.0
	1980	429.8	670.0	377.6	78.8	71.0	0.0
	1985	412.5	627.5	529.5	3.0	0.0	1.5
GA	1975	0.0	4.0	115.6	55.0	0.0	1.2
	1980	0.0	19.0	471.0	390.0	0.0	10.0
	1985	0.0	43.4	559.6	342.2	0.0	55.8
NB	1975	120.4	0.0	405.0	4,500.4	0.0	346.0
	1980	88.5	0.0	856.5	5,550.0	300.0	0.0
	1985	78.0	0.0	873.0	5,555.0	165.0	0.0
TX	1975	594.0	118.0	361.0	3,276.0	1,265.0	2,120.0
	1980	626.1	139.9	379.2	2,333.1	1,229.5	2,260.0
	1985	380.0	96.0	371.0	1,845.0	1,180.0	2,112.0
WA; OR	1975	89.0	237.0	322.0	0.0	208.0	0.0
	1980	145.0	211.0	477.0	236.0	750.0	56.0
	1985	196.0	235.0	525.0	285.0	250.0	50.0
SW[a]	1975	34.2	89.7	78.6	67.0	735.0	332.0
	1980	33.8	125.4	112.5	414.5	345.0	772.3
	1985	42.7	55.6	146.0	320.8	358.6	561.9
WEST[a]	1975	706.5	54.0	752.7	1,378.3	1,065.3	115.0
	1980	718.6	49.8	793.7	3,280.5	1,147.5	845.3
	1985	686.8	51.1	816.6	3,401.8	1,381.0	777.0
KS; OK	1975	39.0	0.0	123.0	2,088.0	516.0	73.0
	1980	11.4	2.1	284.5	2,286.0	716.5	75.5
	1985	14.5	1.0	97.5	2,007.9	798.4	78.1
IA; IL	1975	1.5	0.5	29.2	97.0	0.0	0.0
	1980	4.5	1.0	100.0	270.0	1.0	0.0
	1985	3.5	3.0	314.0	105.0	1.5	0.0
NE[a]	1975	4.2	10.5	44.9	3.2	0.0	0.0
	1980	3.5	15.8	65.6	10.4	0.0	0.0
	1985	5.8	23.8	78.2	22.5	2.0	0.0

[a]For state groupings, see Table 7.5.

Source: *Irrigation Journal*, various issues.

243

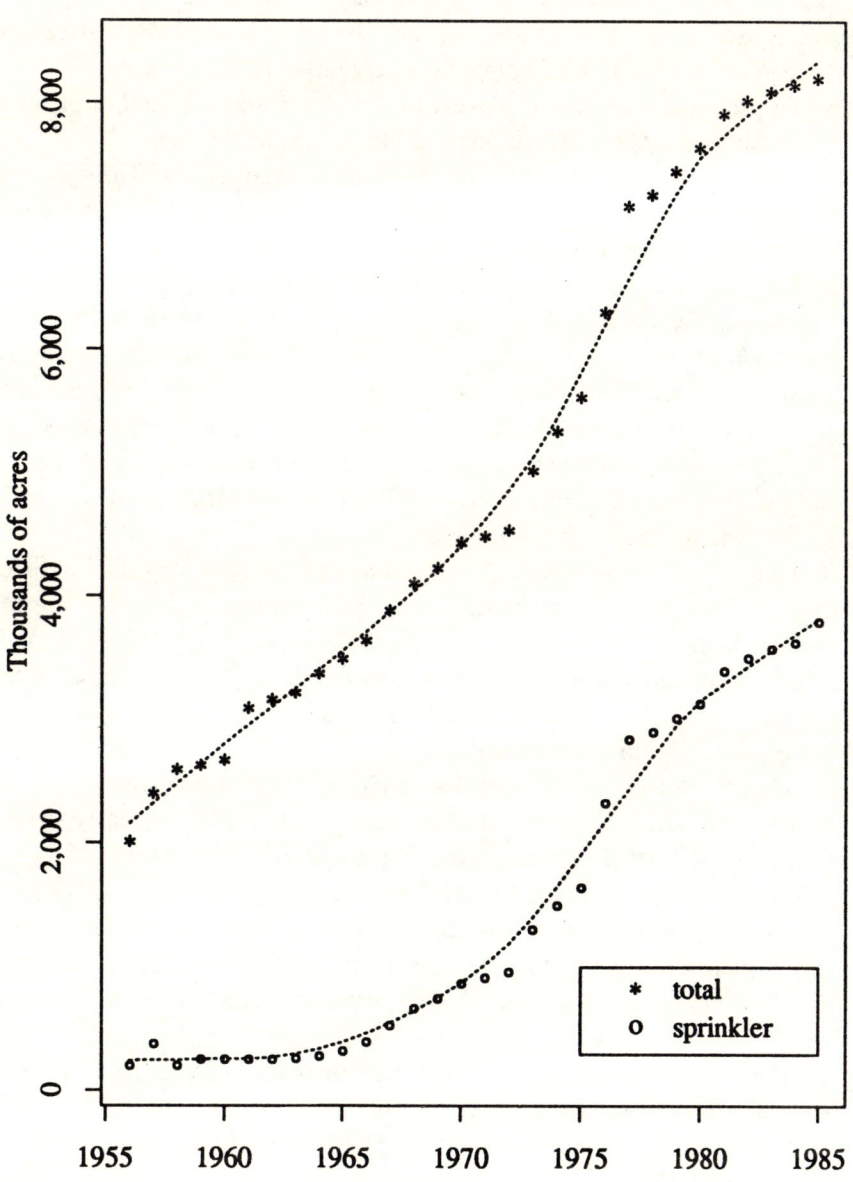

FIGURE 7.1. Nebraska Irrigation, various years

While traditional irrigation technologies were the major form of irrigation of feed grain and oilseed in the 1960s in regions with flat landscape and rather heavy soil (Kansas, eastern Nebraska, and Oklahoma), regions with lighter soils (western Nebraska) and uneven landscape (the West) used sprinkler irrigation to grow these crops in the 1960s. Irrigation (in particular, sprinkler irrigation) is largely responsible for the expansion of apple production in the Northwest in the 1960s and production of other fruits and vegetables in California and Florida during this period.

The 1970s saw a vast diffusion of center-pivot irrigation. The spread of this technology is responsible for much of the growth in sprinkler acreage in this decade. Center-pivot irrigation was used to expand irrigated production of feed grains and oilseed in Nebraska, Kansas, Colorado, the mountain states, and the Northwest. It was also used to irrigate small grains in the mountain states, the Northwest, Texas, and the Southwest. The substantial expansion of sprinkler-irrigated acreage in the 1970s was also manifested in the increase in legume production in the South and all of the West; an increase in sprinkler-irrigated fruits and vegetables in California and Florida; and an increase in sprinkler-irrigated acreage of alfalfa and cotton in California, Texas, and the Southwest.

A more complete view of the diffusion of center-pivot irrigation in field-crop production and its relationship to the behavior of total-irrigated acreage can be seen in Figure 7.1. This figure depicts the dynamics of total-irrigated acreage and sprinkler irrigation (actually, center-pivot irrigated acreage) in Nebraska from 1955 to 1985.

Figure 7.1 indicates that the diffusion of center-pivot irrigation in Nebraska (presented by sprinkler-irrigation acreage) behaves as an S-shaped function of time. This diffusion process has an early period (1957 to 1972) with a slow increase in the spread of the technology. Then it has a takeoff period (1973 to 1977) and, finally, a period of tapering off (1978 to the present). Note that the takeoff period (1973 to 1976) occurred when grain prices were relatively very high, and these high prices provided a strong incentive for the adoption of center-pivot irrigation by Nebraska farmers.

Figure 7.1 demonstrates a substantial increase in irrigated acreage in Nebraska in the late 1950s and the 1960s. This increase is attributed mostly to traditional technologies, and center-pivot sprinkler irrigation increased most in the 1970s.

Drip irrigation was introduced in the United States in the early 1970s. By 1985, it was used on about 1 million acres, and it is currently in use on

1.5 million acres (Riddering). According to Gustafson, the early adoption of drip irrigation occurred in avocado production in San Diego County, California, when the new technology allowed production to expand to rather steep slopes in San Diego County and, later, in Ventura County. The technology has been adopted with other high-volume fruit crops and high-value vegetable crops (fresh market vegetables) mostly in locations with sandy soils, rather uneven lands, and high water costs.

The use of drip irrigation spread very quickly in California in the 1970s, especially after the severe drought of 1977-78. The agricultural recession of the early 1980s slowed this process, but the spread of drip irrigation accelerated in the late 1980s. At present, there are early attempts in using drip irrigation in cotton and tomatoes.

While California was the major adopter of drip irrigation in the 1970s, the use of drip and trickle trickle irrigation increased substantially in the early 1980s in Florida, Hawaii, Georgia, Michigan, and Arizona. Drip irrigation allowed substantial increase in vegetable production on the soils of Dade County, Florida, and expansion of citrus production throughout the state. It increased the output of the peach industries in Florida and cherry and apple production in Michigan and allowed early-season production of vegetables in the Arizona deserts. In Hawaii, the use of drip irrigation has increased sugar production.

There are several manifestations of the increased scarcity of water. One is substantial reduction in the irrigated acreage in Texas between 1975 and 1985 where some groundwater aquifers have been substantially depleted. This has occurred to a lesser extent in Oklahoma and Florida. In other states, the increase in water scarcity is reflected by small changes in irrigated acreage between 1975 and 1985. In almost all states, increased in water scarcity have been largely responsible for the substantial increase in the use of sprinkler irrigation. In some areas, sprinkler or drip irrigation replace traditional irrigation systems and, in others, the use of these water-conserving technologies allow a larger extension of irrigated land base.

A change of technology is one possible way to conserve water. Alternative methods include transition to crops having a lower water requirement and adjustment and improvement of traditional technologies that reduce water use. More complete analysis of water conservation activities in response to increased water scarcity has to incorporate these elements.

Conclusion

This chapter demonstrates that irrigated agriculture has played a major role in U. S. agriculture. Irrigation has extended the product choices available to farmers, extended the utilized land base, and allowed farming in regions that would not have been utilized otherwise.

The importance of irrigation technologies has increased over time. Before World War II, irrigated farming played a major role primarily in the arid West. During the last 30 years, the use of irrigation has spread to many Midwestern states, thus extending their utilized land base and product choice. This spread of irrigation across agricultural regions in the United States affected the markets of many major products including corn, cotton, and even small grains. Moreover, the increase of irrigated acreage has contributed to the substantial increase in the production of oilseed and legumes in the United States.

This 30-year period has seen development of new irrigation technologies that substantially increase irrigation effectiveness. Adoption of modern irrigation technologies is more likely in locations with higher water prices, higher-valued products, and where land and water are of lower qualities. Sprinkler irrigation (in particular, center-pivot irrigation) is responsible for intensive production of corn on the sandy soils of western Nebraska and the production of grains and oilseed on unleveled terrains of the mountain states. Drip irrigation has extended the locations where fruit and vegetable crops can be grown and has substantially increased their yields. These modern technologies may allow the United States to cope with its emerging water scarcity problems.

The importance of developing and adopting new and more efficient irrigation technologies is likely to increase with time since the development of new water resources is very costly. Water reservoirs are declining in quality and quantity, and there are growing demands for water. Thus, we expect a substantial increase in the use of sprinkler and drip irrigation over time, the development of improved computer software to assure improved water efficiency with any pricing, and the development of incentive schemes (water prices and credit support, for example) that will enhance adoption of these modern technologies.

Notes

This chapter was partially supported by the San Joaquin Valley Drainage Program and U. S. Department of Agriculture Grant 58-319-6-00100.

[1] A small number of estimated coefficients are not significantly different from 0 with 95 percent confidence and may lead to some unreasonable estimates. That explains, for example, the unreasonable parameters of the sprinkler equations in New Jersey and Tennessee.

[2] The sum total of acreages irrigated by certain states during certain years may differ between the two tables. These inconsistencies may reflect the diversity of sources used by the *Irrigation Journal* in its survey and omissions of acreages of certain crop categories from Table 7.6.

References

Burt, Oscar R. "The Economics of Conjunctive Use of Ground and Surface Water." *Hilgardia* 36(December 1964):31-111.

Gustafson, C. Don. "History and Present Trends of Drip Irrigation." University of California Cooperative Extension, San Diego, 1979. Mimeographed.

Caswell, Margriet F., and David Zilberman. "The Effects of Well Depth and Land Quality on the Choice of Irrigation Technology." *American Journal of Agricultural Economics* 68(November 1986):798-811.

Cochrane, Willard W. *The Development of American Agriculture: An Historical Analysis*. Minneapolis, Minnesota: University of Minnesota Press, 1979.

Dinar, A., K. C. Knapp, and J. Letey. "Irrigation Water Pricing Policies to Reduce and Finance Subsurface Drainage Disposal." *Agricultural Water Management*, in press, 1989.

Feinerman, E., J. Letey, and H. J. Vaux, Jr. "The Economics of Irrigation with Nonuniform Infiltration." *Water Resources Research* 19(1983):1410-1414.

Hayami, Yujin, and Vernon W. Ruttan. *Agricultural Development: An International Perspective*. Baltimore, Maryland: The Johns Hopkins Press, 1971.

Irrigation Journal. Survey Issues. Tampa, Florida: Brentwood Enterprises, 1950-1985.

Letey, J., A. Dinar, C. Woodring, and J. Oster. "An Economic Analysis of Irrigation Systems." *Irrigation Science*, forthcoming, 1989.

Riddering, T. Personal Communication, 1988.

8

Farm Enterprise Size and Diversification in Prairie Agriculture

William J. Brown

Introduction

The question of farm specialization versus diversification is an important issue facing farmers on the Canadian Prairies. The size of individual farm enterprises must be large enough to capture economies specific to that type of enterprise. Generally, in the past 20 years, generally, farms on the Canadian Prairies have grown to take advantage of these economies of size and in the process have also specialized, mainly into grain production. The downturn in grain prices in the late 1980s demonstrated that perhaps farms on the Canadian Prairies are too specialized.

Diversification is defined here as investing in at least two different assets or enterprises. If the total capital of a farm business manager is invested in one risky enterprise, the realized rate of return is solely dependent on the net return generated by that enterprise. If the farm business manager invests in two nonidentical enterprises (for there to be a low or negative return), both enterprises must have low or negative returns at the same time. Taken individually, both enterprises may be equally risky. However, depending on the correlation between the net returns, investment into both enterprises may reduce the risk over investing in either enterprise alone. The diversification of Prairie crop farms into livestock or other investments may help to improve net returns and reduce risk.

Diversification in the form of the mixed farm of the 1950s, with several enterprises all of which would be too small by today's standards to realize any economies of size, will lower net returns as well as stabilize them. In the 1980s, the substitution of capital for labor has resulted in the cost structure of agricultural enterprises to shift from an emphasis on variable costs (labor) to an emphasis on fixed costs (capital). Farm business managers have had to spread the fixed costs over a larger number of units in order to maintain profits. For the most part, it is not decreasing costs that encourage farm size growth; rather, it is larger incomes which, as long as costs do not increase, translate into increased profits (Miller).[1]

Economically sized and diversified enterprises may not be realizable on the average farm. The management skills and capital required to successfully establish and maintain such a business would be substantial and beyond the ability of most farmers. It may be better for specialized farms to look for diversification off the farm. Off-farm diversification may come in the form of investments in stocks in public companies, government bonds, various forms of joint ventures related to agriculture (such as custom feeding in large cattle feedlots or hog operations), or simply off-farm employment during slack times. The livestock investments should be large enough to take advantage of the economies of size but divisible into units that can be financed by the average farm business. Therefore, these operations will most likely not take place on current crop farms but rather in other locations.

The policy implications of facilitating diversification of Prairie crop farms are manyfold. Governments need to encourage research into the economies of size of the various agricultural enterprises on the Canadian Prairies in order to make certain uneconomically sized enterprises are not encouraged. Research along the lines of this chapter should also be encouraged to make sure that diversification reduces the risk faced by Prairie farms without significantly reducing income. The management skills and education of farmers will have to be improved in order for them to handle investments into a range of enterprises, including bonds and securities. Finally, some form of capital will have to be made available to farmers that wish to diversify their investment portfolios, whether it be on or off the farm.

Objectives

The three major objectives are to:

1. Measure the cost and returns for various sized grain-oilseed, cow-calf, beef feedlot, hog farrowing, and hog finishing operations. The approximate "threshold" size at which the average total costs (ATC) of production stop decreasing dramatically for each farm enterprise type is designated. This result indicates that investments into a particular farm enterprise smaller than the threshold size may prove to be uneconomic or, at the very least, too small to realize economies of size.

2. Measure the net returns from investments in the above-mentioned farm enterprises as well as to measure off-farm investments in stocks and bonds and returns from off-farm employment over the period 1971 to 1987. These net returns are measured as a percentage net return on investment in both nominal and real terms as are net cash returns. The correlation coefficients between the percentage net returns from the farm enterprises and the off-farm investments are calculated.

3. Measure the gains or losses from diversifying specialized farm operations as indicated by increases in net returns and/or reduced variability of net income that could be realized. The diversification options include combinations or portfolios of grains, oilseeds, pulses, and summer fallow in different fixed rotations. To these are added the net returns from cattle and hog enterprises as well as those from such off-farm investments as stocks in public companies and government bonds. Diversification into the various crops can occur on many of the typical specialized grain farms on the Canadian Prairies. Investment into public stocks and government bonds is currently an option open to those grain farmers with the financial means to do so. The investment into cattle and hog operations requires that the size of these operations be large enough to take advantage of the economies of size. Therefore, these investments may take the form of joint ventures such as feeder associations with custom feeding in large cattle feedlots or cooperatively organized hog operations.

Off-Farm Income and Part-Time Farming

Off-farm income has become a more common form of income for farmers on the Canadian Prairies. In 1971, 26 percent, 31 percent, and 34 percent of Saskatchewan, Manitoba, and Alberta farmers, respectively, reported off-farm work. These percentages have grown to 32 percent, 35 percent, and 43 percent, respectively, by 1986. In addition, Prairie farmers have devoted more time to off-farm work. In 1976, the average Prairie farmer spent between 35 to 50 days working off the farm. By 1981, this figure had grown to between 48 and 70 days (Statistics Canada).

Off-farm income can be used to counteract any decrease in net farm income and is, thereby, an excellent form of diversification for crop farmers on the Canadian Prairies. In the past, off-farm income has contributed substantially to total farm family income, especially in times of low net farm income. In 1971, the only year for which such data are available and a low net farm income year, Prairie farmers earned only 39 percent of their total income from net farm income; 40 percent from wages and salaries; and the rest from nonfarm self-employment, investment income, government transfers, or other income (Davey, Hassen, and Lu). If data were available for the rest of the 1970s, which were relatively high net farm income years, it would most likely have shown an increase in the contribution of net farm income to total income. However, if data from the 1986 census were available, it would probably show off-farm income's contribution to total income even higher.

The prevalence of off-farm work is a significant source of income for farm families on the Canadian Prairies showing diversification of making it a legitimate topic for any discussion on farm diversification. The ideal off-farm job should be within the local community to reduce time away from the farm business and family, would coincide with slack periods on the farm, would be consistent and reliable from one year to the next, and would pay a decent wage. Ideally, the off-farm work should be based in an industry that is stable or, at the very least, countercyclical to the agriculture industry; off-farm work would then be available when needed the most, that is, in low net farm income years.

Consistent and substantial off-farm income and the continued substitution of capital for labor have changed the nature of many farm businesses. In many cases, the owner, operator, and manager are no longer full-time farmers; rather, they are permanent part-time farmers. These phenomena may have serious repercussions for economies of farm enterprise size and diversification. Part-time farmers may wish to substitute their more expensive labor (opportunity cost) for even more capital (machinery) per acre than the average farmer. On the other hand, threshold enterprise size may not be as important if off-farm income can compensate net farm income. Part-time farmers may also not want to diversify because of increased time requirements and due to the fact they are already extremely well diversified (off-farm income). Clearly, many of the concepts discussed later apply primarily to full-time farmers. However, as part-time farming becomes more prevalent, studies on economies of size, farm diversification, and the increasing policy implications will have to include this important group.

Economies of Size

In the short run, the farm manager has a fixed number of acres, buildings, and equipment with which to work. The only way output can be expanded is by changing the amount of the variable inputs used. In the long-run all inputs are variable, and the farm manager is able to change the size of the business.

Economies of Size Studies

General Farm Types. Stanton concludes that there are definite economies of size when dealing with the American family farm. However, these soon give way to diseconomies and that, perhaps, a more important determinant of farm size rather than cost structure is the control of risk and uncertainty with respect to production and financing. Other issues relating to farm size, such as diversification and farmer socioeconomic characteristics, are analyzed in Pope and Prescott. A sample of California crop farms were found to be more diversified the larger they became, whereas the smaller, poorly financed farms were more specialized. Both results are contrary to what theory would predict and what other studies have found (White and Irwin; Raup).

Anderson and Powell, in a synopsis of U. S. agriculture, indicate that most long-run average cost (LRAC) curves are L-shaped rather than U-shaped. Relatively small farms can exploit most of the technological cost economies available to larger farms, meaning that LRAC curves tend to be horizontal over a wide range of output. The change in the United States is in the direction of larger farms while the number of smaller farms decline. With both output and costs of inputs measured in constant prices, this suggests that technological change has increased efficiency. Other factors, such as differences in sample of farms between surveys and varying seasonal conditions, could be very important. Overall, Anderson and Powell found that economies of size exist for small- to medium-sized farms and that average cost (AC) curves are nearly horizontal.

Hall and LeVeen discuss issues relating to the structure of agriculture and the survival of the family farm. Smaller farms may be able to survive, but they require more resources to do so; hence, there could be an efficiency cost. However, large farms generally have lower production costs. A relationship exists between farm size and economic efficiency either because there are economies of scale in the physical production function of the farm or because relative prices are such that cost savings result from increasing size. Hall and LeVeen found that the LRAC curve is L-shaped, meaning that production costs decline rapidly with initial increases in size and then decline slowly. Little evidence is found of increasing production costs for very large firms. In spite of higher production costs, small farms are still found to be economically viable. A wide range of farm sizes are found to produce at least enough income to cover all costs, including the opportunity costs of capital.

Economies of size, whereby large farms reduce their costs by spreading fixed machinery and labor costs over more output and land, are evident. Economies of size in volume discounts for purchased inputs are thought to be significant for large farms. However, it cannot be determined if the cost savings is derived from lower input prices or from more efficient uses of the inputs. In either case, the cost advantages associated with purchased inputs did not contribute in any substantial way to the overall advantage of large farms.

Overall, Hall and LeVeen conclude that, while there is a significant technical basis for economies of size, other factors such as management, resource quality, and the overall institutional structure are more important.

Heady states that optimum size will differ between farms depending on the stock of labor and management possessed in each household. The continuance of the family farm as the main structure of agriculture suggests that, if economies of size exist, they soon give away to diseconomies. In addition, Miller, Rodewald, and McElroy demonstrate that much of the gains from economies of size occur at relatively small sizes. Using 1978 data for the Northern Plains area of the United States, farms averaging 232 acres of cropland had 90 percent of the resource return rates of the most efficient farms that averaged 1,476 acres of cropland.

Ehrensaft (1983) concludes that an increase in the scale and degree of concentration in Canadian agriculture is associated with a trend toward new forms of farm enterprises. In 1971 a farm had to have $89,440 or more in sales to rank in the top 1 percent of Canadian farms. In 1981 the percentile limit for the top 1 percent had risen to $400,000. He feels that family farms still constitute the model form of farm enterprise for Canadian agriculture as a whole. However, the tendency is toward nontraditional forms at the upper end of the size distribution. (Dollar figures in this chapter are Canadian dollars.)

Ehrensaft and Bollman (1986) forecast the future of the "traditional" family farm to the year 2000. They are confident that the family farm will still be an important part of the rural landscape by the year 2000. Very large agricultural enterprises, however, are here to stay. Given the high value placed on farming as a way of life, some family farms will struggle quite tenaciously to survive in agriculture, even if it means lower returns and higher risks compared to the larger farms.

Trant discusses farm producers in the past and present. He states that between 1941 and 1981 average farm size continued to increase from 237 acres in 1941 to 463 acres in 1971. Small farms, hobby farms, and farms used for recreation still exist even though farm size is continually increasing over the years. Small farms are a pervasive and persistent group and will probably always be with us (Buttel). With large farms getting larger, farm numbers decreasing, and small farms continuing to persist despite high ATC, it would seem that change would come at the expense of the middle-sized farms. Brinkman and Warley indicate that this is not the case; middle-sized farms appear to be a healthy persistent group of farms.

The Beef Industry. Marshall discusses the size and structure of the livestock industry in Canada. Reconciliation of supply changes with demand shifts forces changes in marginal costs (MC) of production as well as changes in Canada's competitive relationship with other countries. With respect to feedlot services, technological and organizational developments promote growth to a capacity subject only to the limit of economic restraints. In feeder cattle production, expansion creates increasing MC with respect to extending land areas to be used for cow-calf operations. It is noted that any increase in beef herd expansion is encountered by high land values.

Statistics Canada data are summarized by Ehrensaft (1987). He finds that average operating costs follow a curvilinear pattern as farm size increased for most major sectors. Costs are found to exceed gross farm sales for smaller farms, decrease for mid-sized farms, and rise slowly for the largest farms. According to this view, most economies of size are achieved after a relatively modest threshold and then AC remains basically constant over a wide size range.

Between 1966 and 1981, average size of beef feedlots went from 312 to 725 head. The 33 farms with 1,000+ steers in 1966 had 6 percent of the national herd while 93 farms with 1,000+ steers had 14 percent of the herd and average gross farm sales of $3 million in 1981. Low levels of concentration in the beef cow herd (cow-calf enterprises) show no changes over the 1966 to 1981 period. The top 1 percent of farms with beef cows account for only 12 percent of the national herd, while mid-sized farms account for the majority (58 percent) of the national beef cow herd. This proportion has been steady since 1966. In 1966 one needed 23 to 58 cows in an Upper Middle beef cow farm. In 1981 one needed 37 to 99 cows to remain in the same class. Even if relative shares of size classes remain steady, each farm will have to increase its real resources in order to retain the same relative ranking. Previous studies suggest that AC curves would either be L-shaped until one reached the largest size class, with some decline in AC for the 500+ cow herds; or one would observe a modest decline in costs as herd size increased.

Ehrensaft and Bollman also forecast the beef sector for the year 2000. They expect the final stages of feeding animals for meat production to be dominated by the previously defined nonclassic forms of enterprise organization. They also feel that family farms will not disappear from the sector. Some family farms will maintain a position in the sector, but technical conditions, transaction costs, and tax structure tend to favor larger enterprises. Most of these larger enterprises are family farm units that expand to become semimanagerial or independent managerial farms. The minority are agribusiness firms that integrate agricultural production into their enterprise.

Finally, Doll and Orazem conclude the trend is toward fewer and bigger livestock enterprises. Also, a trend toward fewer and larger units in grass-fed cattle and calf operations is apparent, but the degree of change is not as extensive as for grain-fed cattle. Cow herd operations are more difficult to mechanize and automate than feedlots. One of the major problems facing feeders seems to be insufficient volume to effectively use livestock equipment and buildings.

The Hog Industry. Economies of size in the hog production industry in the United States are analyzed rather thoroughly by Van Arsdall and Nelson. They find that large hog operations achieve economies of size over small hog operations. This is done through more intensive use of facilities, somewhat better feed conversion, lower feed costs, and lower unit labor use. The economies of size are large enough that, in a year of low returns, some small enterprises may fail to cover cash costs, whereas large enterprises cover all costs, including capital replacement. The large producers' advantage is less when only short-run cash costs are considered. As the planning period lengthens, so does the large producers' advantage. Hog production is likely to continue to shift toward a smaller number of large, industrialized, and highly specialized operations, increasingly separate from crop production.

Performance varies among producers of all sizes. Large producers fair significantly better on pigs farrowed and weaned per litter, litters farrowed, and pigs weaned per female per year, death losses, and feed conversion rates. They also perform better on four of the five price performance measures—prices received for hogs and prices paid for feeds, rations, and labor costs. Performance varies significantly for all producers on total returns. It is noted that performance varies widely among hog producers of similar size, but variability is greatest among small producers. Overall, however, large size of enterprise alone is no assurance of success.

Producers marketing more than 1,000 hogs a year continue to gain shares in the major hog-producing regions. Large-volume producers may gain economic advantage over small ones through two basic avenues: (1) Producers may have the knowledge and ability to get more output from their physical resources and (2) they may use less costly inputs or get discounts because they buy large quantities. Thus, prices may not be the same for everyone.

As the operations increase from small to large, unit investments change in steps, decrease for a time, then increase before continuing to decline as size of enterprise grows. The unit investment drops as the size of the hog operation increases because of a number of factors, mostly pertaining to size. Investment components are more important to large operations.

Farrow-to-finish producers with large enterprises achieved sizable economic advantages over small enterprises in 1980, 1982, and 1983. Large operations' advantage increased over small in terms of cost versus income when a charge or return was allocated to unpaid labor.

Despite considerable variations among farms, average performance consistently improves as size of hog operations increase. Overall, evidence from this study indicates a continued restructuring of the hog industry to fewer, larger, and more specialized operations.

Trant shows a drop of 65 percent in the number of farms reporting pigs. The number of farms declined 154,528 in 1966 to 55,765 in 1981. The amount of pig numbers almost doubled from 5.4 million in 1966 to 9.9 million in 1981. Also, the concentration of production is increasing in the hog sector. In 1966, 90 percent of pigs were produced by 29 percent of the farms reporting pigs; but, by 1981, this production was concentrated among 18 percent of the farm operations reporting pigs. Thus, Trant concludes that medium to large enterprises will continue to evolve as the dominant size in hog operations.

Wilson and Eidman discuss dominant enterprise size of the swine production industry in the United States. They feel that the size of hog operations is determined by the geographic location of the production unit and the risk attitudes of the producer. The principal structural determinants of the hog production industry are technological change, favorable price ratios (both product and input), labor availability, and urban encroachment. The pattern of dominance across enterprise size is consistent within several geographic locations. The risk-averse individual prefers smaller swine operations while risk-loving agents select larger enterprises. Generally, the larger operations expose the risk-averse individual to high levels of variability which he is not willing to accept.

Wilson and Eidman conclude that, as swine producers become more effective at risk management, medium to large enterprises will continue to evolve as the dominant enterprise size in the hog industry. The medium-sized operations are quite competitive with the larger operations and are in little danger of being forced out of business.

The Threshold Enterprise Size

The current study is not interested in the long-run least cost size of the various agricultural enterprises being investigated. Rather, it is the approximate size at which the LRAC curve ceases to decline as rapidly as it does at smaller sizes. The approximate enterprise size at this point is labeled the threshold size. Enterprises smaller than this size experience severe diseconomies and are most likely unprofitable. Enterprises larger than this size may experience lower costs, but most of the economies of size are captured in the threshold size. The threshold size is by no means fixed from farm to farm and can vary a great deal depending on a number of factors, not the least of which is management ability.

Ehrensaft (1987) has analyzed the National Farm Survey data from 1983. The costs used in this study did not include an estimate for farm labor nor an opportunity cost for land. Figure 8.1 demonstrates that the threshold farm size for all types of farms in the Canadian Wheat Board (CWB) region is around the 75th percentile or $56,000 of gross sales. Since all types of farms are included, it is difficult to translate the gross sales to number of acres or head of livestock. However, this could be interpreted that a full 75 percent of farms in the CWB region are smaller than the threshold size but not necessarily experiencing negative net incomes. Upon close examination of Figure 8.1 and the other figures up to and including Figure 8.6, it can be seen that as much as 25 to 30 percent of some farm types have total costs/sales ratios of greater than one, thereby indicating negative net incomes. How do these farms survive with negative incomes? Undoubtedly, many do not. Other farms do not cover depreciation costs and thereby do not replace machinery but continue to

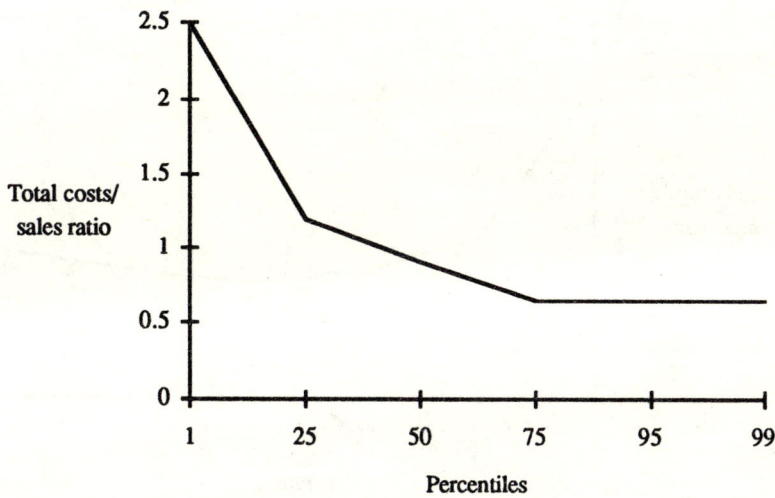

FIGURE 8.1. Total Costs/Sales Ratio: All Farm Types, CWB Region, 1983

Source: Ehrensaft, 1987.

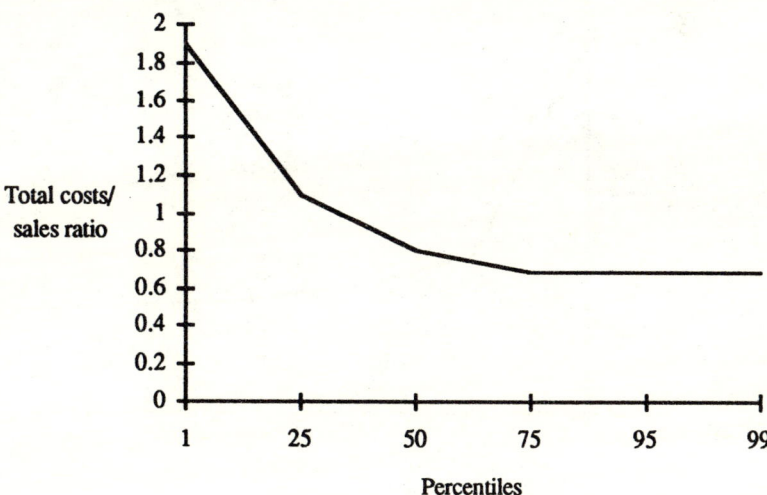

FIGURE 8.2. Total Costs/Sales Ratio: Wheat Farms, CWB Region, 1983

Source: Ehrensaft, 1987.

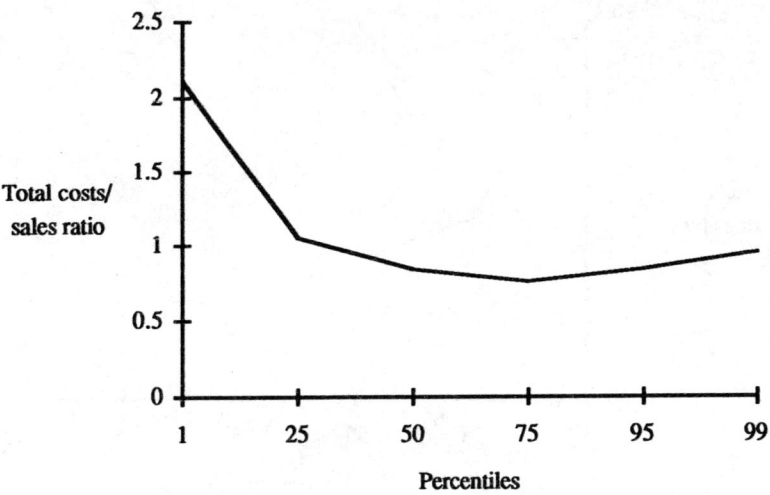

FIGURE 8.3. Total Costs/Sales Ratio: Coarse Grain Farms, CWB Region, 1983

Source: Ehrensaft, 1987.

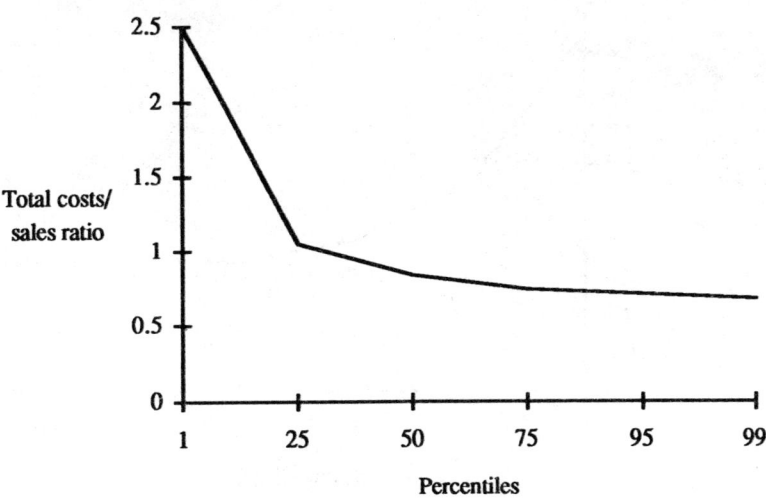

FIGURE 8.4. Total Costs/Sales Ratio: Oilseed Farms, CWB Region, 1983

Source: Ehrensaft, 1987.

261

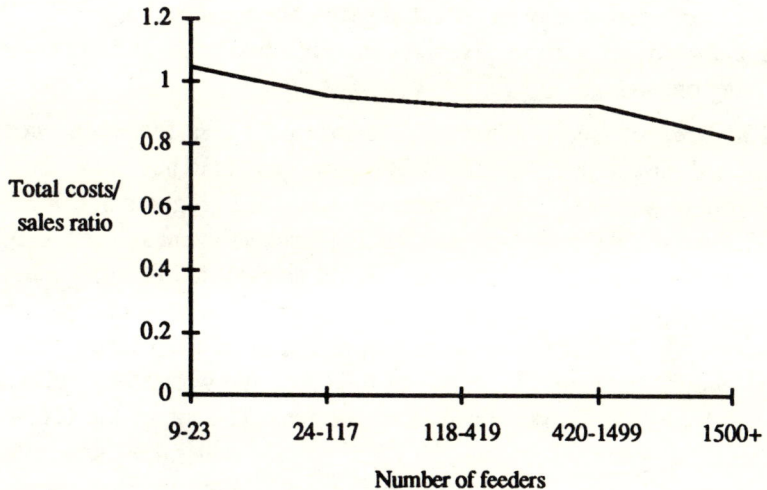

FIGURE 8.5. Total Costs/Sales Ratio: All Beef Types, Canada, 1986
Source: Ehrensaft, 1987.

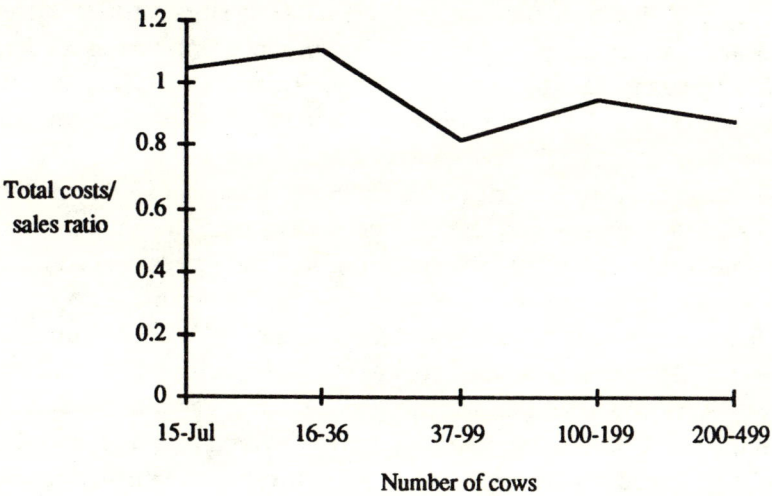

FIGURE 8.6. Total Costs/Sales Ratio: Cow-Calf Enterprises, Canada, 1986
Source: Ehrensaft, 1987.

produce with old machinery. Still others may have off-farm income. Finally, some farms may have had negative incomes in the year the data were gathered but had positive incomes in other years and are thereby surviving on savings.

The Grain-Oilseed Industry. The threshold size for wheat, coarse grains, and oilseed farms in the CWB region appears to be around the 75th percentile or $56,000 of gross farm sales in 1983 (Figures 8.2, 8.3, and 8.4) (Ehrensaft 1987). A grain farm in the black soil zone of the Canadian Prairies—following a $1/3$ fallow, $2/3$ canola-wheat rotation with canola on fallow yields of .476 tonnes per acre and price of $384 per tonne and wheat on stubble yields of .68 tonnes per acre and price of $173.10 per tonne in 1983—would generate the threshold total revenue with 559.1 cultivated acres.[2] Similarly, a grain farm in the brown soil zone of the Canadian Prairies—following a $1/2$ fallow, $1/2$ wheat rotation with fallow wheat yields of .68 tonnes per acre—would generate the threshold total revenue with 951.5 cultivated acres.

Statistics Canada and Canfarm data have also been summarized for grain farms in Saskatchewan (Jensen; Fleming and Uhm). The Jensen study included a charge of $4.00 per hour for operator labor and calculated the residual as a return to land. Therefore, the LRAC curve calculated will be steeper and the diseconomies, associated with sizes smaller than the threshold size, more severe. In 1977 the threshold farm size in the Jensen study appears to be around a total revenue figure of $50,000. In 1977 a grain farm in the black soil zone of the Canadian Prairies—following a $1/3$ fallow, $2/3$ canola-wheat rotation with canola on fallow yields of .635 tonnes per acre and price of $285.73 per tonne and wheat on stubble yields of .871 tonnes per acre and price of $106.57 per tonne—would generate the threshold total revenue with 546.9 cultivated acres. Similarly, a grain farm in the brown soil zone of the Canadian Prairies—following a $1/2$ fallow, $1/2$ wheat rotation with fallow wheat yields of .844 tonnes per acre—would generate the threshold total revenue with 1,111.8 cultivated acres. Applying a similar argument to the 1977 Canfarm data used by Fleming and Uhm, whose threshold total production level is around 350 tonnes produced, would result in a 697.2 cultivated acre farm in the black soil zone and a 829.4 cultivated acre farm in the brown soil zone. Fleming and Uhm did not include an allowance for operator labor nor a residual cost for land. Therefore, the LRAC curve should be lower and less sloped and the diseconomies associated with sizes smaller than the threshold size less severe.

Miller, Rodewald, and McElroy studied the economies of size in field crop farming for several regions of the United States. The threshold size for the Northern Plains region did not change significantly by method of calculation and was as small as 232 acres of cropland, which is significantly smaller than that described in the Canadian studies. Miller, Rodewald, and McElroy show very little additional gains (as stated earlier, only 10 percent) from increasing the size to 1,476 acres of cropland.

The resulting threshold size from the above Canadian studies range between 546 to 697 cultivated acres in the black soil zone and between 829 and 1,111 cultivated acres for the brown soil zone. The studies consistently designate larger farms in the brown soil zone than those in the black soil zone. Unfortunately, they do not predict the evident trend to larger farms, but perhaps this trend is slowing. The threshold sizes may also seem rather small, but it must be remembered that these are not the least cost size of farms. In fact, the literature would indicate that there are still economies, or at least higher incomes to be gained by larger sizes. What is important to remember here is that investments into grain and oilseed farms significantly smaller than these threshold sizes may well be uneconomic.

The Beef Industry. The National Farm Survey done by Statistics Canada in 1986 has been summarized by Ehrensaft (1987). The threshold size of beef feedlots in the CWB region appears to be in the neighborhood of 1,500+ steers (Figure 8.5). The AC curve is almost horizontal, but economies are evident at the larger sized feedlots. U. S. studies with respect to beef feeding have also been summarized (Carter and Schmitz) and have found significantly larger threshold sizes in the neighborhood of 10,000 head. It would appear the U. S. data with respect to cattle feeding are more credible because Statistics Canada does not break down herd sizes larger than 1,500 head. The threshold size for the beef cow herd in Canada, as a whole, appears to be in the 37-99 cow range; however, the data are rather erratic and further economies are evident in the 200-499 cow range (Figure 8.6).

The Hog Industry. To the extent that U. S. data and conditions can be translated to Western Canada, a threshold size in 1983 for a hog farrowing operation and a hog finishing operation was around 3,000 head sold per year in each enterprise. Given a weaning rate of 15 pigs per sow per year, the hog farrowing operation would equal a 200-sow enterprise. Costs continued to decline as size increased up to the maximum measured 10,000 head sold per year in 1983 (Van Arsdall and Nelson).

Economies of Size and Diversification

Pope and Prescott state that the relationship between farm size and diversification is an indicator of trade-offs between risk reduction and possible economies of size in a particular activity. That is, if there are substantial economies of size in a particular activity, one clearly gives up a large expected return in order to insure against risk through diversification. A basic theorem is that, if returns in activities are independently and identically distributed, then diversification is optimal with equal proportions in each activity. Thus, diversification is likely to be optimal for a risk averter. However, large disparities in average returns or resource constraints may provide incentives for specialization. Other variables that may affect diversification choices are net worth, experience of the farm operator, form of ownership (i.e., family farm, corporation), and variables which delineate geographical location and the extent of irrigation, etc.

There is a strong indicator of a positive relationship between diversification and size. Also, there is a negative relationship between diversification and measures of financial "well-being." A farm diversifies to spread risk, and wealthier farmers have fewer incentives to spread risk. Finally, farmer experience or age tends to exhibit a positive effect on diversification. That is, younger or less-experienced farmers are less diversified. Pope and Prescott go on to speculate that younger farmers may be less risk averse, may start small and specialize and diversify later, may be indicative of capital shortages for younger farmers, or, finally, may find it difficult, due to inexperience, to manage diverse activities. Pope and Prescott's discussion would seem to indicate that policies promoting diversification may have less impact among younger, less-experienced farmers.

Gains and Losses from Diversification: The Theory

Mean-Standard Deviation Trade-Off

The most commonly used efficiency criterion is the mean-variance or mean-standard deviation trade-off. When dealing with net income, those alternatives exhibiting the lowest standard deviation of net income for given levels of expected net income, or conversely the maximum level of expected net income for given levels of standard deviation of net income, are said to be on the risk-efficiency frontier of risk-neutral and risk-averse decision makers. The mean-standard deviation trade-off has both strengths and weaknesses. It is an effective means of summarizing data and identifying alternatives having the greatest expected value for a variable for a given level of standard deviation of that same variable. However, in order for it to be technically correct, the net income must be normally distributed or the decision maker's utility must only be a function of mean (expected net income) and standard deviation (variation of expected income). Distributions of alternative net income exhibiting skewness and higher moments are common in agricultural situations (Barry 1980, p. 73). The risk-averse decision maker may choose an alternative that is not on the risk-efficiency frontier when these additional characteristics of the distribution of outcomes are considered. Therefore, efficiency criteria that consider the total distribution of outcomes rather than one or two summary statistics may be preferred. Stochastic efficiency is such a criterion but is beyond the scope of this chapter. A more in-depth discussion of stochastic efficiency can be seen in Brown (1987); Barry (1984); Anderson, Dillon, and Hardaker; and Zentner *et al*.

Portfolio Risk and Diversification

Diversification means investing in at least two enterprises or activities that differ. The rate of return for the resulting portfolio is the weighted average of the returns from the individual enterprises included in it. The total risk of the portfolio depends on the standard deviations and correlations of the individual enterprises included in it. The standard deviation measures the uncertainty as to returns on an individual enterprise or portfolio of enterprises. The correlation coefficient measures the degree to which the returns of one enterprise vary with the returns of another. Diversification reduces risk if the returns from the various enterprises within a portfolio are not highly correlated.

The following example illustrates how diversification affects risk. Take two enterprises, A and C. Both have average returns of 20 percent and standard deviations of returns of 22 percent. If the variation of the returns for the two enterprises was perfectly positively correlated, that is, a correlation coefficient of +1.0, any combination or portfolio of the two enterprises would also have an average return of 20 percent and a standard deviation of returns of 22 percent. However, if the variation of returns for the two enterprises were perfectly negatively correlated, that is, a correlation coefficient of -1.0, any combination or portfolio of the two enterprises would also have an average return of 20 percent and a standard deviation of returns of 0 percent. The standard deviation of the portfolio of enterprises A and B for all possible values of the correlation coefficient is shown in Figure 8.7.

There is a limit to the amount of risk reduction that can take place in any portfolio. Only the nonsystematic risk can be diversified away. Systematic risk, that is, the risk associated with the market portfolio, cannot be diversified away by investing in enterprises from within the market. In addition, it is difficult to find enterprises with negative correlations, especially within the same industry. However, Schall, Haley, and Schachter point out that portfolio risk is reduced, even when the enterprises are positively correlated (+.5), by increasing the number of enterprises (Figure 8.8). In addition, it does not take a great deal of diversification to receive most of the benefits. Investing in more than 10 enterprises reduces risk only slightly. The only risk remaining in a well-diversified portfolio is the market or systematic risk. Therefore, the returns from a well-diversified portfolio are highly correlated (close to 1.0) with the entire market.

Turvey and Driver conclude that the opportunities for diversification within agriculture are limited. In order for farmers to reduce the amount of systematic risk in their market portfolio, that is, the agriculture portfolio, they must make off-farm investments such as securities. By moving outside the agriculture portfolio to a general market portfolio, the systematic risk within agriculture becomes a nonsystematic risk in the general market portfolio. The advantages of off-farm investments are that they have low correlations with the farm sector and can be used to reduce the now nonsystematic risk to zero through diversification. In addition, the liquidity of the capital markets allows for greater flexibility in transferring capital between farm and off-farm uses.

The Capital Asset Pricing Model (CAPM)

The underlying assumptions of CAPM, as outlined in Schall, Haley, and Schachter, are as follows where (1) the investor is concerned only with the return over a single period; (2) the investor has a specific amount of money to invest; (3) the investor likes high expected portfolio return and low standard deviation of portfolio return; (4) the investor has estimates of the expected rates of return and the standard deviations from all portfolios of risky assets; (5) the investor is able to borrow or lend at the same (risk-free) rate of interest; (6) securities are bought and sold in a highly competitive market with no transaction costs (such as brokerage fees); (7) all investors

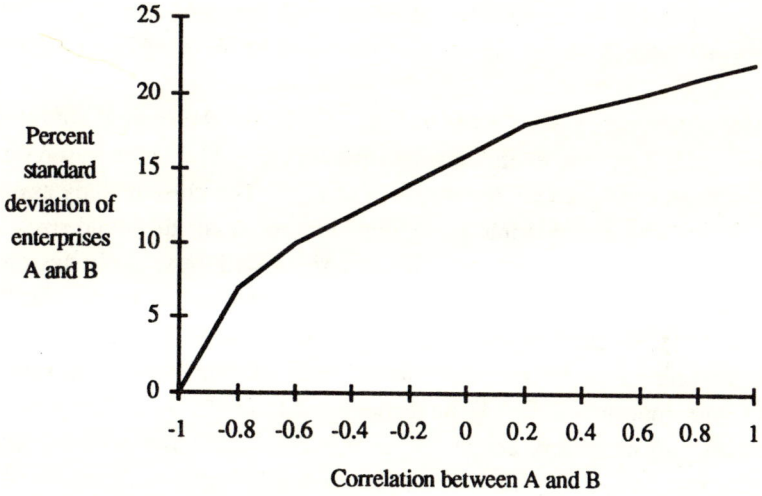

FIGURE 8.7. Portfolio Standard Deviation for Different Correlation Coefficients Between Enterprises
Source: Schall, Haley, and Schachter.

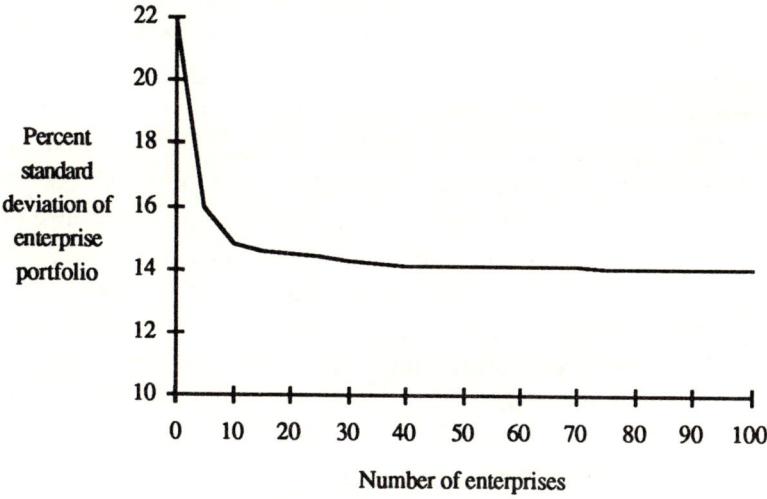

FIGURE 8.8. Portfolio Standard Deviation, Enterprises +.5 Correlated
Source: Schall, Haley, and Schachter.

have the same expectations regarding the future returns from owning securities (same expected rates of return and standard deviations for all portfolios); and (8) taxes do not bias investors in favor of one investment over another.

The underlying assumptions of CAPM hold for the current study. The current study can only estimate what investors should do now based on the data used; as time goes on, the data will change. The current study assumes investments will be made into agricultural enterprises of threshold size. Due to the magnitude of these threshold sizes, these will most likely have to be joint ventures or cooperatives where entrance shares can be standardized between enterprises. The above assumptions (3) and (4) hold in that (3) is a basic assumption underlying the current study, and the means and standard deviations mentioned in (4) have been calculated. It is assumed that investors can enter any enterprise analyzed without any kind of capital constraints. In reality, this may not be true, especially with regard to the magnitude of threshold enterprise sizes. However, as stated above, joint ventures or cooperatives will most likely have to be formed to overcome capital constraints. A competitive market and no transaction costs is somewhat tenuous when dealing with agricultural land but is probably no worse than market securities when dealing with the livestock enterprises, especially if they are part of joint ventures or cooperatives. Assumptions (7) and (8), with regard to future expectations and taxes, also hold.

Schall, Haley, and Schachter point out that the CAPM assumes that properly priced securities or enterprises should provide an expected rate of return equal to the rate of interest on riskless securities (government treasury bills) plus a premium for bearing risk. The risk is measured by the enterprise's beta. The enterprise's beta is equal to the security's correlation coefficient with the market portfolio (e.g., Toronto Stock Exchange (TSE) Index 300) times its standard deviation all divided by the standard deviation of the market portfolio. The beta for the market itself is 1.0 as implied by the definition of beta. Therefore, a beta of 1.0 indicates that the expected rate of return on the enterprise is the expected rate of return on the market. Enterprises with low or negative correlations of returns to the market will have low or negative beta and vice versa. The "security market line" represents the linear relationship between an enterprise's beta and the expected rate of return on that enterprise, that is, the current risk-return trade-off in the market (Figure 8.9). If an enterprise's beta is high, it indicates that the enterprise is associated with high risk. The security market line indicates what rate of return is needed to compensate the

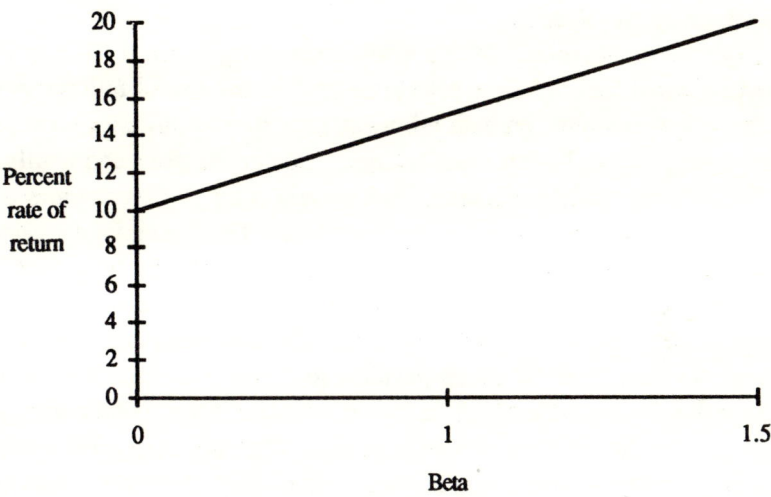

FIGURE 8.9. The Security Market Line

Source: Schall, Haley, and Schachter.

investor for this increased risk. If the rate of return does not meet this rate, investors will not invest until the risk is reduced or the rate is increased. If an enterprise's beta is 0 or less than 0, it indicates that the enterprise is associated with low risk. However, if the rate of return is consistently below the risk-free rate of return (i.e., the rate of return when beta is 0), investors will not invest (they can do better with government treasury bills and have no risk) until the investment cost in the enterprise is reduced and the rate of return increased.

Turvey and Driver used a similar approach to analyze various agricultural enterprises in Ontario. However, they use cash rent for land as their risk-free asset and create a "farm sector portfolio" rather than using the TSE 300 as the market portfolio. Therefore, their beta measures the relative risks of various enterprises within Ontario agriculture, whereas the betas calculated here measure the relative risk of various enterprises as compared to investing in the stock market.

Analysis

Crop Rotation Net Returns

Crop Gross Margins. In this study gross margins from 1971 to 1987 were calculated for the fallow enterprise (that is, the cost of fallowing) and for the following crops (on both fallow and stubble) for the dark brown soil zone: spring wheat, barley, oats, fall rye, flax, canola, peas, and lentils.[3]

The first component in a crop gross margin is output price. Farm price was used for all crops other than spring wheat. The spring wheat price is calculated as follows: The initial payment received each year from the CWB for red spring wheat grades 1, 2, 3, and feed were reduced by charges for transportation to the terminal point, country elevation, and removal of dockage for each year. The final payment received from the CWB for each grade each year was added to the adjusted initial price for the following year to account for the time lag in final payments. This adjusted price received by farmers for each spring wheat grade was further adjusted to reflect the percentage of each grade marketed. The percentage of grade marketings are taken from crop district No. 6 in Saskatchewan, which is in the dark brown soil zone (Ulrich and Furtan). The final result is a weighted farm price received for spring wheat.

The second component in calculating a crop gross margin is yield. Yield data for the crops on both fallow and stubble were based on Saskatchewan Crop Insurance Corporation (SCIC) risk area No. 12, which is in the dark brown soil zone. Note that the dark brown soil zone occurs in both central Saskatchewan and central Alberta. Separate fallow and stubble yield data for 1971 and 1972 were not available, so the average yield for those years was adjusted by the relationship between fallow and stubble yields established through 1973 to 1987. Lentil and pea yields were supplemented by information from Saskatchewan Agriculture (*Specialty Crop Report*).

The final components needed in the calculation of crop gross margins are the direct cash costs which are subtracted from gross income. Direct cash costs were assumed to be the direct operating costs of machinery power and repair as well as those of crop materials. These were obtained from Schoney (1985, 1986, 1987), based on the detailed costs from some 60 to 120 farmers from Saskatchewan. Although this is not a random sample, it is the best estimate of actual production costs presently available in published form. Note, also, that the data used are all from the dark brown soil zone of Saskatchewan; however, cost comparisons with Alberta show larger variations within provinces but across soil zones than across provinces but within soil zones (Alberta Agriculture).[4]

Fallow and stubble cash costs were not available for all crops. Procedures used to estimate these costs are outlined in Brown and Forsberg. The 1985 cash costs were deflated for the period 1971 to 1984 using an index based on the amount expended each year in Saskatchewan on petroleum, diesel oil, and lubricants (machinery power and repair) and fertilizer and other crop expenses (crop materials) (Saskatchewan Agriculture, *Agricultural Statistics*). This index represents inflationary price trends and the shift in agricultural technology between 1971 and 1985 and allows for increased use of fertilizer and chemicals on all crops. Its weakness is that it includes the shift away from fallow and thereby overadjusts the cash costs, particularly of stubble crops. A second indexing procedure, based on the Western Canada farm input price index for machinery and motor vehicle operation and petroleum products (machinery power and repair) as well as crop production expenses (crop materials), was also used (Saskatchewan Agriculture, *Agricultural Statistics*). The results based on this second index were similar and are reported in Brown and Forsberg.

The above method of calculating crop gross margins over time is not ideal. A random sample of producers keeping detailed enterprise records is preferable but is not available. Since costs were deflated by two methods based on rather different assumptions with the results not changing significantly, the approach adopted appears satisfactory. However, no direct relationship between increased expenditures on inputs (fertilizers and chemicals) and increased yields is accounted for. However, the relationship between increased input use and increased yield should be accounted for in the cost and yield data since these data are based on actual producer behavior. The relationship built into the data is really one of yield response to input use and weather conditions.

Correlation Coefficients of Crop Gross Margins. Given the above, gross margins for a number of hypothetical rotations for the period 1971-1987 are calculated. Portfolio theory specifies that individual components within a portfolio exhibiting high positive correlation add to the risk or variance of the portfolio. Gross margins for all crops, both fallow and stubble, were compared with those of wheat on fallow. The correlation coefficients for each crop are shown in Table 8.1.

TABLE 8.1

Correlations of Wheat on Fallow Gross Margins with Alternative Crops, 1971-1985

Wheat on fallow		Dark brown soil zone
Fallow		-0.47
Wheat on stubble		0.89
Barley:	on fallow	0.75
	on stubble	0.53
Oats:	on fallow	0.56
	on stubble	0.31
Durum:	on fallow	0.82
	on stubble	0.77
Canola:	on fallow	0.46
	on stubble	0.48
Fall rye:	on fallow	0.70
	on stubble	0.25
Flax:	on fallow	0.79
	on stubble	0.47
Lentils:	on fallow	0.32
	on stubble	0.13
Peas:	on fallow	0.40
	on stubble	0.49

Source: Computed.

For simplicity, the objective was to have the hypothetical rotations include only one representative from the grain, oilseed, and specialty crop categories. Wheat on fallow and stubble was chosen as the grain because of its dominant area on the Prairies. Barley, oats, fall rye, and durum wheat, on both fallow and stubble, were eliminated from all rotations because of their high correlation coefficients. Canola was chosen over flax as the oilseed representative in the rotations because of its lower correlation coefficient and its greater acceptance by farmers in the past. Lentils was chosen over peas as the specialty crop representative in the rotations because of their lower correlation coefficient.

Hypothetical Fixed Crop Rotations. Hypothetical rotations, as selected, stay constant for the period 1971 to 1987 and are shown in Table 8.2. These rotations were assumed to contain either wheat or canola and no more than 33.3 percent lentils. Note that rotation 11 represents the approximate average crop rotation (or cropland use pattern) for Saskatchewan. The effect on the level and standard deviation of gross margin of reducing fallow intensity and diversifying into canola (oilseeds) and/or lentils (specialty crops) may be calculated by comparing these variables for the 24 hypothetical fixed crop rotations.

In the calculation of the rotation gross margins, two points should be noted:

1. The weightings of the crops in each rotation have been kept constant over time because the objective was to compare the distributions of gross margins from fixed rotations. One or several other rotations in which individual crop weightings change from year to year may well dominate the rotations outlined in Table 8.2.

2. The wheat and canola yields in rotations which include lentils have not been adjusted to compensate for the nitrogen-fixing ability of lentils. This benefit to following crops has not been documented in the literature and may be offset by the increased chances of weed problems in lentils and crops following lentils (Slinkard and Drew).

Saskatchewan Crop Insurance (SCI), Western Grain Stabilization Program (WGSP), and Special Canadian Grains Program (SCGP). The monetary effect on a per acre basis of participation in the SCI, WGSP, and SCGP have been calculated into the net returns of each rotation. Average annual SCIC payments to farmers less premiums for risk area 12 have been

TABLE 8.2

Twenty-Four Hypothetical Crop Rotations Showing Percentage of Each Crop in Each Rotation, 1971-1987

Rotation	Fallow	Wheat Fallow	Wheat Stubble	Canola Fallow	Canola Stubble	Lentils Fallow	Lentils Stubble
				percent			
1	50.0	50.0					
2	40.0	40.0	20.0				
3	30.0	30.0	40.0				
4	20.0	20.0	60.0				
5	10.0	10.0	80.0				
6			100.0				
7	50.0			50.00			
8	40.0			40.0	20.0		
9	30.0			30.0	40.0		
10			33.3		33.3		33.3
11	32.0	28.0	29.0	4.0	4.0		3.0
12	50.0	25.0		25.0			
13	40.0	20.0	10.0	20.0	10.0		
14	30.0	15.0	20.0	15.0	20.0		
15	20.0	10.0	30.0	10.0	30.0		
16	10.0	5.0	40.0	5.0	40.0		
17			50.0		50.0		
18	50.0	25.0				25.0	
19	40.0	20.0	10.0			20.0	10.0
20	50.0	16.7		16.7		16.7	
21	40.0	13.3	6.7	13.3	6.7	13.3	6.7
22	30.0	10.0	13.3	10.0	13.3	10.0	13.3
23	20.0	6.7	20.0	6.7	20.0	6.7	20.0
24	10.0	3.3	26.7	3.3	26.7	3.3	26.7

Source: Computed.

calculated on a per acre basis for 1971 to 1987. These calculated annual acreage benefits or costs have been added to or subtracted from the rotation gross margin for the year in question. The method of calculation used is not rotation specific but does reflect the monetary effect of crop insurance on a per acre basis for the dark brown soil zone.

The WGSP payments to Saskatchewan farmers less producer levies paid were divided by the total marketings of the seven crops included in the program each year to derive an annual per tonne impact of the program (Saskatchewan Agriculture, *Agricultural Statistics*). The per tonne impact was then added to the price of wheat and canola in the appropriate year and calculated into the returns of each rotation. No WGSP impact was calculated for lentils as they are not included in the program.

The SCGP payments on a per tonne basis were added to the price of wheat and canola in the appropriate years. No SCGP adjustment was calculated for lentils as they were not included in the program in 1986 and 1987.

Canadian Wheat Board (CWB) Quotas. The effect of CWB quotas on the rotation gross margins were calculated. The CWB quotas for wheat and canola were gathered from CWB annual reports for the 1971-1987 time period. Quotas were adjusted to account for the level of delivery allowed for all grades of wheat and canola. That is, if one grade of wheat had an open quota and another only 10 bushels per quota acre, the wheat quota that crop year was calculated as 10 bushels per quota acre. A quota acre was considered to be the same as a rotation acre, that is, it included that portion of the rotation acre either seeded to the crops considered (including lentils) or fallowed; perennial forage was not included. The CWB "Bonus Acres" program was included in the calculation from 1982 to 1987.

For years when the production of one crop from a particular rotation was above its quota level and the production of another crop in the same rotation was less than its quota level, quota allocations were adjusted accordingly to allow for the maximum delivery of all crops. Production above the quota level was stored at no cash cost and sold when the quota level permitted. This adversely affected rotation gross margins in low quota years and greatly increased them in subsequent years when quotas eventually increased or became open. This method of calculation is a valid measure of the variability of cash flows (gross margins) resulting from following the 24 fixed rotations during the time period.

Net Return on Investment for the Crop Rotations. The annual net return on investment associated with following each of the hypothetical crop rotations is calculated by the procedure outlined in equation (1).

$$\text{ROIC}_{ij} = \frac{\text{GM}_{ij} - \text{INT}_{ij} - \text{LAB}_{ij} - \text{FC}_{ij} + \text{CAPPL}_{ij}}{\text{BIL}_{ij}} \quad (1)$$

where

ROIC_{ij} = residual return per acre of cropland investment in year i by following crop rotation j

GM_{ij} = gross margin in year i of crop rotation j (see Brown and Forsberg)

INT_{ij} = interest charge on direct costs per acre for $1/2$ year at prime plus 2 percent in year i for crop rotation j

LAB_{ij} = per acre charge for operator labor in year i for crop rotation j

FC_{ij} = per acre charge for the fixed cost in year i associated with rotation j (it includes property taxes, general farm overhead which is 5 percent of direct costs, and a capital recovery charge for machinery and buildings that assumes a 15 percent rate of depreciation and an interest charge of prime)

CAPPL_{ij} = capital appreciation of the land investment per acre in year i as measured by the land value at the end of the year less the land value at the beginning of the year

and

BIL_{ij} = value of the land investment at the beginning of year i as measured by average cropland values for Saskatchewan.

The operator labor charge and the capital recovery charge for machinery and buildings were taken from various cost of production studies over the time period and indexed in those years that data were missing (Brown 1988). Missing labor data were indexed by the hired farm labor price index. Missing machinery and building data were indexed by an 80 percent-20 percent weighting of the machinery and building replacement price indices. Appendix A (Brown 1989) presents the nominal net returns on investment. Appendix B (Brown 1989) presents real net returns on investment by multiplying the nominal net returns by the Consumer Price Index (CPI) for the appropriate year. Appendix C (Brown 1989) presents net cash returns on investment by excluding noncash costs such as interest on operating capital, operator labor, machinery and buildings capital recovery charges (CRC), and noncash income such as capital appreciation.

Cattle and Hog Net Returns

Annual return on investment for each year during the period 1971 to 1987 is calculated for the cow-calf, beef feedlot, hog weanling, and hog finishing enterprises. They are presented on a nominal basis in Appendix A (Brown 1989), real basis in Appendix B (Brown 1989), and cash basis in Appendix C (Brown 1989), based on the following methodology.

Net Return on Investment for the Cow-Calf Enterprise. The return on investment associated with the cow-calf enterprise is calculated by the procedure outlined in equation (2).

$$ROICC_i = \frac{INCC_i - PCC_i + CAPPC_i}{BIC_i} \qquad (2)$$

where

$ROICC_i$ = residual return on investment per cow in year i

$INCC_i$ = income per cow in year i assuming a calf crop of 90 percent and the sale of a 500-pound calf

PCC_i = total costs of production in year i excluding a charge for investment per cow

$CAPPC_i$ = capital appreciation of the investment per cow in year i as measured by the investment value at the end of the year less the investment value at the beginning of the year

and

BIC_i = value of the investment per cow at the beginning of year i.

The total costs of production and investment per cow figures for 1987 are from the Saskatchewan Agriculture (*Farm Business Management Data Manual*). They include grazing, winter feed, bedding, veterinary care and medicine, breeding, machinery, buildings and handling facilities, cow death loss, interest on operating capital, cow replacement, trucking and marketing, and a labor allowance. The 1987 cost of production figures are then indexed back over the time period using the animal production cost index for Western Canada. The calculation of investment per cow includes an allowance for owned pasture, buildings and facilities, and the price of the cow. The 1987 cow price is indexed back over the time period using an index constructed by lagging calf prices by one year. The remainder of the investment per cow consisted mostly of pasture and was indexed back using the change in farmland values in Saskatchewan as the index.

Net Return on Investment for the Beef Feedlot Enterprise. The return on investment associated with the beef feedlot enterprise is calculated by the procedure outlined in equation (3).

$$ROIBF_i = \frac{INCBF_i - PCBF_i}{BIBF_i} \tag{3}$$

where

$ROIBF_i$ = residual return on investment per head in year i
$INCBF_i$ = income per head in year i assuming purchase of a 500-pound calf and the sale of an 1,100-pound finished animal
$PCBF_i$ = total costs of production in year i excluding a charge for feeder investment

and

$BIBF_i$ = value of the feeder investment at the beginning of year i.

The income figures reflect the annual net payouts after producer levies resulting from participation in the Saskatchewan Beef Stabilization Plan (Saskatchewan Beef Stabilization Plan). The total costs of production figures for 1987 are from the Saskatchewan Agriculture, *Farm Business Management Data Manual*. They include feed, bedding, veterinary care and medicine, breeding, machinery, buildings and handling facilities, death loss, interest on operating capital, trucking and marketing, and a labor allowance. The 1987 cost of production figures are then indexed back over the time period using the animal production cost index for Western Canada.

Net Return on Investment for the Hog Weanling Enterprise. The return on investment associated with the hog weanling enterprise is calculated by the procedure outlined in equation (4).

$$ROIHW_i = \frac{INCHW_i - PCHW_i + CAPPHW_i}{BIHW_i} \tag{4}$$

where

$ROIHW_i$ = residual return on investment per sow in year i

$INCHW_i$ = income per sow in year i assuming 13 pigs weaned per sow in 1971 steadily increasing to 17.7 in 1987 and selling a 45-pound weanling pig

$PCHW_i$ = total costs of production in year i excluding a charge for investment in buildings, equipment, and breeding stock per sow

$CAPPHW_i$ = capital appreciation of the investment per sow in year i as measured by the investment value at the end of the year less the investment value at the beginning of the year

and

$BIHW_i$ = value of the investment per sow at the beginning of year i.

The total costs of production and investment per sow figure for 1987 are from the Saskatchewan Agriculture, *Farm Business Management Data Manual*. They include feed, sow and boar replacement, hired labor, buildings and equipment repairs and maintenance, utilities and insurance, veterinary care and medicine, marketing and transportation, interest on operating capital, an operator labor and management allowance, and depreciation. The 1987 costs of production figures are then indexed back over the time period using the animal production cost index for Western Canada. The 1987 investment per sow is also indexed back over the time period. The index used in this case is the building replacement cost index for Western Canada.

Net Return on Investment for the Hog Finishing Enterprise. The return on investment associated with the hog finishing enterprise is calculated by the procedure outlined in equation (5).

$$ROIHF_i = \frac{INCHF_i - PCHF_i + CAPPHF_i}{BIHF_i} \qquad (5)$$

where

$ROIHF_i$ = residual return on investment per head in year i

$INCHF_i$ = income per head in year i assuming the purchase of a 45-pound weanling pig and the sale of a 170-pound dressed weight index 100 finished hog

$PCHF_i$ = total costs of production in year i excluding a charge for investment in buildings, equipment, and weanling pig per head

$CAPPHF_i$ = capital appreciation of the investment per head in year i as measured by the investment value at the end of the year less the investment value at the beginning of the year

and

$BIHF_i$ = value of the investment per head at the beginning of year i.

The income figures reflect the annual net payouts after producer levies resulting from participation in the Saskatchewan Hog Assistance and Rehabilitation Plan (SHARP) (Saskatchewan Hog Assistance and Rehabilitation Plan). The total costs of production and investment per head figures for 1987 are from the Saskatchewan Agriculture, *Farm Business Management Data Manual*. The total costs of production figures include feed, purchase of weanling pig, hired labor, building and equipment repair and maintenance, utilities and insurance, veterinary care and medicine, marketing and transportation, interest on operating capital, an operator labor and management allowance, and depreciation. The 1987 costs of production figures are then indexed back over the time period using the animal production cost index for Western Canada. The 1987 investment per head is also indexed back over the time period. The index used in this case is the Western Canada building replacement cost index for the building and equipment and the price for weanling pigs.

Stocks and Bonds

The return on investments associated with the stocks and bond enterprises can be seen in Appendices A and B (Brown 1989). The stocks are represented by the percentage change in the TSE Index 300 plus the average annual dividend rate on the stocks in question. The bonds are government of Canada bonds with maturity dates greater than 10 years.

Results

Enterprise Mean-Standard Deviation Trade-Off

Nominal and Real Net Returns on Investment. The mean nominal and real net return on investment and standard deviation of these returns for each of the 30 enterprises investigated are plotted in Figure 8.10 (Nominal) and Figure 8.11 (Real). Figures 8.10 and 8.11 demonstrate that, if the 30 enterprises investigated were the only ones available, a visual estimate of the risk-efficiency frontier would run between bonds (Bonds), hog finishing

281

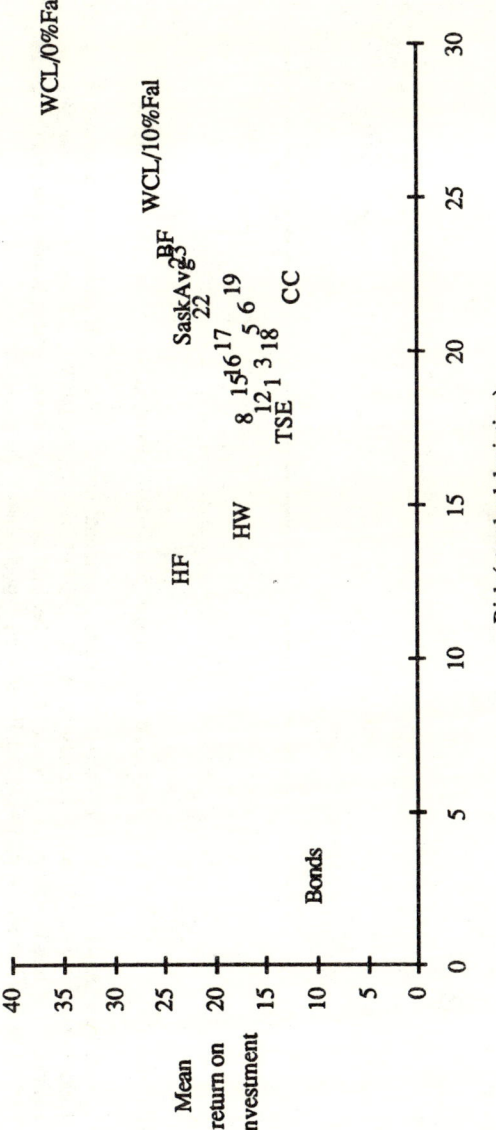

FIGURE 8.10. Mean and Standard Deviation Trade-Off (Nominal Returns on Investment)

Numbers 1-24 are the hypothetical fixed crop rotations presented in Table 8.2. (Note some rotation results are too close for plotting purposes.) SaskAvg = rotation 11; WCL/0%Fal = rotation 10; WCL/10%Fal = rotation 24; HW = Hog weanling; HF = hog finishing; CC = cow calf; BF = beef feedlot; Bonds = government bonds with maturity greater than 10 years; TSE = Toronto Stock Exchange 300 plus an average annual dividend rate.

Source: Brown, 1989.

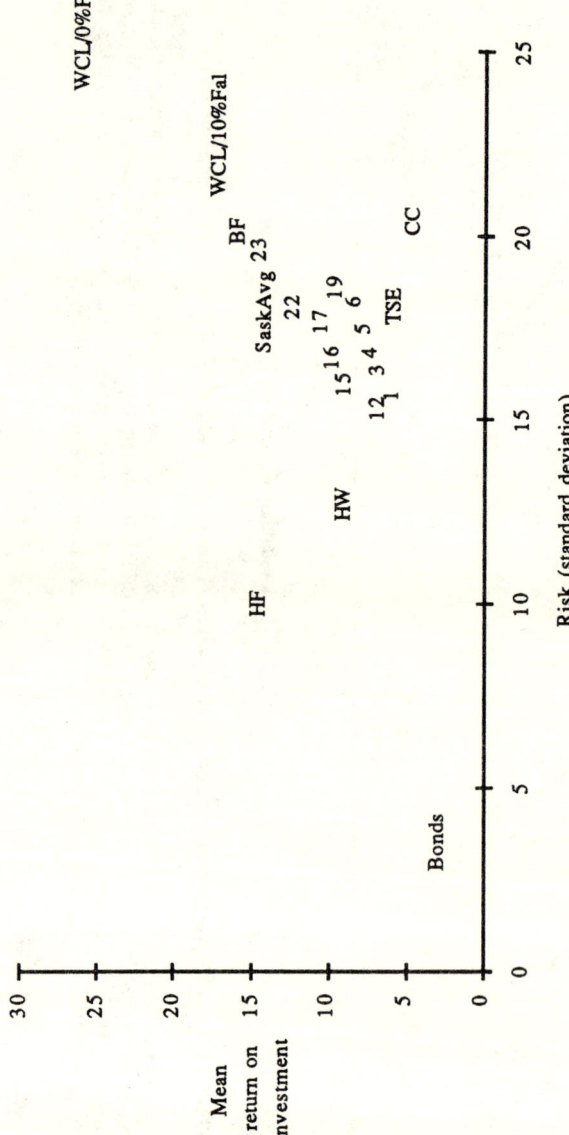

FIGURE 8.11. Mean and Standard Deviation Trade-Off (Real Returns on Investment)

Numbers 1-24 are the hypothetical fixed crop rotations presented in Table 8.2. (Note some rotation results are too close for plotting purposes.) SaskAvg = rotation 11; WCL/0%Fal = rotation 10; WCL/10%Fal = rotation 24; HW = hog weaning; HF = hog finishing; CC = cow calf; BF = beef feedlot; Bonds = government bonds with maturity greater than 10 years; TSE = Toronto Stock Exchange 300 plus an average annual dividend rate.

Source: Brown, 1989.

Farm Enterprise Size and Diversification

(HF) and rotation 10, which consists of 33.3 percent each of wheat, canola, and lentils; 0 percent fallow (WCL/0%Fal). There is little difference between Figures 8.10 and 8.11 other than the real means and standard deviations in Figure 8.11 are smaller.

The three enterprises on the risk-efficiency frontier represent the highest average return on investment for a given amount of risk (standard deviation). All other enterprises have lower returns, higher risk, or both. The most risk-efficient enterprise depends on the risk preferences of the decision maker. Highly risk-averse decision makers would choose bonds. Those with less risk aversion would choose hog finishing. Those decision makers much more willing to take a risk, but are still risk averse, would choose crop rotation 10.

Crop rotation 11 (approximate average Saskatchewan land use: 32 percent fallow, 28 percent wheat on fallow, 29 percent wheat on stubble, 4 percent canola on fallow, 4 percent canola on stubble, and 3 percent lentils on stubble) (SaskAvg) is very close to the frontier and is one of the more risk-efficient crop rotations. It has higher investment returns than most of the other crop rotations and less risk than the crop rotations with large amounts of lentils (23, 24, and 10). It would appear the risk management ability of Saskatchewan farmers is significant. Note also that, other than the addition of lentils to crop rotations, gains from diversification by crop rotation are insignificant as noted by the clustering of deviation.

The location of the other enterprises in the mean-standard deviation trade-off space is also of interest. The TSE does not seem to perform well compared to the other enterprises. However, it may have some advantage in diversification if it has low or negative correlation with the agricultural enterprises. The cow-calf and hog weanling enterprises do not perform as well as their respective finishing enterprises. Part of this is due to the government assistance program payments (Beef Stabilization and SHARP) being allocated to the finishing enterprise. The beef feedlot enterprise (BF) exhibits high returns but also high risk.

Net Cash Returns on Investment. Nominal net cash returns on investment and the standard deviation of these returns have been measured for each of the 30 enterprises and plotted in Figure 8.12. The calculation of net cash returns eliminated noncash costs from the cost side and capital appreciation from the return side. The calculation of returns for Bonds and the TSE were not affected by this adjustment. As can be seen from

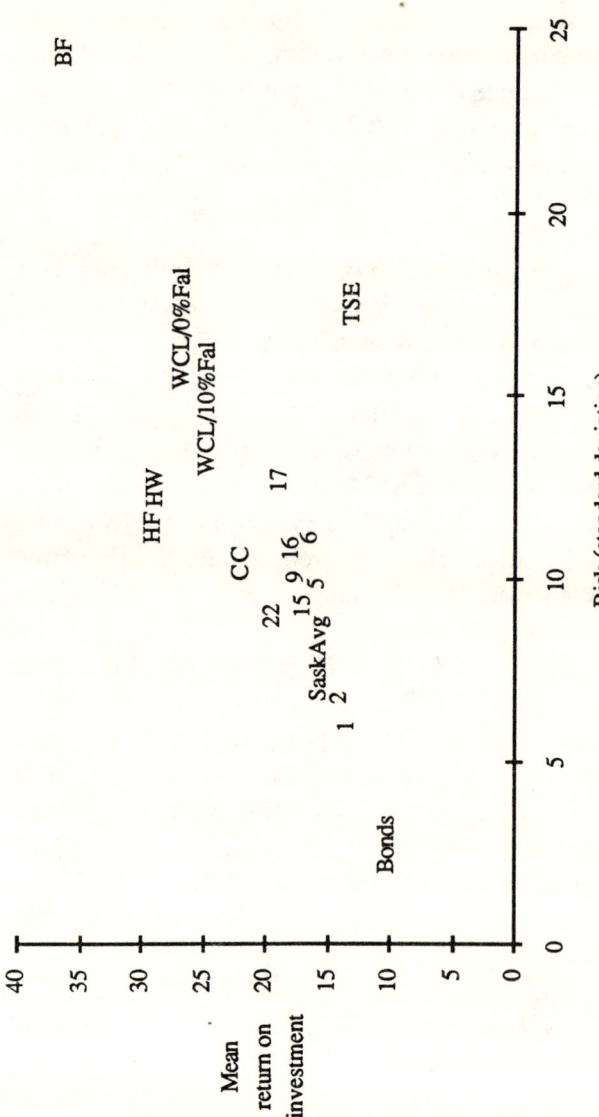

FIGURE 8.12. Mean and Standard Deviation Trade-Off (Nominal Cash Returns on Investment)

Numbers 1-24 are the hypothetical fixed crop rotations presented in Table 8.2. (Note some rotation results are too close for plotting purposes.) SaskAvg = rotation 11; WCL/0%Fal = rotation 10; WCL/10%Fal = rotation 24; HW = hog weanling; HF = hog finishing; CC = cow calf; BF = beef feedlot; Bonds = government bonds with maturity greater than 10 years; TSE = Toronto Stock Exchange 300 plus an average annual dividend rate.

Source: Brown, 1989.

Figure 8.12, this adjustment causes a realignment of the enterprises in the mean-standard deviation trade-off space. The risk-efficiency frontier is more apparent than in Figures 8.10 and 8.11. Bonds and hog finishing (HF) still appear to be on the frontier. The beef feedlot (BF) is also on the frontier, albeit at a higher risk level. However, several of the crop rotations, the cow-calf enterprise (CC), and the hog weanling enterprise (HW) are also very close to the risk-efficiency frontier. The TSE does not seem to be competitive with most agricultural enterprises based on nominal net cash returns.

The adjustment to net cash returns is informative because it presents the picture seen by most farmers. Farmers seldom include noncash costs, such as interest on operating capital, operator labor, and machinery and building capital recovery charges, nor noncash income such as capital appreciation in their calculations. Figure 8.12 demonstrates that, if one is looking only at net cash returns, one sees crop farming as a relatively low-income, low-risk enterprise. In order to increase income, one also must take on considerably more risk either through lentil, hog, or beef production. The remainder of this analysis will investigate which combinations of the 30 enterprises have a chance on increasing income without substantially increasing risk.

Enterprise Correlations

The correlation coefficient is a measure of the mutual relationship between two variables. It ranges between -1.0 and +1.0, with -1.0 meaning perfect negative correlation (the variables move in opposite directions), 0 meaning no correlation between the variables, and +1.0 meaning perfect positive correlation (the variables move in the same direction). The most gains from diversification are realized when enterprises with negative or low positive correlations are combined. The correlations of nominal net returns for the various farm and off-farm enterprises with rotation 11 (SaskAvg) are presented in Table 8.3. Tables 8.4 and 8.5 present the enterprise correlation based on real net returns on investment and nominal net cash returns on investment, respectively. Notice that none of the crop rotations have correlation coefficients of less than +.52 and most are over +.90. Therefore, little gains can be made by crop farmers from diversifying into other crops. The correlation coefficients with the livestock enterprises are lower, all below +.5. Therefore, there are potential gains from crop farms diversifying into livestock enterprises provided their size is large enough. The correlation coefficients for the cow-calf and hog weanling enterprises are very low

TABLE 8.3

Enterprise Correlations Based on Nominal Net Returns on Investment, 1971-1987

Crop rotation	Correlation coefficient	Crop rotation	Correlation coefficient	Crop rotation	Correlation coefficient
1	.99	11	1.00	21	.94
2	.99	12	.98	22	.93
3	.98	13	.99	23	.90
4	.97	14	.98	24	.87
5	.96	15	.97	Cow-calf	.13
6	.94	16	.95	Beef feedlot	.38
7	.93	17	.91	Hog weanling	.40
8	.93	18	.93	Hog finishing	.40
9	.91	19	.91	TSE 300	-.06
10	.85	20	.94	Bonds	-.33

Source: Computed.

TABLE 8.4

Enterprise Correlations Based on Real Net Returns on Investment, 1971-1987

Crop rotation	Correlation coefficient	Crop rotation	Correlation coefficient	Crop rotation	Correlation coefficient
1	.98	11	1.00	21	.93
2	.99	12	.98	22	.91
3	.98	13	.98	23	.88
4	.97	14	.98	24	.85
5	.95	15	.96	Cow-calf	.09
6	.93	16	.93	Beef feedlot	.28
7	.91	17	.89	Hog weanling	.28
8	.91	18	.91	Hog finishing	.22
9	.89	19	.89	TSE 300	-.13
10	.83	20	.93	Bonds	-.87

Source: Computed.

TABLE 8.5

Enterprise Correlations Based on Nominal Net Cash Returns on Investment, 1971-1987

Crop rotation	Correlation coefficient	Crop rotation	Correlation coefficient	Crop rotation	Correlation coefficient
1	.93	11	1.00	21	.69
2	.94	12	.90	22	.66
3	.93	13	.91	23	.63
4	.90	14	.90	24	.61
5	.87	15	.87	Cow-calf	.37
6	.83	16	.83	Beef feedlot	.15
7	.67	17	.79	Hog weanling	.50
8	.69	18	.54	Hog finishing	.06
9	.70	19	.52	TSE 300	-.18
10	.59	20	.68	Bonds	-.73

Source: Computed.

based on nominal and real net returns but increase when based on nominal net cash returns. The beef feedlot and hog finishing enterprises exhibit the exact opposite trend. The correlation coefficients for the TSE and government bonds are all negative with the highest being -.06. Gains from crop farms diversifying into the TSE and/or government bonds may be possible, especially when government bonds exhibit a -.87 correlation coefficient based on real net returns on investment. Bonds and the TSE for the most part exhibit, unfortunately, lower mean net returns than most of the agricultural enterprises measured.

Capital Asset Pricing Model (CAPM) Betas

Beta values were calculated for each of the 30 enterprises and are presented in Table 8.6. The market portfolio was assumed to be the TSE 300 plus an average annual dividend. Beta values for all the agricultural enterprises are extremely low, even for the cow-calf enterprise, when one considers that beta values for many major companies traded on the TSE

TABLE 8.6

Beta Values for Nominal, Real, and Nominal Cash Returns on Investment Enterprise, 1971-1987

Number	Betas: Nominal	Real	Nominal cash
1	-0.005	-0.073	-0.048
2	-0.024	-0.087	-0.059
3	-0.010	-0.076	-0.044
4	0.010	-0.059	-0.024
5	0.036	-0.037	0.002
6	0.058	-0.018	0.023
7	-0.053	-0.082	-0.088
8	-0.078	-0.097	-0.103
9	-0.089	-0.101	-0.103
10	0.125	0.022	0.103
SaskAvg	-0.065	-0.123	-0.071
12	-0.056	-0.100	-0.091
13	-0.067	-0.107	-0.096
14	-0.072	-0.108	-0.094
15	-0.072	-0.105	-0.088
16	-0.068	-0.099	-0.078
17	-0.060	-0.090	-0.064
18	-0.051	-0.129	-0.090
19	0.019	-0.075	0.029
20	-0.052	-0.114	-0.089
21	-0.015	-0.083	-0.055
22	0.021	-0.054	-0.021
23	0.061	-0.021	0.016
24	0.103	0.014	0.055
Cow-calf	0.451	0.449	0.214
Beef feedlot	-0.178	-0.182	-0.157
Hog weanling	-0.006	0.001	-0.030
Hog finishing	-0.022	-0.018	-0.028
TSE 300	1.000	1.000	1.000
Bonds	-0.011	0.033	-0.014

Source: Computed.

Farm Enterprise Size and Diversification 289

usually range between .5 and 1.3 (Schall, Haley, and Schachter). High beta values for agricultural enterprises mean they add risk to a well-diversified portfolio and, thereby, the investor will need a higher rate of return to justify his investment into it. If the rate of return is not there, then one can argue that investors in agricultural enterprises are either not being compensated for the risks they are taking or agricultural asset values are overpriced. A low beta value for an agricultural enterprise means its addition to a well-diversified portfolio lowers the risk, and investors are either willing to accept a lower rate of return on these enterprises or will bid the asset price up. The results presented in Table 8.6 point toward gains from diversification between agricultural enterprises and the TSE. However, Figures 8.10, 8.11, and 8.12 indicate the gainer may well be the TSE investor rather than the farmer.

Portfolio Mean-Standard Deviation Trade-Off

The mean-standard deviation trade-off of a number of combinations (portfolios) of the 30 enterprises are plotted in Figures 8.13 (Nominal), 8.14 (Real), and 8.15 (Nominal Cash). The objective was to see what combinations of rotation 11, which represent the average land use in Saskatchewan (SaskAvg) and the other enterprises, would both increase net returns and reduce risk for the "typical" crop farm on the Canadian Prairies. The size of the agricultural enterprises in the chosen portfolios are assumed to be those that reflect the economies of size investigated earlier. For example, the base portfolio may be a $1\text{-}1/2$ section crop farm following rotation 11 (SaskAvg); portfolio 2 may be the same $1\text{-}1/2$ section crop farm with 20 percent of the investment in a beef feedlot or hog finishing enterprise. The livestock operations may or may not be located on the same farm but should be as large or larger than their respective threshold size. Other portfolios include the base plus an investment in government bonds and/or the TSE 300.

The risk-efficiency frontier would appear to be similar in Figures 8.13 and 8.14 and consist of bonds; hog finishing (HF); and the wheat, canola, lentils, and 0 percent fallow rotation (WCL/0%Fal or rotation 10). The risk-efficiency frontier also would appear to include a portfolio labeled SAWCLHF which is a 33.3 percent equal combination of the SaskAvg or rotation 11, the wheat, canola, lentils, and 0 percent fallow rotation (WCL/0%Fal or rotation 10), and hog finishing (HF).

The only portfolios that both increased net returns and reduced risk compared to the SaskAvg rotation were the above-described SAWCLHF

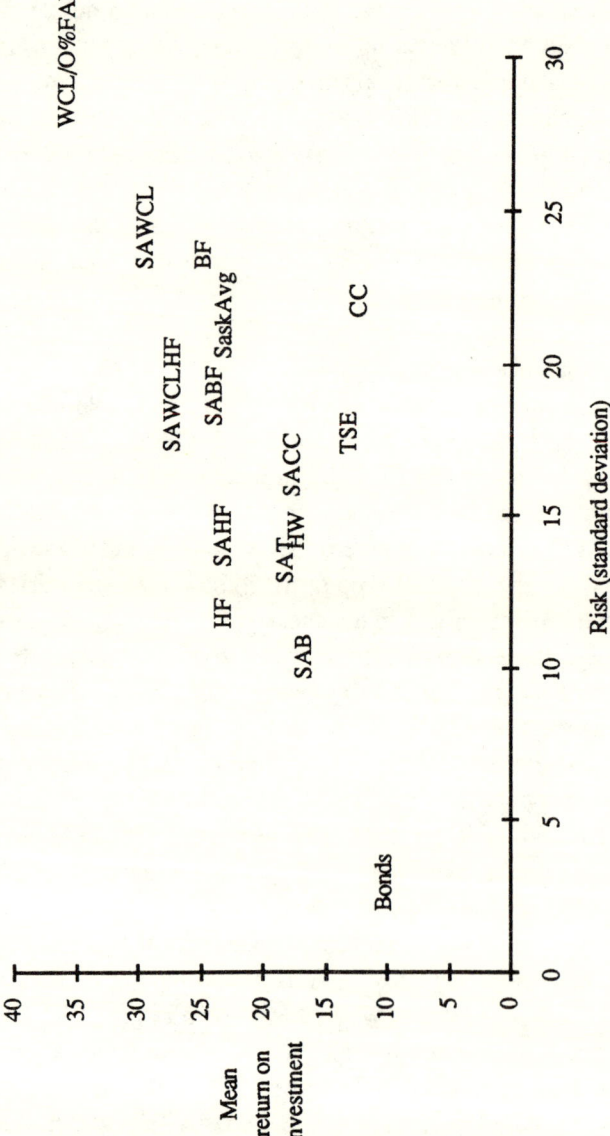

FIGURE 8.13. Mean and Standard Deviation Trade-Off (Nominal Returns on Investment)

SaskAvg = rotation 11; WCL/0%FAL = rotation 11, 33.3% rotation 10, and 33.3% hog finishing; SAHF = 50% rotation 11 and 50% hog finishing; SABF = 50% rotation 11 and 50% beef feedlot; SACC = 50% rotation 11 and 50% cow calf; SAT = 50% rotation 11 and 50% TSE; SAB = 50% rotation 11 and 50% bonds; HW = hog weanling; HF = hog finishing; CC = cow calf; BF = beef feedlot; Bonds = government bonds with maturity greater than 10 years; TSE = Toronto Stock Exchange 300 plus an average annual dividend rate. Source: Brown, 1989.

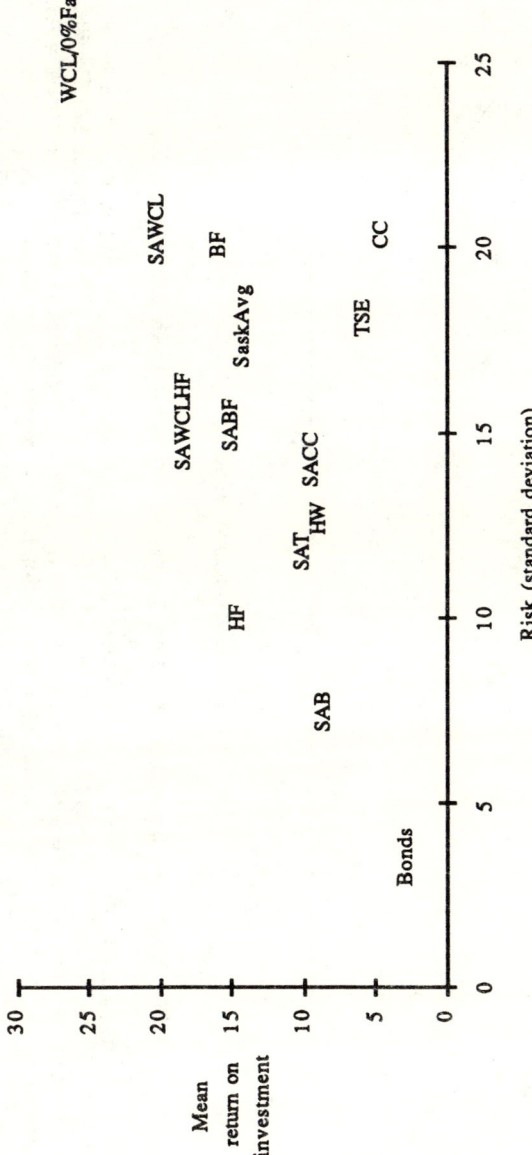

FIGURE 8.14. Mean and Standard Deviation Trade-Off (Real Returns on Investment)

SaskAvg = rotation 11; WCL/0%Fal = rotation 10; SAWCLHF = 33.3% rotation 11, 33.3% rotation 10, and 33.3% hog finishing; SABF = 50% rotation 11 and 50% beef feedlot; SACC = 50% rotation 11 and 50% cow calf; SAT = 50% rotation 11 and 50% TSE; SAB = 50% rotation 11 and 50% bonds; HW = hog weanling; HF = hog finishing; CC = cow calf; BF = beef feedlot; Bonds = government bonds with maturity greater than 10 years; TSE = Toronto Stock Exchange 300 plus an average annual dividend rate.

Source: Brown, 1989.

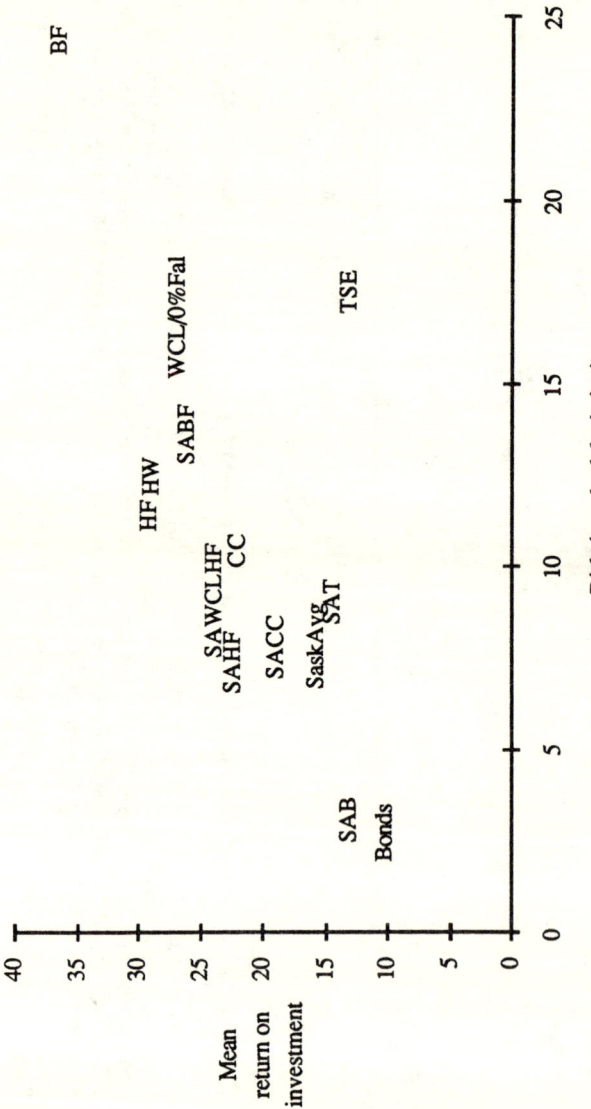

FIGURE 8.15. Mean and Standard Deviation Trade-Off (Nominal Cash Returns on Investment)

SaskAvg = rotation 11; WCL/0%Fal = rotation 10; SAWCLHF = 33.3% rotation 11, 33.3% rotation 10, and 33.3% hog finishing; SAHF = 50% rotation 11 and 50% hog finishing; SABF = 50% rotation 11 and 50% beef feedlot; SACC = 50% rotation 11 and 50% cow calf; SAT = 50% rotation 11 and 50% TSE; SAB = 50% rotation 11 and 50% bonds; HW = hog weanling; HF = hog finishing; CC = cow calf; BF = beef feedlot; Bonds = government bonds with maturity greater than 10 years; TSE = Toronto Stock Exchange 300 plus an average annual dividend rate.   Source: Brown, 1989.

Farm Enterprise Size and Diversification

portfolio, a portfolio consisting of a 50-50 split between SaskAvg and BF labeled SABF, a portfolio consisting of a 50-50 split between SaskAvg and HF labeled SAHF, and the HF enterprise. Undoubtedly, there are numerous other portfolios that increase net returns and reduce risk compared to SaskAvg, but they will most likely consist of combinations of HF and/or WCL/0%Fal and/or BF and/or SaskAvg. Portfolios with larger than 50 percent SaskAvg also increase net returns and decrease risk compared to SaskAvg but are too close to SaskAvg for plotting purposes.

There are several other portfolios in Figures 8.13 and 8.14 that reduce the risk associated with the SaskAvg but also reduce the net returns. The portfolios shown in the figures consist of 50-50 combinations of SaskAvg and the TSE (SAT), bonds (SAB), cow-calf (SACC), and hog weanling (not plotted because very close to SAT). Undoubtedly, there are numerous other portfolios that reduce risk but also reduce net return compared to SaskAvg. However, they will most likely consist of combinations of bonds, and/or the TSE, and/or cow-calf, and/or hog weanling, and/or SaskAvg. Portfolios with larger than 50 percent SaskAvg reduce risk but also reduce net return compared to SaskAvg; however, they are too close to SaskAvg for plotting purposes.

Figure 8.15 plots the same portfolios as outlined in Figures 8.13 and 8.14; only the net returns are measured as nominal net cash returns on investment. The risk-efficiency frontier would appear to consist of a number of portfolios, all defined above and listed here: Bonds, SAB, SAHF, SAWCLHF, HF, HW, and BF. No combinations of SaskAvg and the other enterprises both increase net cash returns and reduce risk as compared to SaskAvg. The only combinations of SaskAvg and the other enterprises to reduce risk and also reduce net cash returns include portfolios consisting of bonds and SaskAvg.

At least one other important point arises from the examination of Figures 8.13, 8.14, and 8.15. Crop farmers can reduce risk (not when calculated on a net cash return basis), and also reduce net return, when they include the TSE in their investment portfolios. However, current investors in the TSE can both increase net returns and reduce risk when investing in most agricultural enterprises. This may have major implications for future investment in agriculture.

Summary, Conclusions, and Policy Implications

Threshold Enterprise

The threshold enterprise size for grain and grain-oilseed, beef, and hog enterprises on the Canadian Prairies was calculated from a number of studies. The resulting threshold size for the grain and grain-oilseed enterprise ranged between 546 and 697 cultivated acres in the black soil zone and between 829 and 1,111 cultivated acres in the brown soil zone. The threshold size of beef feedlots in the CWB region, according to Statistics Canada data, appeared to be in the neighborhood of 1,500+ steers. U. S. studies have found significantly larger threshold sizes in the neighborhood of 10,000 head. It would appear the U. S. data, with respect to cattle feeding, are more credible because Statistics Canada does not break down herd sizes larger than 1,500 head. The threshold size for the beef cow herd in Canada as a whole appears to be in the 37-99 cow range; however, the data are rather erratic and further economies are evident in the 200-499 cow range. To the extent that U. S. data and conditions can be translated to Western Canada, a threshold size in 1983 for a hog farrowing operation and a hog finishing operation was around 3,000 head sold per year in each enterprise. Given a weaning rate of 15 pigs per sow per year, the hog farrowing operation would equal a 200-sow enterprise.

Net returns on investment from 1971 to 1987 were calculated for 24 hypothetical fixed crop rotations, the cow-calf, beef feedlot, hog weanling, and hog finishing enterprises as well as the TSE 300 and government bonds. The net return on investment for the agricultural enterprises was calculated by subtracting total production costs from gross returns plus or minus any change in the value of the capital invested that occurred in the year, all divided by the beginning year value of the investment. The net return on investment associated with the stocks was represented by the percentage change in the TSE Index 300 plus the average annual dividend rate on the stocks in question. The net return on investment associated with bonds was represented by the government of Canada bonds with maturity dates greater than 10 years.

Computer Analysis

The mean and standard deviation of net returns on investment for the 24 hypothetical crop rotations, four livestock enterprises, the TSE 300, and government bonds were plotted in both nominal and real terms. The risk-efficiency frontier contains bonds, hog finishing, and rotation 10 (33.3 percent each of wheat, canola, and lentils; 0 percent fallow). The TSE does not seem to perform well compared to the other enterprises.

Nominal net cash return on investment and the standard deviation of these returns for each of the 30 enterprises were measured and plotted. This adjustment causes a realignment of the enterprises in the mean-standard deviation trade-off space. In terms of the risk-efficiency frontier, bonds and hog finishing still appear. Beef feedlot is also on the frontier, albeit at a higher risk level. However, several of the crop rotations, the cow-calf enterprise, and the hog weanling enterprise are also very close to the risk-efficiency frontier. The TSE does not seem to be competitive with most agricultural enterprises based on net cash returns.

None of the crop rotations have correlation coefficients with respect to crop rotation 11 (SaskAvg) of less than +.52 and most are over +.90. Therefore, little gains can be made by crop farmers from diversifying into other crops. The correlation coefficients with the livestock enterprises are lower, all below +.5. Therefore, there are potential gains from crop farms diversifying into livestock enterprises provided their size is large enough. The correlation coefficients for the cow-calf and hog weanling enterprises are very low based on nominal and real net returns but increase when based on net cash returns. The beef feedlot and hog finishing enterprises exhibit the exact opposite trend. The correlation coefficients for the TSE and government bonds are all negative with the highest being -.06. Gains from crop farms diversifying into the TSE and/or government bonds may be possible, especially when government bonds exhibit a -.87 correlation coefficient based on real net returns on investment.

Beta values were calculated for each of the 30 enterprises assuming the market portfolio to be the TSE 300. Beta values for all the agricultural enterprises are extremely low. A low beta value for an agricultural enterprise means its addition to a well-diversified portfolio lowers the risk, and investors are either willing to accept a lower rate of return on these enterprises or will bid the asset price up.

The mean-standard deviation trade-off of a number of combinations (portfolios) of the 30 enterprises was determined. The objective was to see what combinations of the Saskatchewan average land use (rotation 11, SaskAvg) and the other enterprises would both increase net returns and reduce risk for the typical crop farm on the Canadian Prairies. The size of the agricultural enterprises in the chosen portfolios is assumed to be one that reflects the economies of size investigated earlier. The livestock operations may or may not be located on the same farm. Other portfolios include the base plus an investment in government bonds and/or the TSE 300. The only portfolios that both increased net returns and reduced risk compared to the SaskAvg rotation were a 33.3 percent equal combination of the SaskAvg crop rotation; the wheat, canola, lentils, and 0 percent fallow crop rotation; and the hog finishing enterprise labeled SAWCLHF, a portfolio consisting of a 50-50 split between SaskAvg and BF labeled SABF, a portfolio consisting of a 50-50 split between SaskAvg and HF labeled SAHF, and the HF enterprise.

Economies of Size

It is important that threshold enterprise size be met when crop farmers on the Canadian Prairies are thinking about diversification. This, undoubtedly, will vary from farm to farm due to varying amounts of management ability. The size outlined may well require a larger investment than most farmers are willing or able to raise. Therefore, groups of crop farmers and other investors will have to pool their resources to form joint ventures or cooperatives that can take advantage of the economies of size and provide real diversification for crop farmers on the Canadian Prairies.

Gains and Losses from Diversification

It is clear from the data presented that there is not much to be gained from diversifying within crops on the Canadian Prairies. All of the rotations other than those with large amounts of lentils offered very similar amounts of net returns and risk. Significant addition of lentils, or other specialty crops, raises net returns but also raises risk higher than most Prairie farmers are willing to take. It is significant that the SaskAvg rotation is one of the most risk-efficient rotations tested. It would seem the risk management ability of crop farmers on the Canadian Prairies is substantial.

Diversification into hog finishing and cattle feeding not only increases the net returns of crop farmers but also reduces their risk. However, the economies of size for these enterprises will require most crop farmers to look for joint ventures or cooperatives as a means of entry.

The gains from crop farmers diversifying into bonds or the TSE come only from reduced risk (other than on a net cash return basis) rather than on increased net return. The risk reduction is substantial, however, especially with bonds, and some crop farmers may want to give this special consideration. The real gains from diversification come to the current investor in the TSE. The data demonstrate that this investor can gain increased total net returns by investing in all of the agricultural enterprises investigated, except the cow-calf enterprise. The current TSE investor can also reduce risk by investing in the two hog enterprise types analyzed.

Policy Implications

The major adjustment both policy formulators and farmers have to make when dealing with facilitating diversification of Prairie crop farms is to view the farm as a portfolio of investments. Currently, the investment portfolio of most Prairie crop farms is very specialized. Investment into other crops does not seem to decrease risk. Livestock enterprises offer some gains from diversification, but these must be of threshold size. Therefore, ways must be found to establish these livestock units and encourage crop farmers to participate. In addition, crop farmers need to be encouraged to diversify into bonds and perhaps even stocks rather than continuing to invest in crop agriculture.

The management skills and education of crop farmers will have to be improved in order for them to handle investments into a range of enterprises, including bonds and securities. Finally, some form of capital will have to be made available to farmers who wish to diversify their investment portfolios, whether it be on or off the farm.

Another policy implication that needs to be addressed deals with the apparent gains from diversification that can be made from investing in agriculture by current investors in bonds and the TSE. If the data presented here are correct and investment in agriculture were facilitated by the establishment of equity investment firms or other investment vehicles, there would be a large influx of equity capital into agriculture. This result could have major social and political implications.

Notes

[1] The dynamics of farm growth also have a great deal to do with the eventual success or failure of a farm business. The year chosen to start or substantially expand a farm enterprise can dictate its degree of success.

[2] Yields and prices from Brown and Forsberg.

[3] For more detail on data construction, see Brown and Forsberg 1987.

[4] Crop production costs vary more from soil zone to soil zone than they do from province to province. The 1984 dark brown soil zone variable cash costs used in the paper for wheat on fallow, wheat on stubble, and summer

fallow totaled $93.87 per acre. Similar costs taken from "Costs and Returns for Crop Production in Alberta" (Alberta Agriculture) totaled $104.14 per acre for Stettler-Coronation, which is in the dark brown soil zone and $74.24 per acre for Oyen-Hanna which is in the brown soil zone. There is an approximate 16 percent difference between the two dark brown soil zone costs even though they are in different provinces. There is an approximate 26 percent difference between the Oyen-Hanna costs and those used in the paper and an approximate 47 percent difference between the two Alberta numbers. In addition, the same costs for Barrhead, which is in the black soil zone, exceed those of Oyen-Hanna, by approximately 52 percent (Alberta Agriculture). This magnitude of within-province discrepancy is prevalent in all the provinces because there is more similarity within soil zones than within provinces. The dark brown soil zone was chosen for the study because its cost structure is between that of the brown and black soils.

References

Alberta Agriculture. *Costs and Returns for Crop Production in Alberta.* Agdex 821-42. Edmonton, Alberta: Production and Resource Economics Branch, Alberta Agriculture, 1985.

Anderson, J. R., and R. A. Powell. "Economics of Size in Australian Farming." *Australian Journal of Agricultural Economics* 17(1973):1-16.

Anderson, J. R., J. L. Dillon, and B. Hardaker. *Agricultural Decision Analysis.* Ames, Iowa: Iowa State University Press, 1977.

Barry, P. J. "Capital Asset Pricing and Farm Real Estate." *American Journal of Agricultural Economics* 62(1980):549-553.

Barry, P. J., ed. *Risk Management in Agriculture.* Ames, Iowa: Iowa State University Press, 1984.

Brinkman, G. L., and T. K. Warley. "Structural Change in Canadian Agriculture: A Perspective." Agriculture Canada. Working Paper. Ottawa, Ontario, 1983.

Brown, W. J. "A Risk Efficiency Analysis of Crop Rotations in Saskatchewan." *Canadian Journal of Agricultural Economics* 35(1987):333-355.

_____. "Assumption Behind the Calculation of the On-Farm Economics of Irrigation." Department of Agricultural Economics, Working Paper. University of Saskatchewan, Saskatoon, 1988.

_____. *A Review of the Economics of Farm Enterprise Size and Implications for Farm Diversification.* Discussion Paper No. 360. Ottawa, Ontario: Economic Council of Canada, 1989.

Brown, W. J., and B. H. Forsberg. *A Risk Analysis of Crop Rotations in Saskatchewan.* Technical Bulletin 87.011. Saskatoon, Saskatchewan: Department of Agricultural Economics, University of Saskatchewan, 1987.

Buttel, F. H. "Whither the Family Farm? Toward a Sociological Perspective on Independent Commodity Production in U. S. Agriculture." *Cornell Journal of Social Relations* 15(1980).

Carter, C., and A. Schmitz. *The Economics and Location of Cattle Feeding: Alberta Versus Saskatchewan.* Regina, Saskatchewan: Rural Development and International Affairs Branch, Agriculture Canada, 1983.

Davey, B. H., Z. A. Hassan, and W. F. Lu. *Farm and Off-Farm Income of Farm Families in Canada.* Economics Branch, Publication 74/17. Ottawa, Ontario: Agriculture Canada, 1974.

Doll, J. P., and F. Orazem. *Production Economics: Theory with Applications.* 2d ed. Toronto, Ontario: John Wiley and Sons, 1984.

Ehrensaft, P. "The Industrial Organization of Modern Agriculture." *Canadian Journal of Agricultural Economics* 30(1983):122-133.

_____. *Structure and Performance in the Canadian Beef Sector.* Ottawa, Ontario: Production Analysis Division, Policy Branch, Agriculture Canada, 1987.

Ehrensaft, P., and R. D. Bollman. "Large Farms: The Leading Edge of Structural Change." *Canadian Journal of Agricultural Economics* 33 (Annual Meeting Proceedings 1986):145-160.

Fleming, M. S., and I. H. Uhm. "Economies of Size in Grain Farming in Saskatchewan and the Potential Impact of Rail Rationalization Proposals." *Canadian Journal of Agricultural Economics* 30(1982):1-20.

Hall, B. F., and E. P. LeVeen. "Farm Size and Economic Efficiency: The Case of California." *American Journal of Agricultural Economics* 60(1978):589-599.

Heady, E. O. *Economics of Agricultural Production and Resource Use.* New York: Prentice-Hall, Inc., 1952.

Jensen, K. "An Economic View of the Debate on Farm Size in Saskatchewan." *Canadian Journal of Agricultural Economics* 33(1984):187-200.

Marshall, R. G. "The Size and Structure of The Livestock Industry in Canada 1980." *Canadian Journal of Agricultural Economics* 19(1969):90-100.

Miller, T. A. "Economies of Size and Other Growth Incentives." U. S. Department of Agriculture, Economics and Statistics Service, Washington, D. C., 1979. Mimeographed.

Miller, T. A., G. E. Rodewald, and R. G. McElroy. *Economics of Size in U. S. Field Crop Farming*. Washington, D. C.: U. S. Department of Agriculture, Economics Statistics Service. No. AER-472, 1981.

Pope, R. D., and R. Prescott. "Diversification in Relation to Farm Size and Other Socioeconomic Characteristics." *American Journal of Agricultural Economics* 62(1980):554-559.

Raup, P. "Economies and Diseconomies of Large-Scale Agriculture." *American Journal of Agricultural Economics* 51(1969):1274-1282.

Saskatchewan Agriculture, Economics Branch. *Agricultural Statistics*. Regina, Saskatchewan: Saskatchewan Agriculture, various years.

_____. *Farm Business Management Data Manual*. Update Material: Beef-110.11 (Feb. 1988) & Swine-100.23 (Feb. 1988). Regina, Saskatchewan: Economics Branch, Saskatchewan Agriculture, 1988.

_____. *Specialty Crop Report*. Regina, Saskatchewan: Economics Branch, Saskatchewan Agriculture, various years.

Saskatchewan Beef Stabilization Plan. "Annual Statistics." Economics Branch, Saskatchewan Agriculture. Working Paper. Regina, 1988.

Saskatchewan Hog Assistance and Rehabilitation Program. "Annual Statistics." Economics Branch, Saskatchewan Agriculture. Working Paper. Regina, 1988.

Schall, L. D., C. W. Haley, and B. Schachter. *Introduction to Financial Management*. 2d. ed. Toronto, Ontario: McGraw-Hill Ryerson, Ltd., 1987.

Schoney, R. A. *Costs of Producing Crops and Forward Planning Manual for Saskatchewan*. Bulletin FLB 85-04. Saskatoon, Saskatchewan: Department of Agricultural Economics and Farmlab, University of Saskatchewan, 1985.

_____. *Costs of Producing Crops and Forward Planning Manual for Saskatchewan*. Bulletin FLB 86-04. Saskatoon, Saskatchewan: Department of Agricultural Economics and Farmlab, University of Saskatchewan, 1986.

_____. *Costs of Producing Crops and Forward Planning Manual for Saskatchewan*. Bulletin FLB 87-04. Saskatoon, Saskatchewan: Department of Agricultural Economics and Farmlab, University of Saskatchewan, 1987.

Slinkard, A. E., and B. N. Drew. *Lentil Production in Western Canada*. Publication 413 (Revised). Saskatoon, Saskatchewan: Division of Extension and Community Relations, University of Saskatchewan, January, 1986.

Stanton, B. F. "Perspective on Farm Size." *American Journal of Agricultural Economics* 60(1978):727-737.

Statistics Canada. *Census of Canada and Other Selected Farm Data*. Cat. Nos. 96-107, 96-108, 96-109, 96-110, and 96-111. Ottawa, Ontario: Statistics Canada, various years.

Toronto Stock Exchange. *Toronto Stock Exchange Review*. Toronto, Ontario: Toronto Stock Exchange, various years.

Trant, M. "Farm Producers, Past and Present." *Canadian Journal of Agricultural Economics* 33(1986):122-144.

Turvey, C. G., and H. C. Driver. "Systematic and Nonsystematic Risks in Agriculture." *Canadian Journal of Agricultural Economics* 35(1987):387-401.

Ulrich, A., and Furtan, W. H. *An Economic Evaluation of Producing HY320 Wheat on the Prairies*. Research Bulletin. Saskatoon, Saskatchewan: Department of Agricultural Economics, University of Saskatchewan, 1984.

Van Arsdall, R. N., and K. E. Nelson. *Economies of Size in Hog Production*. Technical Bulletin No. 1712. Washington, D. C.: Economic Research Service, U. S. Department of Agriculture, 1985.

White, T., and G. Irwin. "Farm Size and Specialization." In *Size, Structure and Future of Farms*, edited by G. Ball and E. Heady. Ames, Iowa: Iowa State University Press, 1972.

Wilson, P. N., and V. R. Eidman. "Dominant Enterprise Size in the Swine Production Industry." *American Journal of Agricultural Economics* 67(1985):279-288.

Zentner, R. P., D. D. Greene, T. L. Hickenbotham, and V. R. Eidman. "Ordinary and Generalized Stochastic Dominance: A Primer." Department of Agricultural and Applied Economics. Working Paper. University of Minnesota, St. Paul, 1981.

9

The Effect of U. S. Farm Programs on Diversification

Richard E. Just and Andrew Schmitz

In the late 1920s, President Calvin Coolidge threw down a gauntlet for research on the effect of U. S. farm programs on diversification in agriculture. In vetoing the McNary-Haugen Bill in both 1927 and 1928, he claimed that the introduction of farm-support programs would result in overspecialization of U. S. agriculture. While the philosophy underlying the McNary-Haugen Bill provided the foundation and precedent for farm programs that followed, economists have failed to pick up that gauntlet and investigate in a substantive way the effects of U. S. farm programs on diversification versus specialization. This is in sharp contrast to the research which has taken place on the effect of farm programs on *diversity* in agriculture. Diversity is a term commonly used to refer to the size distribution of farms and is often used synonymously with the term *structure* (U. S. Congress). These studies (e.g., Ball and Heady; Bonnen; Dahl; Krause and Kyle; Madden; Moore) generally show that U. S. farm programs have caused the size of farms to increase and the number of farms to decline and that they have failed to channel benefits to small farmers who need it most.

Diversification versus specialization is also an important element in understanding the structure of agriculture and how it is affected by farm programs. In broad terms, diversification refers to the distribution of resources among crop enterprises. In particular, it is an important consideration in measuring the effects of agricultural policies and in examining the likely impacts of eliminating those programs. This is the context in which diversification is considered here. Both the U. S.-Canada Free Trade Agreement and potential agreements arising from the current Uruguay Round of talks on the General Agreement on Tariffs and Trade (GATT) threaten to reduce the role of farm programs in the United States.

Will the loss in government support for certain heavily subsidized crops lead to a return to greater diversification of farms? This is an important issue because overdiversification is generally associated with a reduction in economic output.

The purpose of this chapter is to examine the effects of farm programs in the United States on diversification in agriculture. It begins by considering the concept of diversification and whether empirical measures of diversification are sufficiently related to other widely researched phenomena to draw ready conclusions. Then some historical data reflecting commodity concentration are examined to see if any clear pattern of effects emerged at the time farm programs were instituted. Since this evidence is inconclusive, the chapter then turns to its own conceptual and empirical analysis. This analysis is somewhat simple and stylized because the investigation of effects of farm programs on diversification is at such a primitive stage.

Definition of Diversification and the Relationship to Farm Size

Given the importance of diversification, it is surprising that so little attention has been paid to how it is affected by farm programs. For example, in an entire volume heralded as "the most comprehensive treatment of farm structure to appear in recent years," the effects of policy on structure in agriculture cover such diverse considerations as credit, labor, technological change, water, taxes, environmental regulations, transportation, input markets, retailing, cooperatives, marketing orders, and rural communities (U. S. Department of Agriculture, p. iii). Yet, the effect of policies on diversification receives only one paragraph of passing comments.

A workable empirical definition of diversification has not been generally agreed upon. The literature uses Herfindahl indexes and entropy coefficients as well as simple measures such as the number of farm enterprises and the maximum proportion of activities devoted to a single enterprise. For example, where A_i is the acreage devoted to crop i and $\Sigma_i A_i$ is total farm acreage, let $p_i = A_i/\Sigma_i A_i$. Then the Herfindahl index is $\Sigma_i p_i^2$, the entropy measure is $\Sigma_i p_i \log(1/p_i)$, the number of enterprises is $\Sigma_i I(p_i)$ where I is a zero-one indicator, and the maximum proportion is $\max(p_1, p_2, \ldots)$. Because of this confusion, Kerr opts to use what he assumes to be the objective of diversification as an empirical measure. That is, he uses the variability of total gross revenue as a measure of diversification. Alternatively, one can use a general multidimensional measure of diversification which includes the entire vector of crop shares $(p_1, p_2, \ldots)$.

Some have suggested a relationship may hold between diversification and farm size. If so, then the large number of studies on farm size effects of policy would also apply to diversification. For example, some have argued that large-scale economies in a single enterprise should cause larger farms to be more specialized. Others have argued that larger farms face more risk and thus have a greater need to diversify. The applicability of such relationships has been investigated by White and Irwin and by Pope and Prescott. White and Irwin use aggregate census data and compare diversification across farm classes. They find that larger farms are more specialized but that the data are not conclusive for all farm types and classes. Pope and Prescott examine data obtained from financial institutions in the San Joaquin Valley of California and find, contrary to White and Irwin, that larger farms are more diversified.

At best, the available results in the literature taken as a whole are inconclusive about the relationship of farm size and diversification. Furthermore, if there is a definite relationship, it clearly appears to be imperfect. For example, while Pope and Prescott find farm size to be one of the two most significant variables explaining diversification, other significant variables are also identified (e.g., experience, net worth, organization structure, and location). Furthermore, all of the variables together only explain 20 to 30 percent of the variation in observed diversification. Thus, a more direct examination of the effect of farm programs on diversification is needed (as opposed to simply using the available literature on the effects of farm programs on farm size to draw implications for diversification effects).

Historical Examination of Diversification

A fitting place to begin the investigation of effects of farm programs on diversification is to examine the data on diversification historically to see whether significant changes took place with the institution of farm programs. To do this, the Herfindahl indexes on crop concentration constructed by Gardner (1987) are used. He computed Herfindahl indexes on commodity concentration in the United States using acreage data across states for the eight largest states in each crop. These indexes were used in another context to explain rates of protection across commodities. The indexes are reported in Table 9.1. It should be noted that these indexes are not computed following the formula given above for the Herfindahl measurement of diversification across commodities but are computed by crop. Nevertheless, they are reflective of changes toward or away from diversification at state aggregate levels.

Farm programs in the United States were instituted with the Agricultural Marketing Act of 1929 with the basic enabling legislation for modern farm programs following the Agricultural Adjustment Act of 1933. Comparing data in Table 9.1 from before and after these changes, however, indicates no clear and consistent changes took place at the time that farm programs were implemented. Some controlled commodities tended to increase in concentration while others tended to decrease. For example, sugar beet concentration declines while sugarcane concentration increases. Concentration increases for peanuts and declines for rice. Only mild changes occur for wheat and tobacco. To the extent that major changes occur for corn and cotton, they occur after the 1950s and are more likely explained by major technological changes that occurred. The same is true for uncontrolled crops. Barley concentration changes little across the 70-year period, and the changes for chicken and dairy cows occur in the 1950s and 1960s and seem to be largely explained by technological innovations. Thus, casual examination of the historical data does not reveal any clear pattern of effects of farm programs on diversification at an aggregate level.

The Conceptual Framework

To address the conceptual problem of how government programs affect diversification, this section examines a simple stylized case with two crops. Diversification is considered at an aggregate level such as would be appropriate for policy issues of production efficiency (ignoring heterogeneity issues). Suppose crop A has a standard government program encompassing a target price, a support price, and an acreage set-aside requirement. Suppose crop B has no government program or an ineffective program. For example, crop A might be corn and crop B might be soybeans (the price support for soybeans has been ineffective except for a

TABLE 9.1

Herfindahl Indexes of Acreage Concentration Across States, Selected Commodities and Years, 1910-1980

Commodity	1910	1920	1930	1940	1950	1960	1970	1980
Barley	.129	.094	.114	.084	.096	.110	.114	.120
Corn	.046	.037	.045	.037	.052	.066	.078	.085
Cotton	.160	.170	.193	.167	.192	.204	.222	.310
Oats	.056	.056	.075	.066	.078	.071	.101	.095
Peanuts	.152	.171	.187	.226	.212	.192	.202	.204
Rice	.413	.325	.365	.320	.263	.242	.242	.246
Sorghum	.288	.289	.324	.280	.347	.297	.290	.229
Soybeans	.598	.193	.088	.096	.117	.097	.083	.061
Sugar beets	.196	.150	.149	.099	.104	.108	.089	.130
Sugarcane	.496	.421	1.000	.845	.796	.735	.519	.458
Tobacco	.196	.202	.213	.234	.221	.223	.236	.229
Wheat	.074	.059	.085	.072	.082	.081	.092	.095
Chicken	.030	.027	.026	.024	.025	.025	.039	.039
Hogs, pigs	.044	.041	.062	.047	.070	.084	.096	.096
Dairy cows	.027	.028	.028	.028	.027	.037	.048	.055
Other cows	.057	.051	.073	.062	.054	.041	.042	.043

Source: Unpublished calculations provided by Bruce Gardner.

small period of time, and soybeans have no target price or set-aside program). The markets for the two crops are represented in Figure 9.1.

In Figure 9.1 (A), the free-market supply of the controlled crop is S_A and the market demand is D_A. The free-market price and quantity are P_A^F and Q_A^1, respectively. Institution of a target price P_A^T results in increasing production to Q_A^3. Imposition of a price support P_A^S then raises the effective consumer price accordingly and results in consumption Q_A^2 with government acquisitions equal to $Q_A^3 - Q_A^2$. To eliminate government acquisitions, an acreage set-aside can be imposed, reducing output from Q_A^3 to Q_A^2. Alternatively, if the government wants to liquidate previously accumulated government reserves, it can impose a more severe acreage set-aside, reducing production further to Q_A^1 so that an amount $Q_A^2 - Q_A^1$ can be dumped on the market without depressing prices below the price-support level. Finally, if the acreage set-aside is set at a stringent level, then acreage can be reduced below the free-market level, say, to Q_A^0 with consumer price above the free-market level at P_A^0. In this case, the deficiency payment $P_A^T - P_A^0$ serves as the inducement for producers to produce under such stringent conditions.

Now consider the market for the uncontrolled crop in Figure 9.1 (B). The market demand for crop B is D_B. Assuming crops A and B are related in supply (they are grown on the same farms), the free-market supply that occurs in market B depends on the government program for crop A. If market A operates as a free market (the target price is below the free-market equilibrium price P_A^F), then the supply in market B is S_B^1, and the market price and quantity are P_B^1 and Q_B^1, respectively. If the target price P_A^T is imposed in market A without an acreage set-aside, then the supply in market B is S_B^3 and the market price and quanity are P_B^3 and Q_B^3, respectively. That is, as a higher target price is offered for crop A, the acreage devoted to crop A is increased and takes away from the acreage devoted to crop B, resulting in a decreased supply of crop B.

Alternatively, if the target price P_A^T is offered for crop A, with an acreage set-aside cutting production back to Q_A^2, then supply in market B is reduced, say, to S_B^2, and the market price and quantity are P_B^2 and Q_B^2, respectively. Finally, if more severe acreage set-asides are imposed on crop A, such as the one cutting production of crop A to Q_A^1, then the supply of crop B is increased to Q_B^1, resulting in market price and quantity P_B^1 and Q_B^1, respectively.

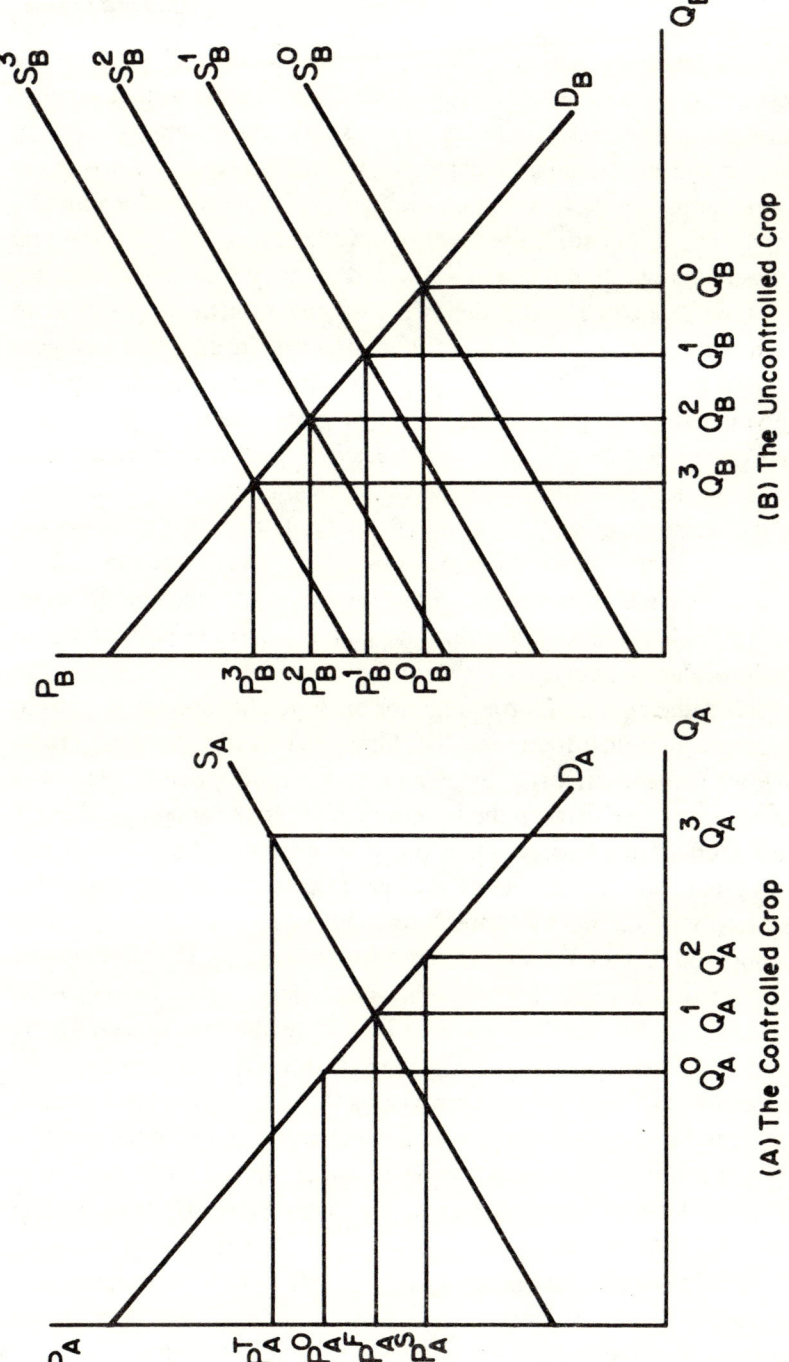

FIGURE 9.1. Conceptual Effect of Farm Programs on Diversification

How do these effects of government programs relate to diversification? The results from the model are inconclusive. Note that the market supplies in Figure 9.1 correspond to individual farm supplies and demands and the amounts produced of each crop correspond to the acreages devoted to the individual crops. In moving among the four production cases denoted by superscripts 0, 1, 2, and 3, the level of diversification between the two crops is changing. If the acreage devoted to crop A is larger than for crop B, then in cases 2 and 3 the effect of government programs is to reduce diversification. In case 1 it is unaffected; in the most stringent case 0, it is increased. Alternatively, if the acreage devoted to crop A is smaller than for crop B, then the effect of government programs is to increase diversification in cases 2 and 3, leave it unaffected in case 1, and reduce it in case 0. Finally, if the acreages devoted to the two crops are about the same (e.g., $Q_A^1 = Q_B^3$ and $Q_A^3 = Q_B^1$), then the government program for crop A may have no effect on diversification. This demonstrates the ambiguous nature of the effect of government programs on diversification and explains why no clear and consistent effects emerge in the results discussed above.

In reality, the effects of farm programs on diversification are somewhat more complicated than represented in Figure 9.1. First, there are cross-commodity demand effects. For example, when the price of crop B is altered because of a change in the government program for crop A, there is a feedback effect that changes the demand for crop A if the two crops are related in demand. Second, there are risk effects. If the provision of a target price is regarded by farmers to reduce the risk associated with producing crop A, then the supply curve for crop A may shift to the right as the target price is implemented. This would also generally be associated with a shift of the supply curve for crop B to the left as acreage is substituted. These two shifts together reflect a decreased need for diversification by risk averse farmers when a less risky activity is provided. Finally, in reality there are many crops competing in supply, some of which have government programs and some of which do not. Of those with programs, the level of subsidization varies so that relative effects may occur among those crops as well.

All of these considerations imply that the effect of farm programs on diversification cannot be resolved except empirically. Even then, it is a complicated issue to take all of the relevant factors into account.

An Empirical Model of the Effect of Farm Programs on Crop Acreage

To begin to address the empirical issue of effects of farm programs on diversification, the rest of this chapter specifies, estimates, and simulates an empirical model of supply and demand for some selected major crops in the United States. The model takes into account all of the considerations in Figure 9.1 as well as the demand interactions discussed subsequently. The model is not complete with respect to all crops that compete in supply and so must be regarded as an abstraction. The model also does not deal directly with the risk effects of government programs because they are not believed to be the primary effects that would occur if farm programs were eliminated.

The model focuses on wheat, corn, sorghum, and soybeans and specifically on the effects of farm programs on the acreage of these crops. This set of crops is chosen because they are believed to capture the most important and widespread effects on diversification that farm programs have. For example, the most important diversification in the corn belt is between corn and soybeans while one of the most widespread cases of diversification in the middle Great Plains is among wheat, corn, and sorghum.

To capture the interactions in demand, the model also includes a livestock component with beef, hogs, and poultry. The grain demand component disaggregates demands by consumption, market inventory, and exports following the specifications of Chambers and Just. Demand for government stocks and the farmer-owned reserve follows the work of Rausser and by Love with somewhat more structure to reflect the qualitative nature of policy instruments. The livestock component follows along lines used by Just (1981) with revisions to incorporate some refinements developed by Rausser and by Love. The grain supply model uses logit equations to represent program participation decisions following the spirit of the work by Chambers and Foster and later empiricized by Rausser and by Love. The acreage equations depart significantly from previous econometric practice and incorporate more structure among important program and market variables in the spirit of the intuitive and conceptual framework developed by Gardner (1988) and Lins. They examine the gains and losses associated with the wheat and corn programs by means of a quantitative graphical analysis of the various policy instruments through which wheat and feed grain commodity polices are administered. Because of the large size of the model, the complete structure and estimates are not reported here but can be found in Just (1989). The discussion here focuses on the general form of the structure so as to provide more understanding of how the model works.

The Crop Supply Structure

The basic form of the acreage equations is as follows. First, acreage in a market free of government programs is assumed to follow

$$A_f = A_f(\pi_n, \pi_a, A_{f,-1}) \tag{1}$$

where

- A_f = free-market acreage of the crop in question
- π_n = anticipated short-run profit per acre from production of the crop in question with free-market price
- π_a = anticipated short-run profit per acre from production of competing crop(s)

and

- $A_{f,-1}$ = lagged free-market acreage (to represent production fixities, etc.).

Profit per acre is defined by price times yield less per acre production cost, e.g.,

$$\pi_n = P_m Y_a - C \tag{2}$$

where

- P_m = market price
- Y_a = expected yield

and

- C = short-run cost per acre.

Unlike most previous models of crop supply and demand, the cost of production is not taken to be exogenous. Following the argument of

Gardner (1984) whereby the price of inputs are bid up to exhaust rents, the cost of production is estimated endogenously as a function of crop prices.

When government programs are voluntary, the nonparticipating component of acreage is assumed to follow equation (1) on the nonparticipating proportion of the acreage so nonparticipating acreage is

$$A_n = (1 - \phi) A_f(\pi_n, \pi_a, A_{f,-1}) \qquad (3)$$

where A_n is nonparticipating acreage and ϕ is the rate of participation in the relevant government program.

The participating acreage is largely determined by program limitations with

$$A_p = B\phi(1 - \theta) - D(G_a) \qquad (4)$$

where

$\quad$ B $\;=\;$ program base acreage
$\quad$ θ $\;=\;$ minimum diversion requirement for participation
$\quad$ D $\;=\;$ additional diversion beyond the minimum

and

$\quad$ G_a $\;=\;$ payment per acre for additional diversion.

The estimating equation for observed total acreage, given the participation level, is obtained by combining (3) and (4):

$$A_t = B\phi(1 - \theta) - D(G_a) + (1 - \phi) A_f(\pi_n, \pi_a, A_{f,-1}) \qquad (5)$$

where $D(\cdot)$ and $A_f(\cdot)$ follow linear specifications.

Determining the level of participation in this framework is crucial. Each farmer is assumed to participate if his/her perceived profit per acre is greater under participation than under nonparticipation ($\pi_p^i > \pi_n^i$). Assuming that individual perceived profits differ from an aggregate by an amount characterized by an appropriate random distribution across farmers, the participation rate can be represented by a logistic relationship with

$$\ln \frac{\phi}{1-\phi} = \phi^*(\pi_n, \pi_p) \qquad (6)$$

where

π_p = the profit per acre under compliance.

Given the qualitative nature of numerous agricultural policy instruments, a conceptually plausible specification of short-run profit per unit of land (producing plus diverted) on complying farms follows:

$$\pi_p = (1-\theta-\mu)\,\pi_z + \theta \cdot G_m + \mu \cdot \max(G_v, \pi_p) \qquad (7)$$

where

μ = maximum proportion of base acreage that can be diverted in addition to minimum diversion

G_m = payment per unit of land for minimum diversion (zero if no payment is offered for minimum diversion)

G_v = payment per unit of land for voluntary diversion beyond the minimum

and

π_p = short-run profit per unit of producing land under compliance.

The latter term suggests no voluntary additional diversion if $G_v < \pi_p$ and voluntary additional diversion to the maximum if $G_v > \pi_p$.

Conceptually, π_p follows

$$\pi_p = [\max(P_t, P_m) \cdot Y_p + \max(P_s, P_m) \cdot \max(Y_a - Y_p, 0)$$
$$+ \max(r_m - r_g, 0) \cdot P_s \cdot Y_a - C] \qquad (8)$$

where

P_t = government target price
Y_p = program yield
P_s = price support

Farm Programs and Diversification 315

r_m = market rate of interest

and

r_g = government subsidized rate of interest on commodity loans under the program (Love).

Equation (8) reflects the complicated relationship through which a participating farmer is entitled to at least the target price on his program yield, at least the (lower) support price on all of his production, and gains an additional interest subsidy on a loan against his stored crop (at harvest time) evaluated at the support price. These benefits must be balanced against the opportunity loss of having to divert some land from production reflected by equation (7).

Once acreage is determined in this framework, it is simply multiplied by yield and added to the carry-in to determine crop supply. Of course, the relationships in (7) and (8) do not necessarily apply exactly. For example, an uncertain anticipated market price may be discounted by a farmer compared to a target or support price which is known with certainty at the time of acreage decisions. Also, not all farmers place their crop under federal loan to take advantage of the interest subsidy. Nevertheless, intuition and experience imply that equations (7) and (8) apply as reasonable approximations and, furthermore, the approximations apply in a global sense. By comparison, the large number of variables with numerous qualitative relationships involved in these relationships suggests significant problems with objective econometric identification and makes the possibility of obtaining even plausible signs remote with estimation of ad hoc or flexible forms.

To illustrate the difference in performance of the approach of simply adding π_p and G_v to equation (1),

$$A_f = A_f(\pi_n, \pi_p, \pi_s, A_{f,-1}, G_v), \qquad (9)$$

compared to that in equations (5) and (6), both were used to estimate acreage response of wheat and of feed grains in the United States over the period 1962 to 1982 and then to forecast acreage in the 1983-1986 period. The results are given in Table 9.2. The results for equation (5) take the participation rate as exogenous whereas the results where the model is specified as equations (5) and (6) include forecasting errors for the participation rate as well.

TABLE 9.2

The Performance of Structural vs. Ad Hoc Models: The Case of U. S. Wheat and Feed Grain Acreage[a]

Model definition (equation)	Period Estimation	Period Forecast	Standard error Within sample	Standard error Post-sample
			million acres	
Wheat				
(9)	1962-1982	1983-1986	4.41	14.90
(5)	1962-1982	1983-1986	3.32	6.21
(5), (6)	1962-1982	1983-1986	b	9.07
Feed Grain				
(9)	1962-1982	1983-1987	1.73	6.40
(5)	1962-1982	1983-1987	6.26	6.38
(5), (6)	1962-1982	1983-1987		5.50

[a]See text for equations which define the various models.
[b]Blanks indicate no within-sample error is computed since the model is derived by combining the estimated equations corresponding to (5) and (6).

Source: Computed.

In the case of feed grains, the ad hoc formulation leads to a much smaller standard error in the sample period than the structural form in (5) even though the structural form performs better than the ad hoc form in ex ante forecasting of the postsample period. The model combining equations (5) and (6) obtains an even lower standard error. In the case of wheat, the structural form fits the sample data better than the ad hoc form and performs substantially better in ex ante simulation.

This superior performance of the structural model carries through when errors in forecasting the participation rate are also considered. The reason the structural form can outperform the ad hoc model even in the sample period is that nonlinearities and kinks in response over a wide range of policy parameters put a premium on global properties of the function. The participation rate over the sample period ranges from 0 (a kink point) to nearly 90 percent in others. As a result, the effects of profits with and without compliance cannot be well represented by a smooth approximating function. Such global structural properties are particularly important for the application of this chapter since the subject is eliminating farm programs or drastically phasing them down.

The Crop Demand Structure

Following numerous previous studies, the demand for crops is broken into food, feed, export, and inventory components for purposes of specification and estimation of a quarterly model. The inventory component is further broken into farmer-owned reserve, government-owned reserve, and market components for crops with government programs. The demand system for a given crop is thus of the form:

$$\begin{aligned}
Q_i &= Q_i(P_m, X_i), & X_i &= \{Q_{i,-1}, Y_c, T_j\}, \\
Q_f &= Q_f(P_m, X_f), & X_f &= \{Q_{f,-1}, F_j, P_j, T_j\}, \\
Q_x &= Q_x(P_m, X_x), & X_x &= \{Q_{x,-1}, E, T_j\}, \\
Q_r &= Q_r(P_m, X_r), & X_r &= \{Q_{r,-1}, P_s, P_r, r_m - r_g, D, T_j\}, \\
Q_g &= Q_g(P_m, X_g), & X_g &= \{Q_{g,-1}, P_s, D, T_j\}, \quad (10) \\
Q_m &= Q_m(P_m, X_m), & X_m &= \{Q_{m,-1}, Q_r, Q_g, r_m, D, T_j\},
\end{aligned}$$

$$Q_{r,t-1} + Q_{g,t-1} + Q_{m,t-1} + A_t \cdot Y_a$$
$$= Q_i + Q_f + Q_x + Q_r + Q_g + Q_m,$$

including the supply-demand identity where

Q_z = quantity demanded (z = i for industry or food, z = f for feed, z = x for export, z = r for farmer-owned reserve, z = g for government stocks, and z = m for market stocks)
P_m = market price
X_z = exogenous variables which determine the relevant demand
Y_a = actual average yield
Y_c = per capita consumer income
T_j = quarterly shift terms
F_j = numbers of various types of livestock on feed
P_j = prices of various types of livestock meat
E = trade weighted exchange rate
P_s = support price
P_r = release price

and

D = shift term reflecting the 1983 PIK program.

The demand system was not estimated in the form of (10) because a system that determines price through an identity equation tends to produce erratic price estimates particularly when demands are inelastic. Alternatively, a demand equation in (10) can be solved for price,

$$P_m = Q_i^{-1}(Q_i, X_i), \qquad (11)$$

and then the identity can be used to determine Q_i. This approach suffers in practice because the coefficient estimates of exogenous variables in the inverted equation are susceptible to spurious correlations with other factors in the system. This can lead to an unreasonably large contribution of these variables relative to other exogenous variables in the system in determining price predictions in practice. The approach used in this study is to solve the system in (10) for a partial reduced-form price equation which is then used to replace one of the demand equations in (10). This partial reduced form equation can be regarded as a convex combination of equations such as (11) which essentially produces a composite price forecasting equation in the sense of Johnson and Rausser where the weights are estimated simultaneously with the coefficients of the price equation. The number of such equations to combine in this manner is roughly determined by the trade-off between the increased forecasting accuracy of combining more forecasting equations and reduced identification as the total number of variables in the composite forecasting equation increases.

To capture the qualitative nature of government market involvement on the demand side, the government inventory demand equation is estimated including a qualitative relationship between market and support price. For example, the government inventory demand for feed grains is

$$Q_g = 0.3873 + 0.5838\ Q_{g,-1} + 39.85 \max\{0, (1.1\ P_s - P_m)\ \phi\}$$
$$(0.38)\quad (9.04)\quad\quad\quad (7.74)$$

$$+\ 20.37\ D - 0.1172\ T_1 + 1.821\ T_2 + 0.5981\ T_3$$
$$(6.90)\quad\ (-0.09)\quad\ (1.33)\quad\ (0.45)$$

$$R^2 = 0.927,\ \bar{R}^2 = 0.919,\ DW = 1.42,\ \text{Sample} = 1973{:}1\text{-}1987{:}3$$

where variables are as defined above and t-ratios are in parentheses (for a complete definition of variables and data sources, see Just 1989). This equation captures the qualitative relationship whereby stocks are not turned over to the government until the market price falls to the government support level but are increasingly turned over as the market price falls below the support (note that only grain produced under voluntary compliance with the program is supported so the market price can fall below the support price). Here the price variable is highly significant as compared to standard cases where a continuous function of market and support prices is used as a term explaining government stocks (in Rausser's work, the price term is a ratio of support price to market price and an implicit t-ratio of 1.48 is obtained in an otherwise similar equation).

The Livestock Supply Model

The livestock supply component accounts for the dynamic nature of breeding herd adjustment and the long lags in breeding and raising livestock to market weight. The basic form of the model for each species is as follows. First, a stock equation is included for the size of the national breeding herd of the form,

$$H_i = H_i(P_c/P_i, H_{i,-1}, r_m, T_j),\quad\quad\quad (12)$$

where

H_i = herd size for species i (e.g., i = cattle)
P_c = price of corn
P_i = price of meat from species i (e.g., beef for i = cattle)

and

T_j = quarterly shift terms.

Next, an equation is included for numbers on feed of the form,

$$F_i = F_i(H_{i,-k}, P_c/P_i, T_j), \qquad (13)$$

where k is the number of quarters required to reach feeding age in species i.

Finally, a meat production equation is included of the form,

$$M_i = M_i(F_i, H_i - H_{i,-1}, P_c/P_i, r_m, T_j) \qquad (14)$$

where M_i is the production of meat from species i. The term, $H_i - H_{i,-1}$, is included to capture the addition to meat production caused by culling breeding herds.

The livestock production model consists of a set of equations similar to (12)-(14) for cattle, hogs, and poultry.

The Meat Demand Structure

The meat demand system is considered independently from the crop demand systems since meats and grains are not very closely related except as grain prices affect meat supply. Each demand equation is estimated in price dependent form with

$$P_i/Y = P_i(P_j/Y_c, P_o/Y_c, C_i/N, T_j)$$

where

Y = per capita income
P_j = prices of other meats (included individually)
P_o = price index for nonfarm prices
C_i = domestic consumption of meat i

and

N = population.

The meat demand system is completed by identities of the form, $M_i + I_i = C_i$, and net import/export equations of the form,

$$I_i = I_i(P_i \cdot E, I_{i,-1}, E, T_j),$$

where I_i is net imports (negative for net exports) and E is trade-weighted exchange rate.

The Results of Policy Simulations

Using the model discussed above, several policy alternatives were simulated to determine the effects of major changes in farm commodity programs on the acreage of wheat, corn, sorghum, and soybeans. The policy alternatives considered are as follows:

1. A reduction of 10 percent in price supports for wheat, corn, and sorghum (with corresponding changes in price controls for the farmer owned reserve).
2. A reduction of 10 percent in both price supports and target prices for wheat, corn, and sorghum.
3. A reduction of diversion requirements by 10 percent for wheat, corn, and sorghum (for example, a 10 percent diversion requirement would be changed to zero).
4. Elimination of the farm programs for wheat, corn, and sorghum.

These various alternatives are investigated by simulating the changes beginning with the 1983 crop year for a period of three years, assuming macroeconomic conditions are unaffected by the changes. While the complete effects of the program changes would take considerably longer to work through the system given the heavy dynamics of the livestock sector, these effects are believed to be representative of the ultimate qualitative effects of farm programs on crop acreages. The results of the simulation are presented in Table 9.3 along with the actual acreages that resulted in 1983-1985.

In these results, the effects of changing the price support are relatively minor because price supports primarily affect the disposition of the crop after production (compare the first three lines of Table 9.3 with the last three lines). Only minor acreage reductions occur for the program crops while a slight increase occurs for soybeans. Only minor effects occur for the case where both target and support prices are reduced. The reason for a small effect here is that diversion requirements prevent much of a response to the change in target price. The only change comes through reduced incentive for participation in the voluntary programs. However, the target prices were sufficiently high to induce high participation. A 10 percent decline in target prices only causes a minor decline in participation.

The results in Table 9.3 show that the major policy instrument controlling acreage in the current program structure is the diversion requirement. In every case, a 10 percent reduction in the diversion requirement causes a 5 percent to 10 percent increase in acreage of the program crops, while the acreage of the nonprogram crop (soybeans) is largely unaffected under the more stringent restrictions on diversion acreage imposed in recent years.

The estimates for the case where all programs are eliminated are more tentative than the others because the estimated equations are used to make predictions far outside the range of observed data. Nevertheless, the results are striking and intuitively plausible qualitatively. The most serious effect is on wheat-growing areas. This can be explained as follows. First, participation in the conservation reserve program is higher in these areas in the case with programs, and this acreage would tend to come back into wheat production with program elimination. Second, the elimination of programs in wheat-growing areas tends to cause an abandonment of irrigated corn production which is replaced by dryland wheat production. Thus, the feed grain acreage does not increase by as much as under a simple reduction in diversion requirements while the wheat acreage increases by substantially more. In the corn belt, on the other hand, the elimination of programs tends to cause a shift toward soybean production (the nonprogram crop) following the diagrammatic analysis of Figure 9.1.

TABLE 9.3

Effects of Changes in Farm Programs for Wheat, Feed Grains, and Soybean Acreage

Policy scenario	Year	Wheat acreage	Feed grains Corn and sorghum acreage	Soybean acreage
			million acres	
Reduce price support[a]	1983	76.40	72.10	63.80
	1984	78.10	97.68	68.31
	1985	73.09	100.78	63.88
Reduce support and target prices[a]	1983	73.36	73.53	64.65
	1984	76.38	97.66	69.26
	1985	73.41	101.41	66.18
Reduce diversion[a]	1983	83.14	78.92	63.39
	1984	83.26	103.04	67.38
	1985	81.16	108.75	62.08
Eliminate programs	1983	98.01	101.28	65.17
	1984	99.91	103.39	71.57
	1985	98.60	106.29	72.98
No change	1983	76.40	72.10	63.80
	1984	79.20	97.80	67.80
	1985	75.60	101.73	63.13

[a]Reduce by 10 percent.

Source: Estimated from the model described in the text.

The Impact on Diversification

Based on the results in Table 9.3, one must conclude at a national level that diversification among wheat, feed grains (corn and sorghum), and soybeans is reduced by reducing diversion requirements or eliminating farm programs altogether. With both of these scenarios, wheat and feed grain acreages rise substantially while soybean acreage changes little. Both the wheat and feed grain acreages are already larger than soybean acreage, so the effect of eliminating the programs is to cause more specialization in the crops which are now program crops. In the context of Figure 9.1, this occurs because acreage reductions tend to follow the stringent case. Eliminating the programs causes acreage increases among the program crops. This is primarily a response to elimination of diversion requirements.

For the cases where price supports and target prices alone are adjusted, the effect is to increase diversification because the acreages of the larger (program) crops are reduced while the smaller acreage crop (soybeans) is increased. Although this effect is negligible, one can conjecture that it could be substantially larger in cases where diversion requirements were not in effect and preventing substantial acreage response. Thus, one must conclude that the effect of phasing down government programs on diversification depends crucially on how it is done. Reducing diversion requirements has the opposite effect as reducing target and support prices.

Of course, determining the effect of government programs on diversification at the national level says little about the effect of government programs on diversification at the individual farm level. These questions are much more difficult to answer empirically because little information on heterogeneity is collected at the farm level. Locally, opportunities for diversification may depend on soil types, weather conditions, market access, etc. To demonstrate these possibilities in a crude way, suppose that the national acreage effects in Table 9.3 occur proportionally across all states and use state data to examine the effects of government programs on diversification at the state level. The two most typical examples applicable to the crops investigated here are the corn-soybean trade-off of the corn belt and the wheat-feed grain trade-off of the middle Great Plains. Here, Iowa is taken to represent the corn belt, and Colorado is taken to represent the Great Plains. Colorado is chosen for the latter case because most of its wheat and feed grain acreage is clearly in the Great Plains while some of the feed grain acreage in states like Nebraska is partially in the corn belt.

In 1985 corn acreage in Iowa was 13.55 million acres and soybean acreage was 8.15 million acres (wheat and sorghum acreage was negligible). If eliminating farm programs has the same proportional impact as in Table 9.3, then this would compare to 14.157 million acres of corn and 9.422 million acres of soybeans, so the proportion of acreage (sum of the two) in corn would decline from 0.6244 to 0.6004. Diversification would increase.

In 1985 wheat acreage in Colorado was 3.45 million acres and feed grain acreage (corn and sorghum) was 1.065 million acres (soybean acreage was negligible). If eliminating farm programs has the same proportional impact on these acreages as in Table 9.3, then this would compare to 4.5 million acres of wheat and 1.113 million acres of feed grains. Thus, the proportion of acreage in wheat would increase from 0.7641 to 0.8017. Diversification would decline.

The main point of these two examples is that diversification is a localized phenomenon. The structure of farm programs is such that effects on diversification can be in either direction depending on local conditions. Data on heterogeneity among farms which contribute to the investigation of these issues are not readily available. Thus, definitive empirical answers are not readily accessible and await some substantial primary data collection efforts at the localized micro level.

Conclusions

It is demonstrated in this chapter that the effects of farm programs on diversification are ambiguous conceptually. Empirical work suggests that relaxing the major determinant of acreage in farm programs—diversion requirements—or eliminating farm programs altogether will reduce diversification at the national level. However, relaxing farm programs through other instruments increases diversification. Furthermore, effects at more local levels and even at state levels may differ considerably, both quantitatively and qualitatively. Thus, the empirical effects of farm programs on diversification are also ambiguous.

The empirical work in this chapter is based on a model with several shortcomings. First, the model only focuses on a few of the major crops. Diversification considerations, particularly at local levels, may involve many other crops. Second, the model ignores one of the main reasons for diversification—spreading risk. Any definitive work on the effect of farm programs on diversification must integrate risk considerations as discussed by Brown in chapter 8 of this book. In this context, such enterprises as livestock activities would have to be included. In addition, in analyzing the impact of farm programs on diversification, one is essentially trying to determine how farm programs affect the efficiency of resource use. There may be other methods more appropriate for this purpose than used in this chapter.

References

Ball, G., and E. O. Heady. *Size, Structure and Future of Farms.* Ames, Iowa: Iowa State University Press, 1972.

Bonnen, J. T. "The Distribution of Benefits from Selected U. S. Farm Programs." *Rural Poverty in the United States: A Report to the President's National Advisory Commission on Poverty.* Washington, D. C.: U. S. Government Printing Office, 1968, 461-505.

Chambers, R. G., and W. E. Foster. "Participation in the Farmer-Owned Reserve Program: A Discrete Choice Model." *American Journal of Agricultural Economics* 65(1983):120-124.

Chambers, R. G., and R. E. Just. "Effects of Exchange Rates on U. S. Agriculture: A Dynamic Analysis." *American Journal of Agricultural Economics* 63(1981):32-46.

Dahl, D. C. "Public Policy Changes Needed to Cope with Changing Structure." *American Journal of Agricultural Economics* 57(1975):206-213.

Gardner, B. G. *Alternative Agricultural and Food Policies and the 1985 Farm Bill,* edited by G. C. Rausser and K. R. Farrell. Giannini Foundation of Agricultural Economics, University of California, Berkeley, 1984.

_____. " Causes of U. S. Farm Commodity Programs." *Journal of Political Economy* 95(1987):290-310.

_____. "Gains and Losses in the Wheat Program." Department of Agricultural and Resource Economics. Working Paper 88-11. University of Maryland, College Park, 1988.

Johnson, S. R., and G. C. Rausser. "Composite Forecasting in Commodity Systems." In *New Directions in Econometric Modeling* and *Forecasting in U. S. Agriculture*, edited by G. C. Rausser. New York: Elsevier North-Holland, Inc., 1982.

Just, R. E. "An Econometric Model of Major U. S. Agricultural Commodities." Department of Agricultural and Resource Economics. Working Paper. University of Maryland, College Park, 1989.

_____. *Farmer-Owned Grain Reserve Program Needs Modification to Improve Effectiveness: Theoretical and Empirical Considerations in Agricultural Buffer Stock Policy Under the Food and Agriculture Act of 1977*. Prepared for Report to the Congress, U. S. General Accounting Office. Washington, D. C.: U. S. Government Printing Office, 1981.

Kerr, W. A. "The Diversification of Prairie Agriculture: Opportunities Arising from Changes in the International Trading Environment." Discussion Paper No. 359. Ottawa, Ontario: Economic Council of Canada, 1989.

Krause, K. R., and L. R. Kyle. "Economic Factors Underlying the Incidence of Large Farming Units: The Current Situation, and Probable Trends." *American Journal of Agricultural Economics* 52(1970):748-761.

Love, H. A. "Flexible Public Policy: The Case of the United States Wheat Sector." Unpublished Ph.D. dissertation, University of California, Berkeley, 1987.

Lins, W. "Gains and Losses from the Corn Program." Economic Research Service, U. S. Department of Agriculture, Working Paper. Washington, D. C., 1988.

Madden, J. P. *Economies of Size in Farming*. U. S. Department of Agriculture, Agricultural Economics Report 107 (February 1967).

Moore, C. V. "Effects of Federal Programs and Policies on the Structure of Agriculture." National Economic Analysis Division, Economic Research Service, U. S. Department of Agriculture (January 1977).

Pope, R. D., and R. Prescott. "Diversification in Relation to Farm Size and Other Socioeconomic Characteristics." *American Journal of Agricultural Economics* 62(1980):554-559.

Rausser, G. C. *Macroeconomics of U. S. Agricultural Policy*. Washington, D. C.: Studies in Economic Policy, American Enterprise Institute for Public Policy, 1985.

U. S. Congress, Congressional Budget Office. *Diversity in Crop Farming: Its Meaning for Income-Support Policy*. Special Study.

Washington, D. C.: U. S. Government Printing Office (May 1985).

U. S. Department of Agriculture. *Structure Issues of American Agriculture*. Economics, Statistics, and Cooperative Service, Agricultural Economic Report 438 (November 1979).

White, T., and G. Irwin. "Farm Size and Specialization." In *Size, Structure and Future of Farms*, edited by G. Ball and E. Heady. Ames, Iowa: Iowa State University Press, 1972.

10

Agricultural Subsidies in Canada: Explicit and Implicit

W. H. Furtan, M. E. Fulton, and K. A. Rosaasen

Introduction

The composition of agriculture is markedly different in the various regions of Canada. Prairie agriculture is dominated by the crop and livestock sectors while, in Ontario and Quebec, the poultry and dairy industries play a much more important role (Table 10.1). The difference in production patterns in these two areas of Canada is matched by differences in the markets for these goods. In the Prairie region, many of the major commodities (e.g., grains and oilseeds) are exported while, in Ontario and Quebec (Central Canada), a much larger percentage of the agriculture production is consumed domestically (e.g., poultry and dairy products). Together, these two regions make up over 90 percent of the value of agricultural production in Canada (Table 10.2).

The contrast in market orientation also reflects a difference in government policy toward these two regions of Canada. As a general rule, Prairie agriculture tends to receive its support in a direct fashion through commodity programs, while Central Canadian agriculture depends more on an indirect transfer of income through price-support mechanisms like supply management.

Crops and livestock have been playing an ever-increasing role in Central Canadian agriculture, while at the same time they have been representing a declining percentage of Prairie agriculture (Table 10.1). Table 10.3 illustrates the components of this shift in the production of livestock. Although cattle and calf numbers on the Prairies have fallen over the 1972-1988 period, the level of production compared to the rest of Canada

TABLE 10.1

Selected Commodity Incomes as Percentage of Gross Farm Cash Receipts, Central Canada and the Prairies Selected Years, 1971 to 1986[a]

Year	Commodity income	Central Canada	Prairies
		percent	
1971	(C + L)/GFR	39	86
	PD/GFR	43	8
1976	(C + L)/GFR	39	87
	PD/GFR	41	7
1981	(C + L)/GFR	43	87
	PD/GFR	37	6
1986	(C + L)/GFR	43	71
	PD/GFR	38	7

[a] C = gross crop receipts.
L = gross livestock receipts.
PD = gross supply-managed commodity receipts.
GFR = gross farm cash receipts.
Source: Statistics Canada, *Agriculture Economic Statistics*.

TABLE 10.2

Percentage of Total Farm Cash Receipts Produced by Central Canada and the Prairies 1978-79 and 1983-84

Year	Central Canada	Prairies
	percent	
1978-79	41.9	50.0
1983-84	41.1	50.4

Source: Agriculture Canada, *Handbook of Selected Agricultural Statistics*.

TABLE 10.3

Total Hogs and Cattle and Calves on Farms, the Prairies, Ontario, and Canada, 1971-1988

Year	Hogs on farms						Cattle and calves on farms					
	Prairies		Central Canada		Canada		Prairies		Central Canada		Canada	
	Total	Share of total	Total	Share of total	Total	Share of total	Total	Share of total	Total	Share of total	Total	Share of total
	thou-sands	percent	thou-sands	percent	thou-sands	percent	thou-sands	percent	thou-sands	percent	thou-sands	percent
Dec. 1 1972	3,400	49.0	3,268	47.1	6,994	100	7,255	58.9	4,738	36.9	12,324	100
Jan. 1 1974	3,280	46.9	3,435	49.1	6,991	100	7,736	57.4	4,871	36.1	13,481	100
1975	2,490	42.2	3,145	53.4	5,895	100	8,283	58.0	5,070	35.5	14,278	100
1976	2,074	38.3	3,404	62.9	5,409	100	8,215	58.5	4,905	34.9	14,048	100
1977	2,107	37.7	3,825	68.5	5,587	100	7,981	58.2	4,806	35.1	13,710	100
1978	2,185	32.8	4,199	63.1	6,653	100	7,391	57.4	4,565	35.5	12,870	100
1979	2,410	29.8	5,340	66.1	8,074	100	6,986	56.7	4,430	35.9	12,328	100
1980	2,815	29.1	6,450	66.6	9,688	100	6,962	56.1	4,489	36.2	12,403	100
1981	2,731	28.6	6,305	66.1	9,544	100	6,850	54.9	4,635	37.2	12,468	100
1982	2,668	26.8	6,660	66.8	9,970	100	6,775	55.7	4,366	35.9	12,163	100
1983	2,678	27.1	6,575	66.5	9,890	100	6,583	55.5	4,291	36.2	11,861	100
1984	3,027	29.3	6,645	64.2	10,346	100	6,485	55.8	4,144	35.6	11,629	100
1985	3,357	31.8	6,545	61.9	10,573	100	6,306	55.7	4,043	35.7	11,330	100
1986	3,158	31.7	6,210	62.3	9,967	100	6,073	55.4	3,940	36.0	10,956	100
1987	3,287	32.9	6,130	61.3	9,996	100	6,138	56.8	3,747	34.7	10,802	100
1988	3,674	34.5	6,380	59.9	10,658	100	6,300	58.2	3,600	33.3	10,818	100

Source: Statistics Canada, *Livestock and Animal Products Statistics*, various issues.

has remained relatively constant. In contrast, while hog production on the Prairies has remained relatively unchanged, the position of this region in terms of total Canadian production has declined in comparison to other regions.

A shift in the pattern of broiler and turkey production has also occurred between the Prairies and Central Canada, with production increasing dramatically in Central Canada during the 1950s and 1960s, while remaining relatively constant in the Prairies (Figures 10.1 and 10.2). This shift was facilitated by changes in technology (e.g., disease control and the use of prepared feeds) which allowed a replacement of the traditional land-based methods of rearing with those methods involving confinement. This allowed for production to occur near the large centers of population. It was also during this time period that the fast food industry started to grow in Canada.

This chapter explores, in some depth, the differences in agriculture in the various regions of Canada and the extent to which they are protected and subsidized by governments. Implications for Canada-U. S. free trade are also discussed. Government transfers to agriculture in different provinces and regions of Canada are examined, with a focus on some of the implications of government policy for conflict within and between regions. Off-farm labor markets are analyzed because they link the structure of agriculture to regional economies. The question of value added is addressed with respect to different commodities and to different provinces. The chapter concludes with a discussion of the implications for agricultural policy and economic development.

Government Transfers to Agriculture

As suggested above, the type of economic protection given agriculture differs between commodities and between regions. As an example, programs like Western Grain Stabilization (WGS), Special Canadian Grains Program (SCGP), and Western Grain Transportation Act (WGTA) provide direct income support to Prairie grain producers, while import and production quotas provide the basis for supply management in the dairy and poultry sectors. However, there are exceptions: the dairy industry also

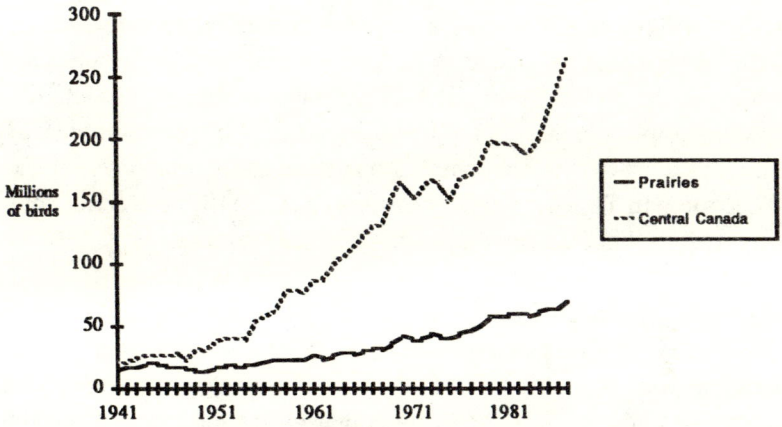

FIGURE 10.1. Broiler Production, 1941-1987, Central Canada and Prairies

Source: Statistics Canada, *Livestock and Animal Products Statistics*, Cat. 23-203.

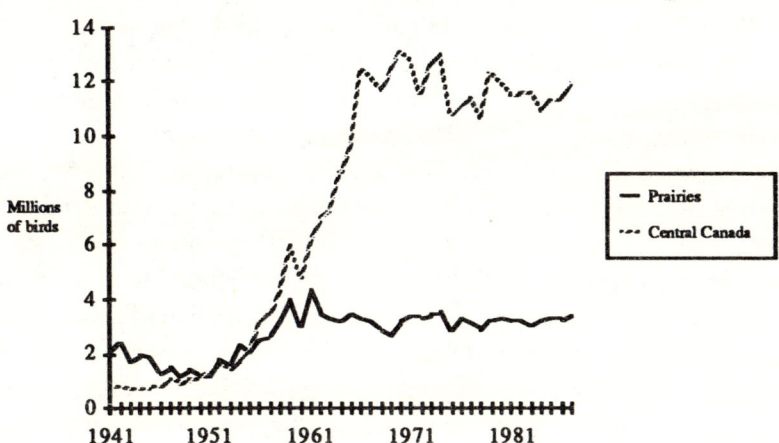

FIGURE 10.2. Turkey Production, 1941-1987, Central Canada and Prairies

Source: Statistics Canada, *Livestock and Animal Products Statistics*, Cat. 23-203.

receives support through deficiency payments, while hog and livestock producers receive deficiency payments through stabilization programs.

The purpose of this section is to document the different types of subsidies in the two regions. The data for this comparison come from a number of sources including the U. S. Department of Agriculture (USDA) and the Organization for Economic Cooperation and Development (OECD). This section will show how these transfers cause conflicts to arise, not only among regions, but also among producers within regions.

Measure of Government Transfers

Estimates of the overall economic effects of selected Canadian agricultural policies on the major farm commodities have been calculated by a number of authors. Barichello finds that by far the greatest transfer to producers is in the dairy sector, followed by wheat and barley (Table 10.4). The $955 million transfer to dairy producers reported by Barichello was also supported in a study by Josling, who calculated the transfer to be $905 million. (Dollar figures in this chapter are Canadian dollars.)

Table 10.5 reports the results of four separate studies on government transfers to producers in the poultry industry. As the results show, there are sizable producer gains. In eggs, producers gain between $38 million and $74 million; for broilers, the estimates range from between $57 million and $94 million. It is important to note that, even though the transfers are considerable, the inefficiencies created were not found to be large. Transfers under supply management in Canada appear to redistribute income, rather than create large distortions in resource use.

A more detailed analysis of the government transfers to producers is required if the impact on a regional basis is to be examined. The measurement of economic protection afforded agriculture in Canada has been carried out by the OECD and the USDA (Goodloe). Both of these studies use the approach of producer subsidy equivalents (PSEs).[1] One of the problems with using the OECD and the USDA data is that they treat Canada as one region. For our purposes, regional differences are required. To get around this problem, we use expenditure data which measure both federal and provincial government transfers to agriculture on a regional basis to determine the amount of support received by each province.

TABLE 10.4

Economic Effects of Selected Canadian Agricultural Policies as Measured by Producer and Consumer Surplus, 1982

Commodity	Economic effects			
	Economy gain	Producer gain	Consumer gain	Taxpayer gain
	million dollars			
Wheat	a	+470	0	-307
Barley	- 3	+246	0	-106
Rapeseed		+ 70	0	- 49
Eggs	- 19	+ 55	- 74	0
Broilers	- 13	+ 57	- 73	0
Dairy	-214	+955	-980	-303

aBlanks indicate less than $1 million.
Source: Barichello.

TABLE 10.5

Economic Effects of Poultry Industry Regulations in Canada, Farm Gate Level, as Measured by Producer and Consumer Surplus, Selected Years, 1981 to 1983

Commodity	Economic effects			
	1	2	3	4
	million dollars			
Eggs				
Economy gain	-19	a	- 0.4	- 5
Producer gain	+55	+45	+38	+74
Consumer gain	-74	-56	-39	- 80
Broilers				
Economy gain	-13		- 5	- 11
Producer gain	+57	+71	+71	+ 94
Consumer gain	-73	-77	-76	-121
Importer gain	+ 4			

aBlanks indicate no data available.
Sources:
    Col. 1: Barichello.
    Col. 2: Arcus.
    Col. 3: Veeman.
    Col. 4: Harling and Thompson.

Agriculture Canada and Finance Canada document the direct and indirect expenditures for agriculture made by the provincial and federal governments for 1981-82 through to 1986-87. What the data do not include is the level of economic support offered in the form of regulations, such as supply management quotas. For a measure of the income transfers under dairy and poultry supply management and two-priced wheat, the PSE estimates reported by the USDA were used (Goodloe). Since the USDA estimates did not include egg production, PSE estimates for eggs were calculated in this report using the USDA's methodology. To obtain estimates for regional economic support of commodities provided under supply management, the total Canadian PSE transfers were allotted to provinces on the basis of production.

A problem exists with the federal expenditure data on the Western Grain Stabilization Act (WGSA). The numbers reported by Agriculture Canada (1986) include the WGSA payout less the producers' contribution to WGSA. To be accurate, the numbers should only include the government's contribution to the WGSA. Two reasons for this are:

1. Farmers will pay back part of this payout in the future, so only contributions to the plan should be included in measuring the level of protection in any given year.
2. Some stabilization plans are contributory (i.e., WGSA) and others are not [i.e., Agricultural Stabilization Act (ASA)].

For these reasons, the Agriculture Canada estimates of federal dollars going to agriculture in each province were adjusted to include only government contributions.

The aggregate economic level of protection afforded agriculture in each province is the sum of federal transfers, provincial transfers, and regulatory benefits. These transfers give an indication of the level of protection on a regional basis. They do not examine the differences in economic protection between commodities within a region.

Estimates of Transfers to Regional Agriculture

The estimates of the level of economic protection given agriculture in the two major agricultural regions of Canada are shown in Figures 10.3 to 10.6.

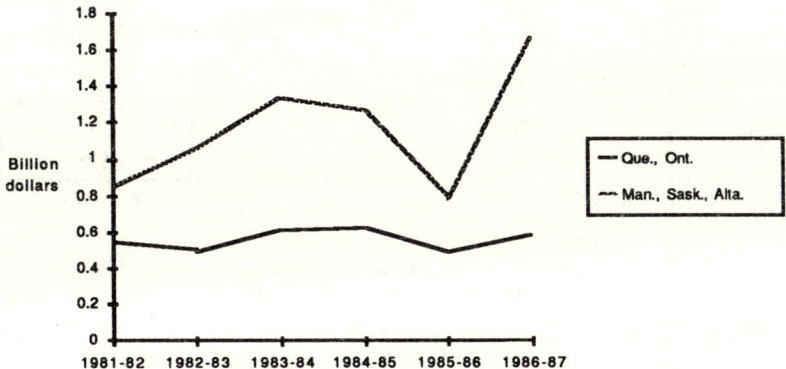

FIGURE 10.3. Adjusted Federal Expenditures, Central Canada and Prairies, 1981-82 to 1986-87

Source: Goodloe.

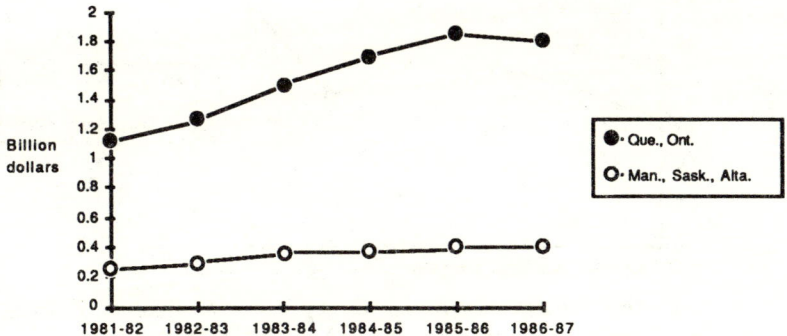

FIGURE 10.4. Indirect Payments to Agriculture, Central Canada and Prairies, 1981-82 to 1986-87

Source: Goodloe.

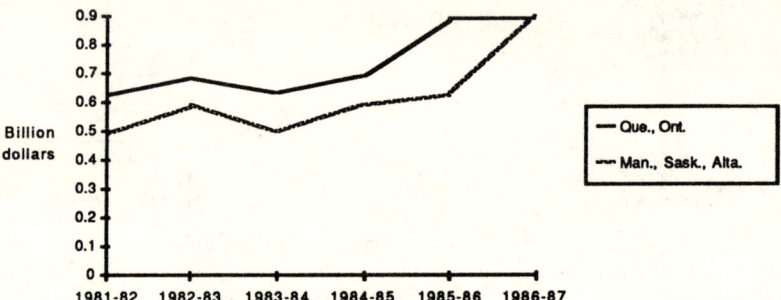

FIGURE 10.5. Provincial Expenditures to Agriculture, Central Canada and Prairies, 1981-82 to 1986-87

Source: Goodloe.

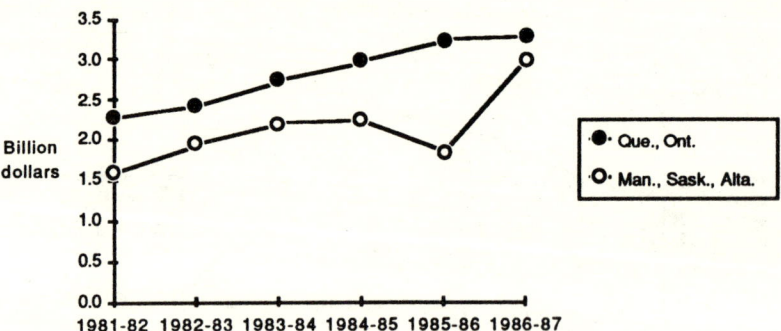

FIGURE 10.6. Total Protection to Agriculture, Central Canada and Prairies, 1981-82 to 1986-87

Source: Goodloe.

The direct federal expenditures in Figure 10.3 show the Prairie region receiving the bulk of the support under this method of transfer. The difference in support between the two regions has widened since 1985 due to WGSA and the deficiency payments (SCGP) made to the grains sector. Since the SCGP is paid on a production basis (Fulton, Rosaasen, and Schmitz), the Prairies, which produce most of the grains, receive the largest share of the support.

The level of economic protection afforded agriculture through quotas or supply management is shown in Figure 10.4. Here, Central Canada receives the bulk of the economic protection since most of the supply-managed commodities are produced in that region. One of the principal objectives of supply management is that each region in Canada be self-sufficient. Therefore, the size of the market, and not necessarily comparative advantage, may be a determining factor in the level of production and, hence, protection. However, it may be that, for the supply management commodities, comparative advantage coincides with market size. Recall that broiler and turkey production shifted to Central Canada well before the national supply-management scheme was introduced in the early 1970s.

The level of provincial support for agriculture is shown in Figure 10.5, while Figure 10.6 shows the total support—federal and provincial expenditures plus the protection afforded through supply management. In aggregate, the data show that Central Canadian agriculture receives more economic protection than does Prairie agriculture. Since the Prairies produce a larger share of Canadian gross farm receipts, it is apparent that agriculture in Central Canada is receiving more protection than its counterpart in the Prairies.

A final comment needs to be made regarding the duration of the economic transfers to farmers in the two regions. In the Prairies, farmers only recently have received the large economic protection detailed above; these will discontinue once product prices improve (Fulton, Rosaasen, and Schmitz). Unless changes are made as a result of the General Agreement on Tariffs and Trade (GATT), however, Central Canada will retain the economic protection it currently enjoys in the supply-management sector. Therefore, the expected dollar value of economic protection is substantially higher for Central Canada than for the Prairies.

Research Expenditures by Agriculture Canada

The rate of return to agricultural research has been documented in Canada by numerous economists (Klein and Furtan). The general conclusion of these studies is that the returns to research are between 30 percent and 60 percent. These expenditures shift the production possibilities frontier to the right and make regions more productive. Over time, this will have a significant impact on the production and income potential of an agricultural region (e.g., canola in the Prairies and corn in Central Canada).

Federal expenditures on agricultural research by provinces are shown in Table 10.6. The expenditures are the largest in Quebec, although the gross receipts of Quebec agriculture are less than those of Ontario and Saskatchewan. Expenditures in Alberta are also higher than in Saskatchewan and Ontario in 1984-85, even though its gross farm receipts are less than the other two provinces. If these research expenditure patterns were maintained over time, the conclusion might be drawn that the production possibilities frontier for Quebec and Alberta would shift outward faster than those in Saskatchewan or Ontario.

It is difficult, however, to analyze research expenditures by province in this fashion because of the spillover effects. That is, canola research done in Manitoba may benefit Saskatchewan farmers, while hog research in Quebec could benefit Ontario farmers. To overcome some of this effect, it is useful to look at regional expenditures. On a regional basis, Central Canada received $51.41 million and $50.80 million in 1984-85 and 1987-88, respectively, while the Prairie region received $53.45 million and $52.27 million, respectively, in the same time periods. These expenditures track the relative level of farm cash receipts by region reasonably close.

An estimate of the commodity research expenditures for 1983-84 are given in Table 10.7. Commodities receive different levels of support, depending upon the region. As expected, wheat research is supported in the Prairies, while dairy research is predominantly expected in Quebec and Ontario. The only major anomaly is horticulture research. Horticulture is the largest expenditure Agriculture Canada makes on research, while its importance in terms of gross farm receipts is less than 10 percent. Also, unlike for other commodities, by far the greatest expenditure is made in Quebec. It is interesting to note that research and development expenditures

TABLE 10.6
Research Expenditures, Agriculture Canada, by Province 1984-85 and 1987-88[a]

Year	Quebec	Ontario	Manitoba	Saskatchewan	Alberta
			million dollars		
1984-85	31.95	19.46	11.97	18.59	22.89
1987-88	26.05	24.75	12.34	18.78	21.34

[a]These numbers do not include research expenditures for institutes.
Source: Agriculture Canada, 1986.

TABLE 10.7
Research Expenditures, Quebec, Ontario, and Western Canada, by Commodity and Activity, 1983-84

Commodity and activity	Quebec		Ontario		Western Canada	
	person years	thousand dollars	person years	thousand dollars	person years	thousand dollars
Animals						
Beef	10	1,180.6	27	959.6	121	5,181.9
Dairy	39	1,216.9	20	3,097.8	22	1,486.4
Swine	8	593.2	23	819.6	31	1,352.1
Poultry	0	0	57	2,010.2	18	586.8
Sheep	10	356.0	30	985.3	6	154.9
Other	0	0	6	179.6	6	229.9
Crops						
Wheat	5	173.5	16	531.8	108	4,014.4
Other cereal	23	784.7	51	1,562.5	89	2,440.3
Oilseed	0	0	30	1,110.2	57	1,867.2
Forage	36	1,153.8	24	701.9	119	3,878.3
Horticultural	70	9,554.3	92	3,284.7	159	6,737.7
Field	16	549.0	38	1,535.2	24	810.8
Research and development	9	357.1	33	3,159.3	18	1,490.9
Protection			97	3,986.3	138	5,108.3

Source: Agriculture Canada, 1983-84.

for horticultural crop research in Quebec are larger than the research for all the other commodities produced in Quebec.

Transportation Policy

Under the WGTA, Prairie grain producers receive a subsidy for shipping grain to export markets. These numbers have been included on income transfers in the above calculations. However, parts of Quebec agriculture also receive subsidies on grain exports, in the form of feed freight assistance, shipped from the Prairies. The concern over the impact of the WGTA on Prairie agriculture, especially livestock production, continues (Fulton, Rosaasen, and Schmitz). If the WGTA were paid to the producer, instead of to the railways, resources on the Prairies would shift from the export of grain to the production of livestock. While the comparative advantage of the Prairies over Central Canada would increase in livestock production, the size of the shift in production is not clear.

In this light, if the WGTA were paid to the producer, it is not clear what the effect would be on the livestock industry in Central Canada. Under the Crow/WGTA, surplus feed grains from the Prairies were shipped to the deficit feed grain areas in Central Canada. Because the cost of shipping the grain was subsidized, an allegiance developed between some farmers in each of the two regions. Since one of the users of western feed grains was Quebec livestock producers, they had a vested interest in seeing the WGTA continued. At the same time, many of the grain producers in the Prairies wished to see Crow/WGTA continued since it provided for lower transportation costs of grain to export position. Recently, the production in Quebec and Ontario has increased to the point where these regions are almost self-sufficient in feed grains (Canada Grains Council). One of the implications of this is that any changes to WGTA are likely to have much less of an effect on that region.

The conflicts that arise within regions because of a transfer to one sector of agriculture are evident in the debate over the method of payment of the benefit under WGTA (the Crow Benefit). Currently, the Canadian government makes a payment to the railways on behalf of farmers for the export of grain from the Prairies. This payment will amount to approximately $31 per tonne in 1989. Livestock producers who purchase feed grains must pay a higher price for their feed grains because of this transfer to grain producers. This reduces the profitability of the livestock industry. In addition, the higher price of grain also makes the processing of this commodity more expensive, thereby having a detrimental effect on this form of diversification.

In Alberta, the provincial government has introduced a program—the Crow Offset Program—designed to mitigate the impact of the Crow Benefit. Under this program, livestock producers who purchase barley as livestock feed receive a payment of $12 per tonne. This program has given livestock feeders in Alberta an advantage over livestock feeders in Saskatchewan, all as a result of one government attempting to offset another government's program.[2]

Transportation policy is clearly one of the major causes of conflict, both within and among regions. Policies that use government subsidies to encourage the export of a product, which in turn hurt other commodities within the same region, will not be successful in the long run.

Trade and Commercial Policy

The Prairies produce commodities that are export oriented. As a result, producers in this region generally support policies that encourage and enhance trade. Producers of supply-managed commodities, on the other hand, are more concerned with the domestic market and the maintenance of trade barriers.

A good example of this is the U. S.-Canada Free Trade Agreement (FTA) which has been largely supported by livestock and grain producers and opposed by producers of supply-managed commodities. The FTA also affects value-added products produced from agricultural production. For example, Canada maintains tariffs on various milk, poultry, bread, and beverage products. If the protection on the value-added products is removed, the price of these commodities will be lowered, reducing producer incomes.

The discussions at GATT reflect a similar set of concerns between producers of supply-managed commodities and other grains and livestock. While all agricultural commodities receive some economic protection that may be trade distorting, grains stand to gain from more liberalized world markets because they are exported. Supply-managed commodities which tax consumers will likely lose economically from liberalized markets. This major difference is reflected in the position of the various farm commodity groups in the move toward liberalizing world trade.

Provincial Labor Markets and Diversification

The well-being of the agricultural economy cannot be measured simply in terms of the income that is generated from farming. While farmers have invested many resources in agriculture, some have also diversified their operations by investing outside the industry. The income earned by these farmers, and hence the economic health of the regions in which they live, is the return earned by all the resources farmers have at their disposal, not just those that are directly invested in agriculture.

One of the most important resources that farmers have invested outside agriculture is labor. Labor is one of the few resources used in agriculture that has a use outside of the industry. In contrast, land and machinery are fixed in agriculture—while the price of these resources will change with market conditions, they do not move to other sectors of the economy. Labor, on the other hand, is relatively more mobile and does move among sectors. For example, a farmer may work in a potash mine, on an oil rig, or in a factory and still operate a farm. The degree of mobility depends on such factors as age and the level of education or training.

The importance of nonfarm income to farm families can be highlighted by examining the total income of farm families and its source. Net farm income was 25 percent of the total farm family income in Canada in 1971—the only year for which such data are available (Table 10.8). For the Prairies, this number is somewhat higher, with 39 percent of farm family incomes derived directly from farming. Nevertheless, even in this region, farming does not constitute the largest source of farm family income.

The impact of off-farm income in the various provinces can be expected to be very different (Table 10.8). The percentages of farms with off-farm work in Quebec, Ontario, Manitoba, Saskatchewan, and Alberta are shown in Figure 10.7. Until 1981, Saskatchewan had the lowest percentage of farms with off-farm work, while Ontario consistently has had the highest percentage. While the percentage of farms reporting off-farm work in Quebec and Ontario has remained relatively constant over the past two decades, this has not been true in the Prairies. All three provinces have seen the percentage of farms with off-farm work increase significantly since 1971, with the greatest increase in Alberta and Saskatchewan. Although the overall trend has been upward, both Saskatchewan and Manitoba experienced a decline in the percentage of farms with off-farm work from

TABLE 10.8

Distribution of Farm Family Income by Source of Income and by Region, Canada, 1971

Source of farm family income	Atlantic	Quebec	Ontario	Prairies	British Columbia	Total Canada
	\multicolumn{6}{c}{dollars per family unit}					
Net farm	579	1,506	1,357	2,149	531	1,620
Wages and salaries	2,939	3,076	5,002	3,227	5,678	3,398
Self-employment (nonfarm)	334	413	320	253	621	331
Investment	222	204	668	362	610	423
Government transfers	925	998	555	515	371	629
Other	34	95	114	49	211	85
TOTAL	5,033	6,292	8,016	5,555	8,022	6,486
	\multicolumn{6}{c}{percent of total income}					
Net farm	12	24	17	39	6	25
Wages and salaries	58	49	62	40	71	52
Self-employment (nonfarm)	7	6	4	5	89	5
Investment	4	3	8	6	7	6
Government transfers	18	16	7	9	5	10
Other	1	2	2	1	3	2
TOTAL	100	100	100	100	100	100

Sources: Davey, Hassan, and Lu.

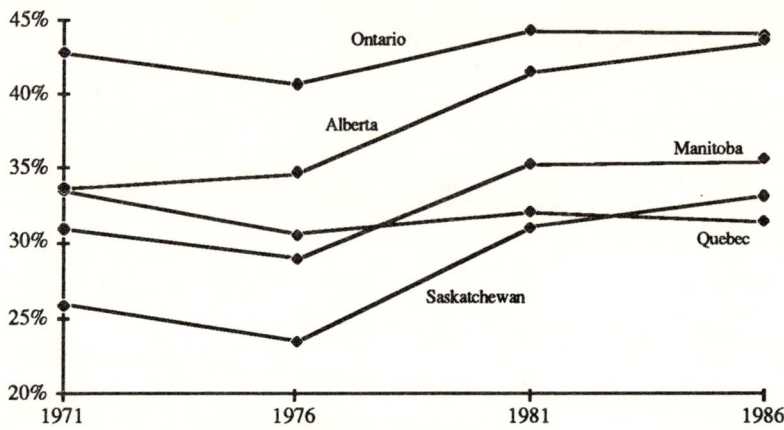

FIGURE 10.7. Percentage of Farms With Off-Farm Work, Central Canada and Prairies, Selected Years, 1971 to 1986

Source: Statistics Canada, *Census of Canada 1986—Agriculture Quebec*, Cat. 96-107; *Agriculture Ontario*, Cat. 96-108; *Agriculture Manitoba*, Cat. 96-109; *Agriculture Saskatchewan*, Cat. 96-110; and *Agriculture Alberta*, Cat. 96-111.

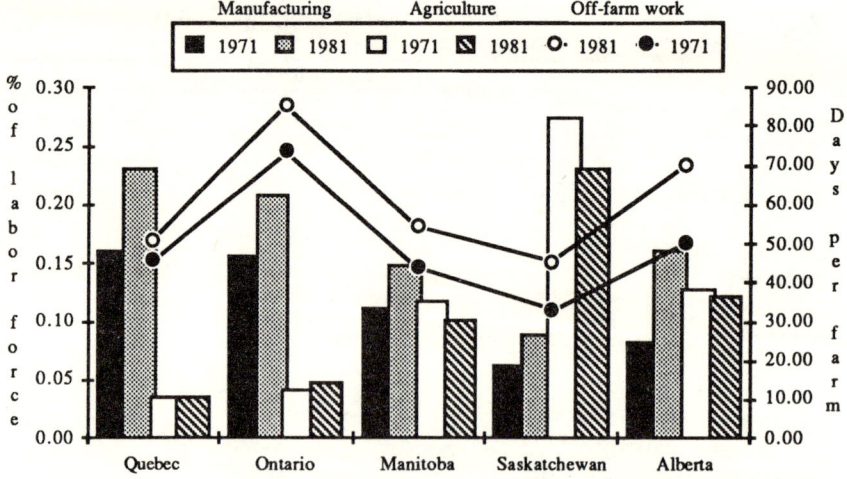

FIGURE 10.8. Labor Force in Manufacturing and Agriculture and Days of Off-Farm Work Per Farm, Central Canada and Prairies, 1971 and 1981

Source: Same as Figure 10.7.

1971 to 1976 (Figure 10.7). This suggests that one of the factors determining the degree to which farmers work off the farm is the financial health of the agricultural industry since this fall in the percentage of off-farm work corresponds with an increase in the profitability of the industry. The relatively slow increase in off-farm work that occurred between 1981 and 1986 may indicate that the health of the nonagricultural sector is also an important factor. During this five-year period, all three Prairie provinces experienced a downturn in both their agricultural and nonagricultural sectors.

The importance of the agricultural and nonagricultural sectors in determining the level of off-farm work can be further highlighted. Figure 10.8 presents the percentage of the labor force in manufacturing and agriculture and the number of days of off-farm work per farm for the provinces of Quebec, Ontario, Manitoba, Saskatchewan, and Alberta. Comparing provinces, the number of days of off-farm work per farm is negatively correlated with the percentage of the work force in agriculture and positively correlated with the percentage of the work force in manufacturing in both 1976 and 1981. Also, when a comparison is made between 1976 and 1981, in all provinces the number of days of off-farm work per farm has increased with an increase in the percentage of the labor force in manufacturing.

The relationship between off-farm employment and net farm income can be further explored by comparing Figures 10.7 and 10.9. Figure 10.9 presents the value of realized net farm income per farm for each of the major agricultural provinces. With the exception of Manitoba, provinces which have smaller percentages of farms reporting off-farm work (Figure 10.7) also tend to have higher levels of realized net farm income per farm (Figure 10.9). This correlation lends support to the point made above that, in provinces where less off-farm income is available, farmers devote more of their effort to farming. The higher level of farm income has not been sufficient to bring farm family income on the Prairies up to the level of Quebec and Ontario (Table 10.8).

Thus, the structure of the nonagricultural economy of a province or region is as important in determining the financial health of the agricultural economy as is the structure of the agricultural industry itself. Steeves draws a similar conclusion. He argues that rapid urbanization in the 1950s and 1960s in provinces outside the Prairie region has meant that most farms in

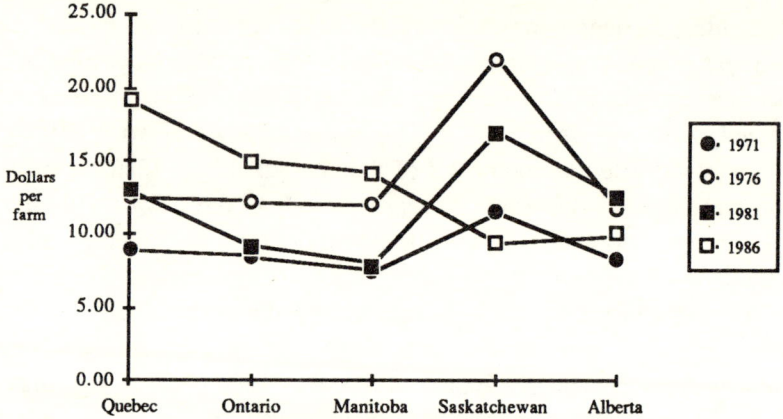

FIGURE 10.9. Real Realized Net Farm Income Per Farm, Central Canada and Prairies, 1971, 1976, 1981, and 1986 (1981 dollars)

Source: Statistics Canada, *Agriculture Economic Statistics*, Cat. 21-603.

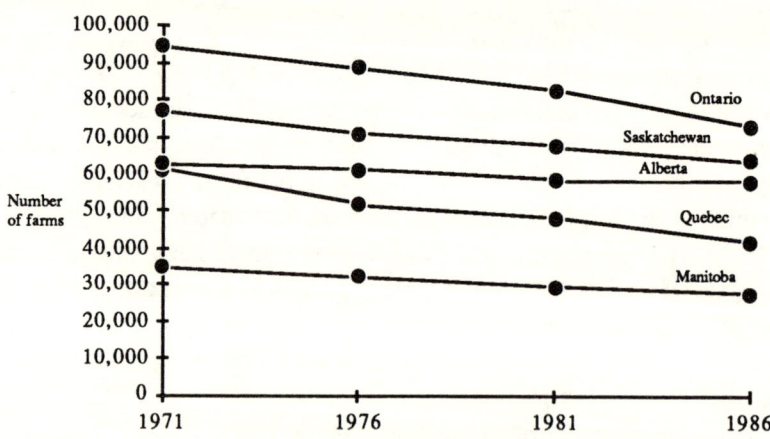

FIGURE 10.10. Number of Farms in Quebec, Ontario, Manitoba, Saskatchewan, and Alberta, Selected Years, 1971 to 1986

Source: Same as Figure 10.7.

these areas are now within commuting distance of a substantial market community. This has allowed farmers to continue farming, though often with a much smaller operation, and still commute to urban employment. In the Prairies, however, this possibility is less likely. Greater distances and much less urbanization have resulted in far fewer opportunities for off-farm employment in urban areas.

The lower level of off-farm employment in the Prairie region, combined with the smaller percentage of manufacturing jobs, suggests fewer opportunities for farmers who wish to move out of agriculture altogether. This notion, in fact, seems to be supported. As Figure 10.10 illustrates, the percentage reduction in the number of farms was the largest in Quebec and Ontario—provinces that have a considerably greater proportion of their work force in the manufacturing sector. This same argument is made by Steeves who concludes that a larger proportion of agricultural producers east of the Prairies have been forced out of agriculture due to greater nonfarm opportunities.

The Structure of the Canadian Economy

The linkages between the health of the agricultural industry and the nature of the nonagricultural market take on added importance when the overall structure of the Canadian economy is examined. One of the striking features of Canada is the large economic disparities that exist between the various provinces and regions of Canada. Ontario and British Columbia have historically had higher personal per capita income than the Prairies and Quebec which, in turn, have had higher personal per capita income than the Atlantic Region (Economic Council of Canada).

A number of reasons for these disparities in income have been suggested, including differences in resource levels; a poor industrial structure; and differences in labor quality, level of technology adoption, and the level of aggregate demand (including both export and domestic). For Canada as a whole, the latter three appear to be the most important in explaining the level of productivity, the primary factor in determining the level of income in each region of Canada.

When the Prairie region is examined by itself, however, the forces influencing the lower levels of income appear to be somewhat different. For instance, a poor industrial structure appears as a major factor in lowering labor productivity in Manitoba, Saskatchewan, and Alberta. More specifically, the predominance of goods-producing industries (which tend to have lower productivity and incomes associated with them) in these provinces has contributed to the relatively lower levels of income for the Prairie region.

While differences in labor quality and technology adoption are less important for the Prairie region than they are for the rest of Canada in explaining economic disparities, the level of aggregate demand, which is strongly influenced by the degree of urbanization, does appear to play a prominent role. In particular, a resource-type urban system has a relatively low level of economic interaction among its components, with shipments out of the region being much more important than shipments within the region. One of the consequences of this type of aggregate demand pattern (i.e., export-based rather than locally or regionally oriented) is a lower level of income. Among the Prairie provinces, Saskatchewan and, to a lesser extent, Manitoba appear to fall into this category. The Maritime provinces also exhibit this type of urbanization structure.

Alberta and British Columbia, on the other hand, are more correctly classified as transformation regions. Since the transformation of raw materials usually requires several steps, often physically located some distance apart, these regions experience a much higher degree of interdependence and complementarity than do the resource regions.

Finally, fabrication industries characterize the economies of Quebec and Ontario. These industries are often directed at the production of finished goods and require a large number of inputs. The result is that economies based on these industries are highly integrated and competitive. As would be expected, the more integrated a region, the higher its level of per capita income.

The structure and urbanization of the Canadian economy becomes important when it is related to the problem, as described above, facing agriculture. Historically, there has been a continual pressure to move labor out of agriculture, a sector with a relatively low level of income. In Saskatchewan and Manitoba, this out-migration of labor has gone hand in hand with an out-migration from the province. The implication is that, at least for these provinces, there are insufficient employment opportunities in the nonagricultural sector.

There is reason to believe that this out-migration of people from regions like Saskatchewan and Manitoba will not have any effect on the overall level of income within the province. Out-migration from a region simply reduces the size of the nonprimary trading industries and sectors (Copithorne). Since the general wage level in competitive economies is determined in this sector, and the health of this sector depends in part on the level of urbanization, the movement of people out of the region is likely to have, at best, no impact at all on the wage levels in the province or region.

Thus, movement of people out of a primary sector such as agriculture in a heavily resource-based region is not expected to increase the overall level of income within the region. Instead, the relatively low overall income level in such provinces compared to the other provinces tends to encourage an out-migration of labor. This, in turn, reduces the overall level of economic activity in the region, which has a further detrimental impact on wage levels in the region. It also leads to fewer opportunities for off-farm employment. This has led some observers to comment that a true diversification policy for Prairie agriculture requires an expansion of the manufacturing sector in rural areas in such forms as automobile plants in order to facilitate off-farm jobs.

Value Added in the Food Industry

The agricultural sector in Canada encompasses more than just the production of primary products. It also includes the processing of the raw products into finished products as well as the wholesaling and final retailing of the finished products. One of the goals of diversification has been to increase the amount of processing activity undertaken in Canada. The purpose of this section is to examine the trends that have occurred in value-added processing in Canadian agriculture. One of the conclusions reached is that growth in value added in the meat and grains industries has been much less than in the dairy and poultry industries.

There are many types of agricultural processing activities undertaken in Canada. These include the processing of meat, meat products, and poultry products; the manufacture of fluid milk; and the production of bread and other bakery products, cereal grain, flour, vegetable oils, and animal feed. One of the more important indicators of the degree to which these processing activities have contributed to diversification is value added. Value added in an industry represents the difference between the value received when the final good is sold and the value of the materials and supplies (including fuel and electricity) that had to be purchased to produce the final good. An important component of value added is wages and salaries since one way that value is added is through the use of labor to transform the raw material to a finished form.

Table 10.9 presents the level of value added in constant 1981 dollars contributed by four of the most important of the manufacturing sectors in the food industry. Two of the manufacturing industries process products from primary agricultural sectors that are governed by supply management (poultry and dairy), while the other two industries process products from sectors that are market oriented (meat and meat products and bread and other bakery products).

One of the most notable features of the data in Table 10.9 is the substantial growth in value added in the poultry and dairy industries as opposed to the meat and bakery industries. While the increase in some provinces is greater than in others, all regions of Canada have seen an increase in the level of value added in the poultry and dairy industries. In some regions and for some commodities (such as poultry and dairy in Ontario and Quebec and poultry in Alberta), the increase in the level of value added is quite substantial—400 percent to 500 percent in real terms over the period 1955-1985.

In contrast, the increase in the level of value added in the meat and meat products industry and the bread and other bakery products industry has been much less significant. The larger percentage increase in value added occurred in the meat and meat products industry, where Alberta, Ontario, and Quebec saw value added increase by approximately 60 percent. Some provinces actually saw the real level of value added decrease over the period 1955-1985. This was true, for example, in the bread and other bakery products industry in Manitoba and Saskatchewan.

While more research on this question is needed, at least part of the explanation of the difference in changes in value added among the industries can be attributed to the nature of the demand for the products. As noted in the Introduction, poultry and dairy products are sold primarily to the domestic market, grains are sold primarily for export, and livestock products are sold to both markets. This pattern of demand is correlated with the growth in value added, suggesting that a reliance on export markets has reduced Canada's ability or desire to invest in value-added processing.

TABLE 10.9

Value Added in the Major Food Industries by Province and the Prairies, 1955-1985

Year	Ontario/Quebec	Manitoba	Saskatchewan	Alberta	Prairies
		thousands, 1981 dollars			
Meat and Meat Products Industry					
1955	421,092	89,831	33,408	98,092	221,331
1960	422,474	90,336	35,900	112,488	238,723
1965	514,965	72,420	38,782	141,366	252,568
1970	555,485	77,147	50,768	167,106	295,021
1975	726,603	103,532	59,618	230,361	393,511
1980	802,294	78,968	67,258	116,187	262,413
1985	682,784	92,602	55,272	160,444	308,318
Poultry					
1960	43,1874,218	4,429	5,754	14,401	
1965	70,4456,543	4,625	9,637	20,804	
1970	104,925	6,964	4,657	10,479	22,101
1975	160,181	12,516	6,010	17,853	36,380
1980	162,704	16,832	a	28,586	
1985	223,589	17,949		31,937	
Dairy					
1955	331,669				
1960	518,920				
1965	597,858	35,817	34,205	69,735	139,757
1970	712,361	35,727	35,979	74,188	145,894
1975	706,632	43,528	38,884	90,356	172,768
1980	818,384	41,947		92,018	133,965
1985	1,119,890	44,463			
Bread and Other Bakery Products					
1955	392,515	29,358	20,462	34,727	84,546
1960	453,682	33,118	22,284	41,657	97,059
1965	523,785	33,817	24,284	42,032	100,132
1970	507,149	25,871	18,923	41,938	86,732
1975	576,647	28,242	18,581	46,138	92,961
1980	490,602	27,259	18,159	57,466	102,884
1985	564,159	24,943	9,570	41,412	75,926

[a]Blanks indicate data unavailable because of confidentiality.
Source: Statistics Canada, *Food Industries*, Cat. 32-250B.

Canada's value-added activities are among the lowest of the high-income countries; during the period 1980-1982, roughly 70 percent of farm exports were exported in unprocessed form (Schmitz). As Table 10.10 illustrates, Canadian exports of flour have actually been declining over the period 1950-1988. The Grains 2000 report makes a similar point when it notes that Canada's share of the $26 billion U. S. market for bread and related products (cookies and crackers, cereal breakfast foods, and pasta and noodles) was less than half of 1 percent.

Part of the reason for the poor performance by Canada in expanding its export of value-added items relates to the domestic policies of importing nations. Variable levies by these countries, for instance, discourage the Canadian processing of products for export. A case in point is canola. The Japanese, who are large importers of canola from Canada, allow canola in its raw form to enter Japan duty free, whereas variable levies are placed on imports of canola oil (Carter, McCalla, and Schmitz; Schmitz). The result is the reduction of the crushing margins for canola in Canada, thereby forcing Canada to export its canola in an unprocessed form.

Value added in the agricultural processing industries has been shifting among the regions within Canada (Table 10.9). In the meat and meat products industry, there has been a shift in value added from Manitoba to the other Prairie provinces and to Ontario and Quebec. More specifically, while value added in Manitoba has remained relatively constant over the period 1955-1985, it has increased by 50 percent to 70 percent in the other provinces.

In the poultry industry, the shift in value added appears to be away from Saskatchewan and toward Alberta, Manitoba, Ontario, and Quebec. Although value added in the Saskatchewan poultry industry has increased over the period 1960-1985, it has increased at a faster rate in the other provinces. This is particularly the case in Alberta, which has had the fastest growth in value added in the poultry industry.

In the dairy industry, both Manitoba and Saskatchewan appear to have lost ground to the other provinces in terms of the amount of value added. While the lack of data for Saskatchewan makes conclusions hard to draw, both it and Manitoba appear to have had only a small increase in value added over the years, especially as compared to Alberta, Ontario, and Quebec. The lack of data for Alberta for 1985 makes it difficult to determine whether

TABLE 10.10

Canadian Exports of Wheat Flour, 1949-50 to 1987-88

Year	Exports thousand tons	Year	Exports thousand tons
1949-50	1,243.2	1969-70	733.8
1950-51	1,521.9	1970-71	676.2
1951-52	1,390.8	1971-72	672.7
1952-53	1,537.7	1972-73	635.6
1953-54	1,258.6	1973-74	511.6
1954-55	1,105.5	1974-75	510.0
1955-56	1,088.6	1975-76	647.6
1956-57	912.8	1976-77	734.5
1957-58	1,099.0	1977-78	548.7
1958-59	1,010.4	1978-79	552.2
1959-60	1,006.2	1979-80	454.7
1960-61	971.1	1980-81	508.5
1961-62	869.6	1981-82	348.1
1962-63	742.0	1982-83	302.4
1963-64	1,494.4	1983-84	352.5
1964-65	858.4	1984-85	470.0
1965-66	1,037.6	1985-86	371.8
1966-67	866.8	1986-87	429.5
1967-68	671.9	1987-88	342.4
1968-69	670.1		

Source: Statistics Canada, *Agriculture Economic Statistics,* Cat. 21-603.

it has been able to keep pace with Ontario and Quebec over the last five-year period.

A shift in value added from Manitoba and Saskatchewan to other provinces in Canada also appears to be occurring in bread and other bakery products. Over the period 1955-1985, both provinces have seen a reduction in the value added originating from that industry. On the other hand, Alberta, Ontario, and Quebec have experienced an increase of between 20 percent and 40 percent.

Although there are exceptions, the overall conclusion appears to be that Manitoba and Saskatchewan have generally been losing value added when compared to either Alberta or Central Canada. The shift in value added from these two provinces is consistent with the structure of the Canadian economy discussed above. Both Saskatchewan and Manitoba have resource-type urbanization structures. Such economies have a relatively low level of economic interaction among their components, as well as a low level of local demand. Alberta, Ontario, and Quebec, on the other hand, have much more integrated economies and are much more dependent upon local demand. Hence, it is not surprising that these provinces would see the largest increase in value added since their economies are generating a demand for processed products, a demand that can be met by further value added in that province.

The relative loss of value added in Manitoba and Saskatchewan implies that these economies will be more resource based than they were before. Such an economic structure makes it difficult for these regions to generate off-farm employment opportunities. As well, it further encourages the out-migration of people from these areas since relative wages in these regions cannot keep pace with those in regions that are more economically integrated (Copithorne).

The loss in value added in agriculture in these two provinces is in contrast to what has been occurring at the level of the overall provincial economy. Stabler and Howe argue that, between 1974 and 1979, service exports increased more rapidly that did exports of goods. This result is all the more interesting when it is realized that, during this period, the value of raw material exports (grain, potash, uranium, oil, and gas) was increasing rapidly as a result of commodity price inflation. Thus, service exports appear to have the ability to initiate growth in a regional economy. One of the implications is that the failure of agriculture to develop its service sector (e.g., its value added) will further leave agriculture behind in terms of growth.

Summary and Conclusions

Agriculture in Canada is very diverse with different regions producing different products, serving different markets, and operating under different types of agricultural policies. However, the pattern of production in the various regions of Canada has not been static. Crops and livestock have been playing an increasing role in Central Canadian agriculture while representing a declining portion of Prairie agriculture. Poultry production has increased dramatically in Central Canada since the 1950s; the same pattern has occurred in Alberta since the late 1970s.

The income transfers that have been made to agriculture have differed between regions. Western Canada, on average, receives most of its support directly through commodity programs like WGSA. Central Canadian agriculture, on the other hand, tends to receive its transfers more indirectly through mechanisms like supply management. While total support for Prairie agriculture has increased substantially since the late 1970s, it still does not match that received by Central Canada. It is expected that, as commodity prices recover, the level of support going to Western Canadian agriculture will be reduced to former levels.

The importance of off-farm income in the two regions of the country is also very different, with farmers in Central Canada receiving a much greater percentage of their total income from off-farm employment. However, in both regions of Canada, nonfarm income makes up a significant proportion of the total income received by farmers. This makes the health of the farm economy highly dependent upon the health of the overall provincial economy and dependent upon the extent to which employment opportunities are spread throughout the provinces. Indeed, the tremendous importance of off-farm income to agriculture suggests that a very important part of any agricultural policy must be an industrial policy that creates employment opportunities in rural communities which, in turn, can be beneficial to many sectors in the region.

Value added in the agricultural industry in Canada was shown to be relatively low compared to elsewhere in the world. This was particularly true for the livestock and grains sector where value added has experienced very little growth since the 1950s. As might be expected from the discussion of the structure of primary agricultural production, value added also differs substantially from province to province. For example, Quebec, Ontario, and Alberta have seen increases in value added relative to Saskatchewan and Manitoba. The low level of value added overall and the relative decline in provinces like Saskatchewan and Manitoba are disturbing, given the importance of off-farm income to the health of the farming sector and the fact that services (of which value added is one) appear to be a factor in the growth of the Prairie provinces.

Discussed are a number of implications for agricultural policy and diversification of the industry. The most obvious conclusion is that, because of the diversity of agriculture across Canada, it is very difficult to establish policies that will not cause conflicts either within a region or among regions. The existence of such conflicts has historically led to additional government policies. A good example of this is transportation policy, where subsidies on the movement of grain out of Western Canada have spawned policies that attempt to offset some of the damage the subsidies have allegedly caused.

The dismantling of agricultural policies that are being discussed as part of the GATT negotiations are also likely to affect the various regions of Canada in different ways. For instance, a movement to freer trade in international agriculture has the potential to increase the welfare of Prairie grain producers (Carter, McCalla, and Schmitz). At the same time, however, this movement toward less protection is likely to adversely affect poultry and dairy producers (primarily located in Central Canada) as the supply management system is dismantled.

The impact of these international developments will also have impacts beyond the farm level. Proponents of freer trade and the removal of transportation subsidies argue that this will mean greater potential for diversification in the Prairie region for both grains and livestock. While the higher price for grain that results from the transportation subsidy is no doubt having a detrimental impact on livestock production and the further processing of grains and oilseeds, the magnitude of this effect is largely unknown. The analysis in this chapter, however, may provide some clues. It was pointed out that the growth in value added in the grains and livestock industries has been much less than that in the poultry and dairy industries.

The Crow Benefit only became a significant factor after 1975, suggesting that prior to that time the pattern of value added was not being influenced to any great extent by government policy (Fulton, Rosaasen, and Schmitz). This lends some support to the notion that removal of the subsidy may not have as large an effect as many assume it will.

The analysis in this chapter also suggests that the dismantling of supply management may not have as large an impact on value added and diversification in the poultry and dairy industries as some believe. Recall that the rapid expansion of poultry production occurred before the introduction of national supply management—the same is true of value added in this sector and in the dairy industry. While the presence of supply management has undoubtedly influenced the degree of diversification, it may be that the nature of these commodities (e.g., the demand is largely domestic) has as much of an impact in determining the potential for value added as does agricultural policy.

Notes

[1] For a discussion of the strengths and weaknesses of the PSE measure, see McClatchy.

[2] For a more general discussion of how governments use agricultural policy as a way of responding to the policy decisions made by other governments, see Fulton.

References

Agriculture Canada, Policy Branch. *Handbook of Selected Agricultural Statistics*. Ottawa, Ontario: Agriculture Canada, December 1986.

Agriculture Canada, Research Branch. *Summary of Budgets by Branch Sub-Objectives and Regions*. Ottawa, Ontario: Agriculture Canada, 1983-1984.

Arcus, P. L. *Broiler and Eggs*. Technical Report No. E13. Ottawa, Ontario: Economic Council of Canada, 1981.

Barichello, R. R. "Government Policies in Support of Canadian Agriculture: Their Costs." U. S./Canada Agricultural Trade Research Consortium Meeting, Airlie House, Virginia, December 16-18, 1982.

Canada Grains Council. *Statistical Handbook '87.* Winnipeg, Manitoba: Canada Grains Council, 1988.

Carter, C., A. McCalla, and A. Schmitz. *Canada and International Grain Markets: Trends, Policies, and Prospects.* Ottawa, Ontario: Economic Council of Canada, 1989.

Census of Canada. *Agriculture.* Ottawa, Ontario: Minister of Supply and Services, 1986.

Copithorne, L. *Natural Resources and Regional Disparities.* Ottawa, Ontario: Economic Council of Canada, 1979.

Davey, B. H., Z. A. Hassan, and W. F. Lu. *Farm and Off-Farm Incomes of Farm Families in Canada.* Economics Branch, Publication 74/17. Ottawa, Ontario: Agriculture Canada, 1974.

Economic Council of Canada. *Living Together: A Study of Regional Disparities.* Ottawa, Ontario: Economic Council of Canada, 1977.

Fulton, M. "Canadian Agricultural Policy." *Canadian Journal of Agricultural Economics.* Proceedings of the 1986 Annual Meeting of the Canadian Agricultural Economics and Farm Management Society, 34(May 1987):109-126.

Fulton, M., K. Rosaasen, and A. Schmitz. *Canadian Agricultural Policy and Prairie Agriculture.* Ottawa, Ontario: Economic Council of Canada, 1989.

Goodloe, C. A. *Government Intervention in Canadian Agriculture.* U. S. Economic Research Service, AGES871216, 1988.

Grains 2000. *The Road Not Taken: An Opportunity for the Canadian Grains and Meat Industry.* Ottawa, Ontario: Agriculture Canada, 1988.

Harling, K. F., and R. L. Thompson. "The Economic Effects of Intervention in Canadian Agriculture." *Canadian Journal of Agricultural Economics* 3(1983):153-176.

Josling, T. *Intervention and Regulation in Canadian Agriculture: A Comparison of Costs and Benefits Among Sectors.* Institute for Research on Public Policy, Technical Report E/12. Ottawa, Ontario: Economic Council of Canada, 1981.

Klein, K. K., and W. H. Furtan, eds. *Economics of Agricultural Research in Canada.* Calgary, Alberta: University of Calgary Press, 1988.

McClatchy, D. *The Concept of Producer Subsidy Equivalents: Some Considerations With Respect to Its International Negotiability.* Ottawa, Ontario: Agriculture Canada, 1987.

Organization for Economic Cooperation and Development (OECD). *National Policies and Agricultural Trade.* Paris: OECD, 1987.

Schmitz, A. *Canada's Agriculture in a World Trade Context.* Ottawa, Ontario: Agriculture Canada, 1984.

_____. "Trade in Primary Products: Canada, the United States, and Japan." In *U. S.-Canadian Trade and Investment Relations with Japan*, edited by R. M. Stern. Chicago, Illinois: The University of Chicago Press, 1989.

Stabler, J. G., and E. Howe. "Service Exports and Regional Growth in the Postindustrial Era." *Journal of Regional Science* 28(1988):303-315.

Statistics Canada. *Census of Canada: Population and Labour Trends.* Cat. 92-920. Ottawa, Ontario: Minister of Supply and Services, 1981.

_____. *Agriculture Economic Statistics.* Cat. 21-603. Ottawa, Ontario: Minister of Supply and Services, September 1986.

_____. *Census of Canada, 1986: Agriculture Alberta.* Cat. 96-111. Ottawa, Ontario: Minister of Supply and Services, various years.

_____. *Census of Canada, 1986: Agriculture Manitoba.* Cat. 96-109. Ottawa, Ontario: Minister of Supply and Services, various years.

_____. *Census of Canada, 1986: Agriculture Ontario.* Cat. 96-108. Ottawa, Ontario: Minister of Supply and Services, various years.

_____. *Census of Canada, 1986: Agriculture Quebec.* Cat. 96-107. Ottawa, Ontario: Minister of Supply and Services, various years.

_____. *Census of Canada, 1986: Agriculture Saskatchewan.* Cat. 96-110. Ottawa, Ontario: Minister of Supply and Services, various years.

_____. *Food Industries.* Cat. 32-250B. Ottawa, Ontario: Minister of Supply and Services, various years.

_____. *Livestock and Animal Products Statistics.* Cat. 23-203. Ottawa, Ontario: Minister of Supply and Services, various years.

Steeves, A. "The Dissociation of Occupation and Residents: A Study of Multiple Job-Holding Among Canadian Farm Operations." *Canadian Review of Sociology and Anthropology* 17(1980):145-168.

Veeman, M. M. "Social Costs of Supply-Restricting Marketing Brands." *Canadian Journal of Agricultural Economics* 30(1982):21-36.

Index

Alberta Crow Offset Program, 128–129, 343

Bakery products, 352, 356. *See also* Flour milling industry
Beef
   and economies of size, 255–256, 261(tables), 263, 294
   net returns on, 277–278, 283–293, 295–297
   processing, 102, 103(table), 104–105(table), 111
   U.S.-Canadian trade in, 31, 113, 114(table), 128
   See also Livestock industry
Biotechnological industry, 120, 131–133
Brewing industry, 139–157, 141(fig.), 143(figs.), 148(table), 151(table), 152(table), 181–182, 183(n5)
   See also Malt

Canada
   agricultural domestic policy, 116–121, 124, 128, 161–162, 329, 332–343, 335(tables), 357–359. *See also* Farm programs
   economy, 349–351, 356
   regional differences, 329–332, 330(tables), 331(table), 333(figs.), 337(figs.), 338(figs.), 339, 344, 347, 354–359. *See also* Prairie region; Provinces
   and trade, 25, 77, 78–80, 80(table), 141–142, 182, 343. *See also* Trade, U.S.-Canadian; Trade barriers; Trade disputes; United States–Canada Free Trade Agreement
Canadian Wheat Board (CWB), 67, 117–118, 166, 177, 275
Canola, 64, 68–69, 70, 73, 354
CAP. *See* Common Agricultural Policy
Capital Asset Pricing Model (CAPM), 266–269, 287–289, 288(table)

CAPM. *See* Capital Asset Pricing Model
Chemicals, 226
Common Agricultural Policy (CAP), 18, 20–21, 22, 79
Corn, 29, 311, 321–325. *See also* Grains
Crop acreage, 311–317, 316(table), 321–325, 323(table)
Crop gross margins, 270–277, 272(table), 297–298(n4)
Crop mix
   and diversification, 83–84, 127
   and government assistance, 32–33, 34
   irrigation and, 195–196, 201–204, 202(fig.), 203(table), 216, 227
   markets for, 113
   See also Crop rotations
Crop rotations, 270–277, 272(table), 274(table), 283, 285–293, 294–297
CWB. *See* Canadian Wheat Board

Dairy industry, 332–334, 352, 354–356, 359
Diversification
   definitions of, 3–4, 39–42, 195, 217(n4), 249, 304–305
   and economies of size, 5, 66, 249–251, 253, 264, 296, 305. *See also* Economies of size
   and government policy, 6, 32, 38–39, 303–304, 305–310, 307(table), 309(fig.), 324–326, 358–359. *See also* Farm programs; Regulation; Trade barriers
   and income variability, 48–66. *See also* Income
   irrigation and, 187–188, 195–207, 216, 227, 235. *See also* Irrigation
   off-farm, 250, 251–252, 351. *See also* Income, off-farm
   portfolio risk and, 265–269, 283, 285–293, 286(tables), 287(table), 288(table), 295–297, 325–326. *See also* Risk

and trade liberalization, 66-75, 78, 79-84. *See also* Trade, international, liberalization of; United States-Canada Free Trade Agreement
and value-added activities, 89, 351. *See also* Value-added activities
See also Crop mix; Crop rotations

EC. *See* European Community
Economics
 exchange rates, 116
 of irrigation, 190, 198, 207-215, 209(table), 210(table), 212(tables), 213(table), 217, 217(n2), 217-218(n7), 218-219(n15), 225, 227-235, 228(table), 230-231(table), 232-233(table). *See also* Irrigation
 and trade law, 26-27, 30. *See also* Trade barriers
 See also Investment; Prices; Return on investment; Risk
Economies of size, 5, 66, 249-251, 254, 255-264, 259(tables), 260(tables), 261(tables), 294, 296, 297. *See also* Farms, size of
EEP. *See* Export Enhancement Program
Employment
 off-farm, 250, 251, 346(figs.), 347-349, 351, 356, 357. *See also* Income, off-farm
 in Prairie region, 88, 349, 350
 and value-added activities, 97, 108
Environment, 215-216, 219(n19), 222, 226, 227, 235
European Community (EC)
 subsidies and, 96, 122
 and trade disputes, 14-22, 68, 157
 and trade liberalization, 75, 79
 and wheat flour exports, 160
European Community Treaty, 12-13
Export Enhancement Program (EEP), 161
Exports, 2, 80(table), 92, 92(table), 95-97, 99(table), 100, 106, 108, 124, 343, 354, 355(table), 356. *See also* Trade, international; Trade, U.S.-Canadian

Farm programs
 Canadian, 3, 6, 26-27, 113, 126, 129, 134, 190, 273-275, 278, 280, 329, 332-342, 335(tables), 337(figs.), 338(figs.), 357. *See also* Canada, agricultural domestic policy
 and crop acreage, 311-317, 321-325, 323(table)
 and crop demand structure, 317-319
 livestock supply model, 319-320
 and meat demand structure, 320-321
 United States, 161, 303-304, 305-310, 309(fig.), 324-326
 See also Subsidies
Farms
 family, 253-254, 256
 number of, 348(Figure 10.10)
 size of, 253-255, 264, 297(n1), 303, 305. *See also* Diversification; Economies of size
Flour milling industry, 158-181, 158(fig.), 159(fig.), 164(table), 165(table), 167(table), 169(figs.), 171(fig.), 172(fig.), 175(table), 176(table), 178(table), 180(table), 182, 354, 355(table). *See also* Bakery products; Pasta industry
FTA. *See* United States-Canada Free Trade Agreement

GATT. *See* General Agreement on Tariffs and Trade
General Agreement on Tariffs and Trade (GATT)
 effects on Canada of, 130, 339, 343, 358
 and patent rights, 133
 and trade disputes, 8-11, 14-22, 15(table), 16(table), 30, 31, 121
 and trade restrictions, 24-25, 45, 75, 82, 157
 and U.S. farm programs, 303
Government assistance. *See* Farm programs; Subsidies
Grains
 and economies of size, 260(Table 8.3), 262-263, 294
 and value-added activities, 124-127, 358. *See also* Bakery products; Flour milling industry; Pasta industry
 and WGTA, 342-343
 See also Corn; Malt; Wheat

Hog industry, 64, 113, 114(table), 128, 332, 334
 and economies of size, 256-258, 263, 294
 net returns in, 278-293, 295-297

Index 365

processing, 102–107, 104–105(table), 106(table), 111
U.S.-Canadian trade disputes over, 23–27, 33–34, 123
See also Livestock industry

Income
 distribution, 214–215, 334, 349–351
 and diversification, 39–42, 48–66, 250
 and the free trade agreement, 72, 74–75, 81
 and government policies, 4–5, 38–39
 and international trade liberalization, 77–78, 79–80, 82–84
 negative, 258–262
 net farm, 348(Figure 10.9)
 off-farm, 251–252, 262, 344–349, 345(table), 357, 358. *See also* Employment
 and risk-efficiency, 264–265, 285
 variability, 46–48, 49–54(table), 56–61(table), 63(fig.), 65(table), 83(table), 84–85(n1)
 See also Return on investment
Investment, 250, 251, 266–269, 296, 297. *See also* Return on investment
Irrigation
 and diversification, 187–189, 195–201, 216, 264
 future of, 207–216, 216–217, 239–240
 in Prairie region, 189–195, 191(fig.), 192(fig.), 193(table), 194(table), 197(fig.), 199(tables), 200(table), 201–207, 202(fig.), 203(table), 205(table), 206(table), 209(table), 210(table), 212(tables), 213(table), 216, 217(nn 2, 3), 218(nn 8, 10, 14), 219(n19)
 technology, 222–223, 224–235, 228(table), 230–231(table), 232–233(table), 235–245, 246
 in the U.S., 222–224, 227, 235–246, 237(table), 241(table), 242(table), 243(fig.)

Japan, 31, 34

Labor, 350–351, 356. *See also* Employment
Livestock industry, 329–332, 331(table)
 and diversification, 127–130, 285–287, 294–296, 297, 358

 and irrigation, 196–201, 197(fig.), 199(tables), 200(table), 204
 and U.S. farm programs, 311, 319–321, 334
 value-added activities, 352
 and WGTA, 342–343
 See also Beef; Dairy industry; Hog industry
Locational factors
 for farms, 257–258, 264, 332
 for industry, 109–111

Malt, 93–97, 96(table), 98(table), 110
 markets for, 112, 115, 118–119, 125
 See also Brewing industry
Management
 farm, 254, 258, 264, 297
 and risk, 258, 283
Manufacturing sector
 and diversification, 7, 351, 352
 See also Short-line machinery; Value-added activities
Market expansion
 barriers to, 33, 34. *See also* Trade barriers
 and the free trade agreement, 73, 81, 127
 and industry locational factors, 110–111
 potentials for, 82, 95–97, 111–115, 125, 131, 134, 218(n11)
 See also Trade, international; Trade, U.S.-Canadian
Market structure
 and the beer trade, 147–149
 and crop demand structure, 317–319
 and the flour trade, 162–170, 164(table), 165(table), 167(table), 169(figs.)
 and meat demand structure, 320–321
 and value-added activities, 352
 See also Economics; Prices
Mean-standard deviation trade-off, 264–265, 280–285, 281(fig.), 282(fig.), 284(fig.), 289–293, 290(fig.), 291(fig.), 292(fig.), 295–296, 297
Minerals, 30–31

New crops. *See* Crop mix
Nontariff barriers. *See* Trade barriers

OECD. *See* Organization for Economic Cooperation and Development

Oilseed, 246, 260(Table 8.4), 262–263, 294
Organization for Economic Cooperation and Development (OECD), 76–78

Pasta industry, 97–102, 99(table), 101(table), 110–111
  markets for, 113, 115, 125
PFRA. *See* Prairie Farm Rehabilitation Administration
Pork. *See* Hog industry
Portfolios. *See* Diversification
Potash, 28–29, 34
Poultry, 332, 333(figs.)
  and the free trade agreement, 69, 85(n4)
  and government policy, 334, 335(Table 10.5)
  markets, 114(table)
  and value-added activities, 352, 354, 359
Prairie Farm Rehabilitation Administration (PFRA), 190
Prairie region
  economy of, 37–39, 46–48, 66, 88–89, 329–332, 344, 347–351
  and government policy, 339, 342, 343
  and irrigation, 187–195, 191(fig.), 192(fig.), 193(table), 194(table), 201–207, 202(fig.), 203(table), 205(table), 206(table), 207–217, 209(table), 210(table), 212(tables), 213(table)
  and labor out-migration, 350–351, 356
  See also Canada, regional differences
Prices
  and crop gross margins, 270, 271
  flour, 170–171, 171(fig.)
  government policy and, 23–24, 29, 42–45, 97–100, 117–119, 125, 179, 315, 317–322, 324, 339, 357
  irrigation and, 211, 213(table), 215, 223, 225
  and product dumping, 28–29
  and specialization, 1
  variability in, 33, 38, 46–66, 49–54(table), 56–61(table), 63(fig.), 65(table), 77–78
  See also Economics; Income; Market structure
Processing activities
  of agricultural products, 89–95, 91(table), 97–102, 118
  and diversification, 88, 351
  and irrigation, 196, 201
  locational considerations, 109–111
  meat, 102–107, 103(table), 104–105(table), 128, 130
  See also Value-added activities
Protectionism
  Canadian regional, 336–339
  and the flour market, 172–177, 172(fig.), 182
  and market access, 38, 41
  and trade liberalization, 68, 75, 78–79, 343
  See also Trade barriers
Provinces
  and agricultural domestic policy, 336, 338(Fig. 10.5), 343
  and trade barriers, 145–147, 148(table), 150, 157, 182
  See also Canada, regional differences; Prairie region

Rail rates. *See* Transportation
Regulation
  environmental, 235
  health, 73, 123
  trade, 9–14, 24–25. *See also* General Agreement on Tariffs and Trade; Trade, international; Trade barriers
Research and development
  in biotechnology, 131–133
  government expenditures on, 6, 35, 239, 340–342, 341(tables)
  in irrigation technologies, 225
  for red meats, 130
  and value-added activities, 120–121, 124, 134
Return on investment
  and CAPM betas, 266–269, 287–289, 288(table)
  and crop rotations, 270–277, 294–295, 296
  and enterprise correlations, 285–287, 286(tables), 287(table)
  and livestock, 277–280, 294–295, 296
  and mean-standard deviation trade-off, 264–265, 280–285, 281(fig.), 282(fig.), 284(fig.), 289–293, 290(fig.), 291(fig.), 292(fig.), 295–296, 297
  and stocks and bonds, 280, 294–295, 296–297
Risk
  and farm size, 305

Index 367

and portfolios, 265–269, 267(figs.), 269(fig.), 271–277, 280–293, 294–297, 310, 325–326
See also Return on investment

Saskatchewan Beef Stabilization Plan, 278
Saskatchewan Crop Insurance (SCI), 273–275
Saskatchewan Hog Assistance and Rehabilitation Plan (SHARP), 280
SCGP. *See* Special Canadian Grains Program
SCI. *See* Saskatchewan Crop Insurance
SHARP. *See* Saskatchewan Hog Assistance and Rehabilitation Plan
Short-line machinery, 107–108, 111, 120, 130–131
  markets for, 113–115, 131
Sorghum, 311, 321–325
Soybeans, 311, 321–325
Special Canadian Grains Program (SCGP), 6, 332, 339
  and crop rotation, 273–275
Special Import Measures Act, 11
Specialization
  versus diversification, 1, 249–250, 303, 305, 324
  in hog industry, 257
  and international trade, 4
  See also Diversification
Stocks and bonds, 280–285, 287, 289, 293, 294–297
Subsidies, 85(n7)
  and diversification, 6–7
  effects of, 96, 119, 122, 160–161, 166, 223–224
  and the free trade agreement, 68, 179, 180(table), 182, 188, 216
  and rail rates, 6–7, 31, 126, 128–129, 134, 342–343, 358–359
  and trade disputes, 9–14, 20, 21–22, 23–27, 29–30, 32–35
  See also Farm programs
Subsidies and Countervailing Duties Code, 9, 11–12, 13

Tariffs. *See* Trade barriers
Technology
  improvements in, 110, 125, 239, 306, 332

irrigation, 222–223, 224–235, 228(table), 230–231(table), 232–233(table), 235–245, 246
regional disparities in, 350
See also Biotechnological industries
Trade, international
  in beer, 140–142, 141(fig.)
  disputes over, 14–22, 15(table), 16(table), 18(table), 19(table), 31, 35(n3). *See also* Trade disputes
  and diversification, 4, 40–42
  in flour, 158–166, 158(fig.), 159(fig.)
  and government policy, 3, 37–39, 42–45, 121–122, 129–130. *See also* Regulation
  liberalization of, 75–80, 82–84, 124, 343, 358
  See also Exports; Trade barriers; Trade, U.S.-Canadian
Trade, U.S.-Canadian, 108
  barriers to, 100, 106, 122–124, 127, 145–147, 148(table), 177–178, 178(table)
  in beer, 142–157, 143(figs.), 183(n5)
  disputes, 8–14, 22–35, 35(n2), 106
  in flour, 177–181
  and the free trade agreement, 66–75, 81, 100, 123–124, 127, 139–140, 178–181, 343
  See also United States–Canada Free Trade Agreement
Trade barriers
  effects of, 42–45, 43(fig.), 121–124
  and the free trade agreement, 66–67, 69–72, 73, 74, 81, 100, 127, 177–179, 178(table), 343
  provincial, 145–147, 148(table), 150
  trade disputes about, 20, 21, 31–32, 106. *See also* Trade disputes
  See also Protectionism
Trade disputes, 3, 8–32, 15(table), 16(table), 18(table), 19(table)
  U.S.-Canadian, 8–14, 22–35, 35(n2), 106
Transportation
  and industry location, 109–111, 124–125, 170
  rates, 6–7, 31, 89–90, 116–117, 126, 128–129, 134, 332, 342–343, 358–359

United States
  agricultural domestic policy, 161, 303–304, 306. *See also* Farm programs

and beer trade, 140–141
irrigation in, 222–224, 227, 235–246, 237(table), 241(table), 242(table), 243(fig.)
and trade law, 10–11, 25, 122–124, 160–161
See also Trade, U.S.-Canadian
United States–Canada Free Trade Agreement (FTA)
brewing industry exceptions, 150–157, 181–182
impact of, 4, 27, 45, 66–75, 81–82, 100, 123–124, 127, 134, 139–140, 178–181, 182, 188–189, 216, 303, 343
as negotiating tool, 121
United States Trade Agreements Act of 1979, 10–11
United States Trade and Tariff Act of 1984, 10
Urbanization, 350, 351

Value-added activities
agricultural processing, 93, 94(table), 97, 98(table), 100, 107, 124–127, 139, 351–357, 353(table)
and diversification, 2–3, 41–42, 64, 88–89, 109–124, 133–134, 351, 357–359
government assistance and, 6, 34
and trade liberalization, 83–84, 343
See also Processing activities

Water, and irrigation, 214–215, 219(n18), 223, 224–227, 238–240, 245, 246
Western Diversification Initiative, 3, 113
Western Grain Stabilization Act (WGSA), 126, 332, 336, 339, 357
and crop rotation, 273–275
Western Grain Transportation Act (WGTA), 126, 129, 332, 342. *See also* Transportation
WGSA. *See* Western Grain Stabilization Act
WGTA. *See* Western Grain Transportation Act
Wheat
Canadian processing of, 89–93, 91(table), 94(table), 109–110, 118. *See also* Bakery products; Flour milling industry; Pasta industry
and diversification, 1, 62–64, 217(n5)
and economies of size, 259(Table 8.2)
exports, 92, 92(table), 112
and farm programs, 311, 321–325
trade in, 158, 161, 177, 180–181, 183(n11)
See also Grains

Yields
and crop gross margins, 270, 271
and irrigation, 211, 222–223, 225